Henry Piddington

The Sailor's Horn-book for the Law of Storms

Being a Practical Exposition of the Theory of the Law of Storms

Henry Piddington

The Sailor's Horn-book for the Law of Storms
Being a Practical Exposition of the Theory of the Law of Storms

ISBN/EAN: 9783337192297

Printed in Europe, USA, Canada, Australia, Japan

Cover: Foto ©Andreas Hilbeck / pixelio.de

More available books at **www.hansebooks.com**

THE SAILOR'S HORN-BOOK

FOR THE

LAW OF STORMS:

BEING

A PRACTICAL EXPOSITION OF THE THEORY
OF THE LAW OF STORMS,

AND

ITS USES TO MARINERS OF ALL CLASSES,

IN ALL PARTS OF THE WORLD,

SHEWN BY

TRANSPARENT STORM CARDS AND USEFUL LESSONS.

BY HENRY PIDDINGTON,

PRESIDENT OF MARINE COURTS, CALCUTTA.

FIFTH ENLARGED AND IMPROVED EDITION.

"Wherein, if any man, considering the parts thereof which I have enumerated, do judge that our labour is to collect into an art or science that which hath been preturmitted by others as a matter of common sense and experience, he judgeth well."
BACON—*De Aug. Scient.*

"Perhaps this storm is sent with healing breath
From neighbouring shores to scourge disease and death?
'Tis ours on thine unerring Laws to trust,
With thee, great Lord! 'whatever is, is just.'"
FALCONER—*Shipwreck, Canto II. 884.*

WILLIAMS AND NORGATE,
14, HENRIETTA STREET, COVENT GARDEN, LONDON;
AND
20, SOUTH FREDERICK STREET, EDINBURGH.
1869.

TABLE OF CONTENTS.

I. DESIGN OF THE WORK.
II. PUBLICATIONS BY THE AUTHOR ON THE LAW OF STORMS.

PART I.
HISTORY AND THEORIES.

	PAGE	PARA.
HISTORY OF THE SCIENCE	1	1
Dr. Blane's account of the Barbadoes Hurricane	1	2
Captain Langford, Col. Capper, Horsburgh, Romme, Franklin	2	3-7
Professor Farrar. Mr. Thom	3	8-9
Redfield	4	10
North American Review, Dove, Brande, and Brewster	4	11
Reid	5	13
Espy, Thom, Piddington	6	16
EXPLANATION OF THE WORDS LAW OF STORMS	7	16
Theory and Law explained . . *Note*	7	
Motions of the wind in Storms	8	16-17
Applications of the Law	8	19
Storm, Gale, Hurricane	9	20
Words confounded and new one proposed. Straight-lined and Circular winds	9	20
Willy-waws of the Straits of Magellan, and Pamperos of La Plata	10	20
Storm Wave	12	21
Currents	12	21
Cards	12	22
Track	13	23
Turning and veering, example	13	24
Storm Disk	16	25
Centre or Focus	16	26
VARIOUS THEORIES as to the motion of Winds in Cyclones	16	27
Of Redfield and Reid	16	27
Espy, Hare	16	28
Redfield, Thom and Espy	17	29
VARIOUS THEORIES as to the causes of Cyclones	17	30
Redfield § 31, Reid § 32, Espy § 33	18	31-33
Hopkins	19	34
Dr. Thom	20	35-36

	PAGE	PARA.
L'Abbé Rochon's description . *Note*	20	
Whirls of Wind; Kaemtz: instances .	21	36
Sir John Herschel . . .	21	37
Piddington	22	38
Volcanoes . . .	22	39

PART II.
Tracks and Rates—Stationary Cyclones—Sizes—Dividing Cyclones.

	PAGE	PARA.
Tracks of Cyclones explained . .	24	40
Winds on Shore and at Sea . .	24	41
Average Tracks of Cyclones, and data required		
Reid, Redfield and Dove's views . .	25	42-44
Diagram of Reid's Theoretical tracks .	27	
Tracks of West Indies, Carribean Sea, Gulf of Mexico, Coast of North America and North Atlantic	28	45
Straight-lined tracks . . .	28	45
Curved tracks . . .	29	46
Towards the Coasts of Europe .	31	47
Cyclones crossing the Atlantic . .	31	48
Azores	32	49
Madeira	32	50
Cape de Verds	33	50
Breaking up of Atlantic Cyclones .	34	51
Eastern Coast of South America .	34	52
Southern Atlantic . . .	36	53
Off the Cape	37	54
Natal Coast and Madagascar . .	39	55
Southern Ocean, Cape to Van Diemen's Land .	40	56
H. M. S. *Havannah* . . .	40	56
Southern Indian Ocean, References to Chart	42	57
Belle Poule's Hurricane . .	46	57
Keeling or Cocos Isles . . .	47	57
Storm Track, Mauritius Cyclones .	48	58
Mosambique Channel . .	49	59
N. W. and West Coast of Australia .	50	60-61
Arabian Sea	51	62-63
Bay of Bengal . . .	54	64
Andamans to Madras . . .	54	65
Coast of Ceylon . . .	54	66
Table of References, Bay of Bengal, and Arabian Sea	55	66
Andaman Sea. . . .	56	67

CONTENTS.

	PAGE	PARA.
China Sea	57	68
Curved tracks, China Sea	58	68
Palawan Passage	59	68
Table of References	59-61	68
Java and Anamba Sea	62	68
Banda, Celebes, and Sooloo Seas	62	69
Timor Sea	63	70
South Coast of Australia	64	71
Bass' Straits	65	71
Northern Pacific and Loo-Choo Sea	65	72
Shanghae and Behring's Straits	67-68	72
Bonin Islands	69	72
Marianas, Carolinas and Radack Islands	71	72
Southern Pacific	73	73
New Caledonia	76	73
New Zealand	78	74-75
Van Diemen's Land to Cape Horn	79	76
Eastern Coast of Australia	80	77
Patagonia	81	78
West Coast of South America. Don Juan de Ulloa	82	79
To California	84	80
European Seas, Nelson; Malta	85	81
British Channel	88	82
National Research	*Note* 88	
AVERAGE RATES AT WHICH CYCLONES TRAVEL	89	83
West Indian and North American Cyclones	90	84
Southern Indian Ocean	90	85
Mosambique Channel	90	86
Arabian Sea	90	87
Bay of Bengal	90	88
Andaman Sea	91	89
Coast of Ceylon	91	90
China Sea	91	91
Pacific Ocean	91	92
STATIONARY CYCLONES	91	93
Albion's Cyclone	92	93
Burisal and Backergunge Cyclone	92	94
Bahamas	92	95
Azores, Tercera	93	96
SIZES OF CYCLONES	93	97
West Indies	94	98
Southern Indian Ocean	95	99
Arabian Sea	95	100

	PAGE	PARA.
Bay of Bengal	95	101
China Sea	96	102
CONTEMPORANEOUS, PARALLEL, AND DIVIDING CYCLONES	96	103
Parallel Cyclones	96	103
In opposite hemispheres	96	103-104
Contemporaneous	97	104-105
Meeting of Cyclones, Charleston Tornados	97	106
Hail Storms	98	107
Diverging Cyclones	99	108
Dividing Cyclones	99	109

PART III.

PRACTICAL APPLICATIONS—BEARING OF CENTRE—CYCLONE AND COMPASS POINTS—INCURVING—PROPER TACK TO LIE TO ON—ROADSTEADS AND RIVERS.

	PAGE	PARA.
General practical application	101	110
Three cases of management for every ship	102	111
Avoiding a Cyclone	102	112
When involved in one	103	113
Profiting by a Cyclone	104	114
BEARING OF THE CENTRE OF A CYCLONE	104	115
Directions to find it	104	116
Table of bearings of Centre	105	117
Method from the Nautical Magazine	106	118
WIND POINTS AND COMPASS POINTS	106	119
Table	107	
PROBABILITY OF THE INCURVING OF THE WINDS IN A CYCLONE	108	120
Charles Heddle's storm, and figure	109	121
Effects of incurving of winds. Redfield's estimate	110	122
Thalia's Log	111	124
Other examples	111-113	124-125
Indication from birds	113	127
Flattening-in of Cyclones	114	128
Ascertaining the track of a Cyclone	115	129
By projection	116	130
From the shift	119	133
Examples	121	134
PROPER TRACK TO LIE TO ON. REID	122	136
Admiral Graves' Disaster	125	137
Scudding or heaving to; special examples of the rule	126	138
Considerations	126	138

CONTENTS.

	PAGE	PARA.
Diagram	129	141
Extracts from imaginary Logs	128	141
Men of War and Merchantmen	131	143
USE OF THE HORN CARDS	132	144
Practical Lessons	132	145
Special examples. West Indies	135	147
New Orleans to the Havannah	137	149
Northern Atlantic	138	150
Bermuda Vessels, Reid	138	150-151
Great Western Steamer and Atlantic Storms	140	152-153
Coasts and Seas of Europe	141	154
Entrance of the Channel	142	155-156
Narrow Seas of Europe	143	157
The Black Sea	144	157
Southern Indian Ocean	144	158-160
Mauritius, Mr. Thom	146	160
Sailing round a storm	147	161
Orient, Maria Somes	148	162
Earl of Hardwicke	148	164
Sailing parallel to, or before the Cyclone track	150	165-166
Arabian Sea	151	167
Bay of Bengal	152	168-171
Rules for management in	154	172
China Sea	155	173-174
Golconda, Thetis and *Pluto*	157	175-176
Northern and Southern Pacific	158	177
RECAPITULATION	159	179
SHIPS PREVENTED FROM RUNNING INTO THE CENTRE	159	180
PROOFS OF THE ACCURACY OF RULES	160	181-182
IN ROADSTEADS, RIVERS, ETC.	162	184
Examples, West Indies	163	185
Madras Roads	164	187
Ships *Baboo*, and *General Kydd*	165	187
Capt. Millar, *Lady Clifford*	166	188
Example for Rivers	166	189

PART IV.

STORM-WAVE AND CURRENT—INUNDATIONS—CROSS SEA—
NOISES—ELECTRICITY—SQUALLS—EARTHQUAKES.

	PAGE	PARA.
STORM-WAVE AND STORM-CURRENT	168	193
Explanation, Storm-wave	169	193

	PAGE	PARA.
Coasts of Cornwall and Devon	170	195
Electric Rise, Ramsgate	172	196
On the coasts of Italy	172	197
Storm-currents explained	173	198
Instances of both	174	199
West Indies	176	200-204
Southern Indian Ocean	179	205
Coasts of Australia	179	207
Bay of Bengal	180	209
CHINA SEA	181	210
The Channel and H. M. S. *Repulse*	Note 185-7	215
Great Liverpool	186	*Note*
Bass' Straits	188	216
Caution	189	217
Nortes of the Gulf of Mexico	189	218
Loss of the *Tweed* Steamer	189	219
INUNDATIONS FROM THE STORM-WAVE	190	220
In the West Indies	192	222-223
Raz de Marée	192	*Note*
Barbadoes	193	225
Bermuda	193	226
New York	194	227
East Indies, Coringa	195-7	228-230
Ingeram	196	229
Cuttack and Balasore	197	231
Burisal and Backergunge	198	232
False Point	198	233
Laccadives	199	234
St. Petersburgh	199	235
PYRAMIDAL AND CROSS SEAS	200	236
Lt. Archer, Mr. Thom, Capt. Rundle, H. M. S. *Serpent*	201	237-241
Captain Sproule, *Magellan*	202	241
SWELL FELT AT A DISTANCE	203	242
NOISE OF CYCLONES	205	244
Instances	207	245
PASSAGE OF THE CENTRE	207	246
Instances	207-8	247-249
Flaws and gusts, Oscillation	208-9	250
Clouds at the passage of the Centre	210	252
SIZES OF THE CENTRAL CALM SPACE	211	253
Instances	212	253-254

	PAGE	PARA.
ELECTRICITY AND ELECTRO-MAGNETISM	213	256
Instances of Electric action	213-15	257
West Indies	215	259
Bay of Bengal	216	260
Electric action	216-17	260-62
Kaemtz's Electricity of Storms	217	264
Electro-Magnetism	219	265
ARCHED SQUALLS AND TORNADOS	219	266
Straits of Malacca, Bengal	219	267-268
Pamperos of Buenos Ayres	221	269-271
Tornados, Coast of Africa	222	272-274
EARTHQUAKES	225	275
Instances of them	225	276-280
At what time occurring	228	281

PART V.

BAROMETER AND SIMPIESOMETER—MEASURING THE DISTANCE OF THE CENTRE—SEASONS, AT WHICH CYCLONES OCCUR, WHIRLWINDS AND SPOUTS—COMMENCEMENT AND BREAKING UP OF CYCLONES.

	PAGE	PARA.
BAROMETER AND SIMPIESOMETER	229	282
Nelson's use of the Barometer	230	283
Cause of their motions	231	284-285
Fall explained: Redfield's experiment	233	286
Experiment varied. *Making* a Hurricane	234	286
Good Barometers and Improved Simpiesometer	234	287
RISE OF THE BAROMETER BEFORE CYCLONES	235	288
Oscillations, Barometer and Simpiesometer	236	289
Instances	236	290-296
Flashes in Barometer tube	239	297
Barometer not rising	240	299
Barometer in following storms	240	301
Rise of Barometer before the Storm is over	241	303
BAROMETER AS A MEASURER OF THE DISTANCE OF THE CENTRE	242	305
Question, views of Mr. Thom	243	306
Data for the Problem	244	308
Standard Observations	245	310
Barometrical Chart	246	311
Table of Rates of fall	248	314
Remarks	249	315

CONTENTS.

	PAGE	PARA.
Examples	249	316
Results, Plate and Table	250	317-318
Scale of distances	251	320
Precautions in applying the rule	253	321
Explanation of Table of Examples	255	322
Allowances to be made	256	323
Table of Examples and results	258	
Notes on the table of examples	260	327
Excessive falls of the Barometer	265	328
Table of them	264	
Height of Cyclones above the ocean	265	329
Kaemtz, Espy, Redfield	266	330-332
Horn Card for a disk	268	333
Cyclone a disk, not a column	269	334
Disks on Barometrical Chart	269	335
Disks seen through: Instances	270	336
Banks of Clouds	272	337
Instances of these	272	338
Banks of clouds surrounding ships: Instances	274	339
Signs of approaching Cyclones	276	340
Table of them, Celestial and Atmospheric	278	
—— Terrestrial	281	
Remarks on the Table of signs	283	341
Red Sky	283	341
Instances of it	284	343-345
—— at night	288	346
Stars, appearance of them	290	347
Green Light	292	347
Clouds	293	348
Lightning, Instances of	293	349
Aurora Borealis—Lightning	293	349
Explanation	294	350
SEASONS AT WHICH CYCLONES OCCUR	295	351
Table of them	296	352
Whirlwinds and Water-spouts	296	353
Theory explained	297	354
Simple forms	297	355
Indian dust whirlwinds	297	356
Instances, India, Africa	298	357-362
Land whirlwinds	301	363
Fine weather whirlwinds	305	364
Instances of them	305	365

CONTENTS.

	PAGE	PARA.
Bull's-eye squalls, *Kandiana's* & *St. George's squalls*	307	367
Electric whirlwind	308	367
Passing through water-spouts	309	368
Ships involved in spouts, Dampier	309	369
Father Piancini	310	370
Water-spouts, Capt. Howe's ; Franklin	311	371-372
In Storms	312	372
Instances	313	373-374
Become Storms (*Vautour's*)	316	374
Noises in them	317	376
Boiling agitation of the sea	317	377
M. Peltier's work on spouts	318	378
His views of them	318	379
Experiments	318	381
Table of spouts	319	382
Water-spouts, probably whirlwinds	320	383
Great spout on Teneriffe	320	384
COMMENCEMENT AND BREAKING UP OF HURRICANES	322	385
Instances, *Algerine* and *Vernon*	322	386-387
Considerations how they commence	324	388
Commence at the level of the ocean or in the atmosphere	324	389
Difficulty as to streams of air	325	391
Are they descending vortices or disks?	325	392
Theory of this, and conditions	325	393
Conditions fulfilled	327	394
Relation of Redfield's theory of the Barometer to this	328	395
St. Domingo Hurricane of 1508	329	396
Remarks on it	329	397
Other instances: Lifting up on the rear	330	397
Settling down of Cyclones	331	398
Termination of Cyclones	332	399
By rising up	332	399
Law of Rotation not accounted for	334	400
Constancy of revolution, and evidences derivable from it	334-336	401-404
Solar spots. Analogy with Cyclones	336	405
Observations wanting	337	406
Objection to the rotatory and progressive theory answered	337	407-409
Commencement of Cyclone accounted for	338	410
Facts supporting the theory	339-342	412-417

PART VI.

DIRECTIONS FOR STUDY—POINTS OF INQUIRY—MISCELLANEA
—CONCLUSION.

	PAGE	PARA.
DIRECTIONS FOR STUDYING THE SCIENCE	342-344	418-422
POINTS OF INQUIRY AND DIRECTIONS for Observers	344	423
Before a Cyclone	344	424
During it	347	425
At the close	348	426
Directions for Observers	348	427
How to forward Notes	349	429
Charts of Tracks required	350	430
Data required	350	431
Observers on shore	353	432
MISCELLANEA AND ADDITIONS. The Log Book	354	433
French Log Books: Lorimer's letters	355	435
Effects of Cyclones on the Compasses	356-360	435
Treacherous moderating of the wind	360	436-438
Etymology of "Hurricane"	361	439
"Tyfoon"	362	440
Utility of knowing these names	362	441
Surgeons and Passengers	363	442
Tornado-Cyclones near and in the Channel	363	443
Barometer signals to ships	364	444
at Barbadoes	365	445
Nortes of the Gulf of Mexico	366	445
Revolving winds true Cyclones but not of Hurricane force	367	445
at Calcutta	368	445
Nelson and Villeneuve's Gale	368	445
Tracks; New Caledonia	370	445
Cyclones of Iceland	370	445
Hail in its relation to Electricity	372	445
Electric Telegraph warning of the approach of Cyclones	375	445
CONCLUSION.		
Medina on the Variation	376	446
Burroughs on Mercator	376	447
Difficulties with some classes of Sailors	376	448
Claims of the Science on all classes	377	449

PREFACE TO THE THIRD EDITION.

DESIGN OF THE WORK.

WHAT I propose in this work is, to explain to the seaman, in such language that every man who can work a day's work can understand it, the Theory and the Practical Use of the LAW OF STORMS for all parts of the world; for this science has now become so essential a part of nautical knowledge that every seaman who conscientiously desires to fulfil his duties, from the Admiral of a great fleet down to the humble Master of a West India or Mediterranean trader, must wish to know at all events what this new science is, of which he hears it said, that it teaches how to *avoid* Storms—teaches how best to *manage in Storms* when they cannot be avoided—and teaches how to *profit by Storms!* A man who thoroughly understands all this must have as great professional advantages in that respect, over one who does not, as our fleets and ships of the present day, when scurvy is almost unknown, have over those of the days of ANSON, when whole crews were swept off by it.

To enable the plainest ship-master, then, clearly to comprehend this science in all its bearings and uses, and as far as our present knowledge extends, is my first object; and on this account I have endeavoured to make the work as clear and as brief as possible; though as still intended for all classes of mariners, the scientific as well as the unlearned, and intended also while it explains, to suggest, to ask for, and to urge further inquiry, I have not passed over the more scientific part while striving to give to the mariner of all classes a cheap book of the results of all the researches up to the present time; together with the principles of the science so clearly indicated, that he might be fully satisfied as to them, and, if he pleased, look farther to investigate their origin and effects, and furnish those who are prosecuting the research with additional materials.

Every one who has mixed much with sailors, and every educated man who reflects on his early life amongst them, must have remarked, from the naval officer to the coasting skipper, not only the many instances in which misdirected, imperfect, or limited education in youth has dulled the faculties for study; and thus all times and opportunities in after life have been neglected; but with too many also, alas! even where in youth or manhood inclined to any self-improvement, the truth of Falconer's beautiful lines, which, though written of classical literature, are not less true of scientific pursuits, and of the professional improvement dependent upon them—

> ". . . for blasted in the barren shade
> Here, all too soon, the buds of science fade!
> Sad Ocean's genius in untimely hour
> Withers the bloom of every springing flower:
> Here fancy droops, while sullen clouds and storm
> The generous climate of the soul deform."

And this I trust will be my apology, if any be needed, for preferring the familiar terms of common sailor-language where I could use them as I think with better effect, to more scientific forms of expression. I have thought, in a word, that the work might lose much of (I trust I may say) its national utility, if written only for the state-rooms of science; and thus I have preferred to seat it at her cabin-table—claiming only in this respect from those who might wish it otherwise, a moment's reflection on how large a class of our brother-sailors there is, and always must be, who though worthy and most valuable men, have wanted the inestimable advantages of a good education; and might be repelled from the study by the sight of "hard words" and the sound of scientific phrases, however familiar such may be to our ears.

In the style of printing, and by the frequent use of italics, as well as by a copious Index and detailed Table of Contents, I have endeavoured, as far as possible, to assist the reader to a quick and clear comprehension of what is said; for I know that many forget that "a hurricane at sea is like a battle in a campaign; an important, but unfrequent occurrence, for which it is wise to be well prepared;"* and rarely looking at works like the present till they want assistance from them, are thus very liable to mistakes in a moment of anxiety. If such, however, will but give it one fair perusal at leisure, they may perhaps, recollecting with quaint old THOMAS FULLER that "the winds are not only wild in a storm, but even stark mad in a hurricane,"† find that a little study in fine weather may save a world of labour and mischief in bad.

I should perhaps apologise for the apparent egotism of quoting so often my own Storm Memoirs, but in truth they form, as will be seen by the list of them in the next page, a considerable portion, as to bulk at least, of the evidence, new and corroborative, upon which our science rests; and often furnish details and exact data which I could not elsewhere obtain.

Writing in Calcutta, I have been unable to consult many books of voyages, and especially the works of the older and foreign navigators

---

* Foreign Quarterly Review for April, 1839: supposed to be written by Captain BASIL HALL.
† FULLER'S "HOLY STATE." "*The Good Sea Captain.*"

which I could have desired to examine or refer to, and my labours are continued under difficulties and even against positive discouragements which would be hardly credible if told; but the reader will nevertheless, I trust, find that he has a tolerably complete work for all really useful purposes, and that he has in this Third Edition of the SAILOR'S HORN BOOK, a very considerable addition to his former stock of knowledge on the subject.

H. P.

CALCUTTA, *May*, 1859.

Publications by the Author on the Law of Storms, and upon Miscellaneous Nautical Subjects.

	pp.	Charts & Diag.
1.—*First* Memoir. "Hurricane in the Bay of Bengal, June, 1839." Journ. Asiat. Society of Bengal, Vol. VIII.	45	5
2.—*Second* Memoir. "Coringa Hurricane of November, 1839, with other Storms." J. A. S. Beng. Vol. IX.	45	6
3.—*Third* Memoir. "Cuttack Hurricane of April and May, 1840." J. A. S. Beng. Vol. IX.	46	3
4.—*Fourth* Memoir. "The *Golconda's* Tyfoon in the China Seas, September, 1840." J. A. S. Beng. Vol. X.	11	1
5.—*Fifth* Memoir. "Madras Hurricane, of May, 1841, and Whirlwind of the *Paquebot des Mers du Sud*." J. A. S. Beng. Vol. XI.	19	1
6.—*Sixth* Memoir. "Storms of the China Sea, from 1780 to 1841." J. A. S. Beng. Vol. XI.	100	1
7.—Two editions of "Notes on the Law of Storms in the Indian and Chinese Seas, presented to, and printed by the Government of India, for the use of the Expedition to China."	42	4
8.—Notes on the Law of Storms in the Indian and China Seas, arranged for, and presented to Rushton's Directory	8	
9.—*Seventh* Memoir. "Calcutta Hurricane, 2nd—3rd June, 1842." J. A. S. Beng. Vol. XI.	24	2
10.—*Eighth* Memoir. "Madras and Arabian Sea Hurricanes, 22nd October to 1st November, 1842." J. A. S. Beng. Vol. XII.	61	2
11.—*Ninth* Memoir. "Pooree, Cuttack, and Gya Storms, of October, 1842." J. A. S. Beng. Vol. XII.	42	1
12.—*Tenth* Memoir. "Madras and Masulipatam Storm, 21st to 23rd May, 1843." J. A. S. Beng. Vol. XIII.	45	1
13.—*Eleventh* Memoir. "Storms in the Bay of Bengal and Southern Indian Ocean, 26th November to 2nd December, 1843." J. A. S. Beng. Vol. XIV.	66	1
14.—*Twelfth* Memoir. "The *Briton* and *Runnymede's* Hurricane in the Andaman Sea, Nov. 1844." J. A. S. Beng. Vol. XIV.	23	1 .
15.—*Thirteenth* Memoir. "The *Charles Heddle's* Storm off the Mauritius." J. A. S. Beng. Vol. XIV.	33	2
16.—*Fourteenth* Memoir. "Bay of Bengal, Ceylon and Arabian Sea Storms of 29th November to 5th December, 1845." J. A. S. Beng. Vol. XIV.	40	1
17.—*Fifteenth* Memoir. "*Cleopatra* and *Buckinghamshire's* Hurricane, Malabar Coast, 16th—18th April, 1847." J. A. S. Beng. Vol. XVII.	30	1
18.—*Sixteenth* Memoir. "Hurricane of the *Maria Somes*, Southern Indian Ocean, March, 1846." J. A. S. Beng. Vol. XVII.	15	1

19.—*Seventeenth* Memoir. "Cyclones of the China and Loo Choo Seas, 1842 to 1847, and of the Northern Pacific Ocean." J. A. S. Beng. Vol. XVIII. . . 73 . 3
20.—*Eighteenth* Memoir. "Cyclone of the Bay of Bengal, 12th—14th October, 1848." J. A. S. Beng. Vol. XVIII. . . . . . . 84 . 3
21.—*Nineteenth* Memoir. "H. M. S. *Jumna's* and other Cyclones in the Southern Indian Ocean, January to April, 1848." J. A. S. Beng. Vol. XIX. . . 40 . 1
22.—On Storms of Wind in Tartary. J. A. S. Beng. Vol. XIX. . . . . . . 7 . 0
23.—*Twentieth* Memoir. "April Cyclone of the Bay of Bengal." J. A. S. Beng. Vol. XX. . . 75 . 1
24.—*Twenty-first* Memoir. "Cyclone of H. M. S. *Fox* in the Bay of Bengal." J. A. S. Beng. Vol. XXI. . 25 . 1
25.—"Horn Book of Storms for the Indian and China Seas." Three Editions, each . . . 35 . 2
26.—On the Geometric measurement of the Barometric Waves in a Cyclone. J. A. S. Beng. Vol. XXII. . 6 . 1
27.—The SAILOR'S HORN BOOK for the Law of Storms for all parts of the world. 4th Edition . . 408 . 18
28.—CONVERSATIONS ABOUT HURRICANES . . 109 . 27
29.—*Twenty-second* Memoir. Tornadoes and Cyclones in the Bay of Bengal, &c. J. A. S. Beng. Vol. XXIII. 43 . 1
30.—*Twenty-third* Memoir. *Precursor's* Cyclone. J.A.S. Beng. Vol. XXIII. . . . . . 45 . 1
31.—*Twenty-fourth* Memoir. Sunderbund Cyclone. J. A. S. Beng. Vol. XXIV. . . . 65 . 1
32.—On the Cyclones of the Black Sea . . 22 . 1

Miscellaneous.

pp.

1.—Queries on the Deviations of the Compass on board of Steamers. Submitted to the Right Honourable The Lords Commissioners of H. M. Admiralty 3
2.—Method of Obtaining Warning of Spontaneous Combustion of Coal on board ship. H. M. Admiralty . . . 12
3.—Steamers can't stop for soundings! Naut. Mag. . . 2
4.—On a Spontaneous Combustion of Coal wetted with salt water, on the Ship *Sir Howard Douglas.* Jour. As. Soc. Bengal . 4
5.—On the Rates of Chronometers as influenced by the Local Attraction, and by Terrestrial Magnetism. Jour. As. Soc. Bengal . 17
6.—On the Comparative Action of the Marine and Aneroid Barometers. Jour. As. Soc. Bengal 9
7.—Letter to the Marquis of Dalhousie on the Cyclone Wave in the Sunderbunds 20
8.—Separate Report on the state of the Hooghly, Sept. 1854, folio 20
9.—On the silt of the waters of the Hooghly, 1st Note . . 4
10.—On clearing the River channel 12
11.—Memorandum addressed to the Government of India on the Emigration of Coolies and Transport of Troops from India, 28th October, 1846 2
12.—On the silt of the Hooghly, second Memoir . . 14

DIRECTIONS TO THE BINDER.

Table of Contents to follow the Title.

Place the Barometrical Chart fronting page 246, and fold it along the double line, below the word Centre at top.

A table of References to be placed fronting Charts II, III, and IV.

The Charts to be placed at the end of the book.

THE
SAILOR'S HORN-BOOK.

PART I.

BRIEF HISTORY OF THE SCIENCE OF THE LAW OF STORMS.—
2. WORKS ON STORMS.—3. EXPLANATION OF THE WORDS USED.
—4. VARIOUS THEORIES NOTICED AS TO THE MOTIONS, AND
BRIEFLY AS TO THE CAUSES, OF STORMS.

The word Cyclone proposed in the first edition of this work as a fit name, whereby to designate hurricanes, tyfoons, or any circular blowing gale, having been generally approved of, is now adopted throughout as a recognised term.

1. IT is perhaps more than a century ago since the first accounts were published of ships which had scudded in a hurricane for a day or more and yet found themselves "nearly in the same place as when the gale began," and the oldest seaman has heard instances of this strange, and then inexplicable circumstance; as well as of ships, which in lying to, had the wind either veering all round the compass, or shifting suddenly, with or without a calm interval, to an opposite point of the compass, and blowing harder than before, or of ships which, though but a short distance from each other, had furious hurricanes, dismasting and bringing them near to foundering from opposite points of the compass, and yet veering differently with both: but no author before the commencement of the last century appears to have thought of accounting for this, and most perhaps believed that it never could be accounted for.

2. In Dr. BLANE's Account of the hurricane of 1780 at Barbadoes (Trs. Roy. Soc. Edinburgh, Vol. I. p. 85), it is noticed that "the wind blew all round the compass, a circumstance which distinguishes the hurricane from all other gales within the tropics," and then, a little further on, that—

"The ships which put before the wind during the hurricane were not carried with the velocity which might have been expected from the violence of it. A merchant ship with the crew on board was driven from her anchors at Barbadoes, all the compasses were broken, and after tossing about for two days and two nights, the people found themselves at the mouth of Carlisle Bay, the very point whence they set out, at a time when they supposed themselves 100 leagues from it."

3. The earliest published account we have of these Storms being distinctly considered as whirlwinds is by Captain LANGFORD, in the "Philosophical Transactions" for 1698, in a paper on the West Indian hurricanes, which he calls "whirlwinds." He describes the veering of the wind, advises putting to sea with the Northerly winds to get sea-room, and come back with the S. Easterly ones "when *his* fury is over," describes their progression! and gives some limits for them. He has of course also a theory to account for their origin.

There are, indeed, frequent passages to be met with in the older navigators, and travellers, which directly or indirectly speak of the violent storms of the tropics as "whirlwinds." It is remarkable that DON JUAN DE ULLOA in 1743 fully describes rotary storms and shifts of wind on the Pacific Coast of South America, as will be seen in an extract subsequently given, but does not seem (in the English translation, which may be an abridgment) to have heard of, or to consider them as whirlwinds.

4. In a work published in 1801,* Colonel CAPPER, speaking of the Madras and Coromandel Coast hurricanes, says—p. 61,

"All these circumstances properly considered, clearly manifest the nature of these winds, or rather positively prove them to be whirlwinds whose diameter cannot be more than 120 miles, and the vortex seems generally near Madras or Pulicat."

And again, after describing some on the Malabar Coast and in the Southern Indian Ocean, he says—p. 64,

"Thus then it appears that these tempests or hurricanes are tornadoes or local whirlwinds, and are felt with at least equal violence on the sea coast and at some little distance out at sea."

5. In HORSBURGH's Directory, first published in the early part of the present century, he alludes to storms, particularly those of the China Sea, as being *rotatory,* i. e. great whirlwinds.

* Observations on Winds and Monsoons.

PART I. § 8.] *History ; Romme, Franklin, Farrar.* 3

6. A French author, ROMME, in a work of much research,* (published in 1806) describes the Tyfoon of the China Sea about the Gulf of Tonkin as a "*tourbillon*" (whirlwind); and again speaking of the gales of the Mosambique Channel, he distinctly says, that during the N. E. monsoon the heaviest tempests are felt there, and that "*then the winds change to whirlwinds* (tourbillons) with a high sea, cloudy sky, and heavy rain.*"* Vol. II. pp. 106, 132. The same author describes those of the Gulf of Mexico, p. 45, as hurricanes and whirlwinds. He also describes very fully the veering of the winds in the hurricanes of the Bay of Bengal, pp. 148 to 150, but does not, when speaking of these last, call them distinctly whirlwinds.

7. Hence we see that, up to the first ten years of the present century, all that appears known and published of tropical storms, and hurricanes was, that they were often great whirlwinds. FRANKLIN had indeed shewn in 1760 that the N. E. storms of the American Coast came from the S. W., but he did not prosecute the inquiry farther.

8. Professor FARRAR of the University of Cambridge, New England, in an account of the Boston Storm of 23rd September, 1815, printed in the American Philosophical Transactions, and reprinted in the Quarterly (Brande's) Journal of Science of 1819, p. 102, uses the following remarkable expressions. The italics are mine.

"I have not been able to find the *centre* or the limits of this tempest. It was very violent at places separated by a considerable interval from each other, while the intermediate region suffered much less. Its course through forests in some instances was marked almost as definitely as where trees have been cut down for a road.† *In these cases it appears to have been a moving vortex and not the rushing forward of the great body of the atmosphere.*"

He then goes on to describe the extent of the storm and the veering of the wind, and its veering in opposite directions at Boston and at New York, as well as the difference of time between the maximum violence at the two places, so as to leave no question about this Storm's having been a true Cyclone ; but he did not generalize the facts, and thus missed the honour of Mr. Redfield's discovery !

* Tableaux des Vents des Marées et des Courans.

† This was probably the furious vortex at the centre, which was there a true tornado, for the storm by Professor FARRAR's statement was about 200 miles in diameter.

9. I overlooked in the first edition the following notice from Mr. THOM'S work.

Professor MITCHELL, "On the proximate cause of certain winds," published in 1831,* expresses his opinion that "the phenomena of winds and storms are the result of a vortex or gyratory movement generally of no great extent, especially in the region of the atmosphere where they prevail." Although (adds Mr. THOM) it is not possible to agree altogether with this view of the subject, it sufficiently conveys the theory of circular action in stormy disturbances of the atmosphere.

10. Fortunately for science, however, Mr. WILLIAM REDFIELD, of New York, had his attention drawn to the subject in the course of his professional pursuits as a Naval Architect, and in 1831 published in the "American Journal of Science," a valuable paper, the first of a numerous series which have since followed, in which he demonstrated not only that the storms of the American Coast were whirlwinds, but, moreover, that they were *progressive* whirlwinds, moving forwards on curved tracks at a considerable rate, and were traceable from the West Indies, and along the Coast of the United States till they curved off to the Eastward between the Bermudas and the Banks of Newfoundland. He also gave many excellent practical rules for the management of ships, so as to avoid, or at least to diminish, the chances of damage from rotatory storms, and some valuable remarks on the Barometer as a guide.

11. In an able article on the Law of Storms in the "North American Review," for April, 1844, is the following, which is not only part of the history of our Science, but moreover a well merited tribute to the labours of the gentleman named in it—and I do not mean here to speak with less respect of those whose views have, I think, been proved erroneous. No one can have undertaken independent researches in any branch of the physical sciences without soon feeling that mankind owe much more than is usually supposed to those who have missed their road; they are the gallant soldiers who have fallen; the successful are the victors who survive to reap the honours of the combat.

"About the time that the American philosopher REDFIELD was employed in his earlier labours of observing and collecting the observations of others, a similar inquiry was in progress in Germany. On Christmas eve, 1821, after a long conti-

---

* "Edinburgh New Philosophical Journal," by Jameson, Vol. XI.

nuance of stormy weather, the barometer sank so low in Europe, that the attention of meteorologists was strongly drawn to the circumstance. Mr. BRANDE, having obtained the registers kept at this time in various places, came to the conclusion that, during this storm, the winds blew from all points of the compass towards a central space, where the barometer was, for the moment, at its lowest stand. The conclusion was disputed by Professor DOVE, of Berlin, who subjected the observations in the possession of BRANDE, as well as others, to a new examination, and made it appear, that an explanation of all the phænomena was afforded by the assumption of one or more great rotary currents, or whirlwinds, advancing from the Southwest to the Northeast.* Before this discussion was known in the United States, Mr. REDFIELD, by an independent course of investigation, had arrived at the result we have already announced, and his opinion is fortified by facts and cases so numerous and well authenticated, as fully to justify the distinction which Sir DAVID BREWSTER has accorded to him in the following language, 'The theory of rotary storms was first suggested by Colonel CAPPER, but we must claim for Mr. REDFIELD the greater honour of having fully investigated the subject, and apparently established the theory upon an impregnable basis.'"†

12. Mr. REDFIELD was followed in 1838 by Lieut.-Col. REID of the Royal Engineers, who published the valuable and well known work usually called "Reid on the Law of Storms," in which he not only fully confirmed Mr. REDFIELD's views, but added most extensively to the proofs of them by investigations of the West Indian Hurricanes, and those of the Southern Indian Ocean, and moreover by proving farther what REDFIELD had already announced theoretically, that in the Southern Hemisphere the Storms revolve in a contrary direction to those in the Northern one.

13. The great step of bringing the science to full practical use was also made by Colonel REID, who shewed that safe rules for scudding or lying to in a hurricane, might be deduced from the theory, and that farther, when obliged to lie to, ships should, to avoid being taken aback by the veering of the wind heading them off, choose a particular tack to lie to on, according to the side of the path of the Storm on which they were; and lastly that ships might often, by means of this knowledge, *profit* by storms by sailing *round* instead of *through* them; and the theory, like so many other sciences, then became from a speculative view, or *theory*, of which the uses had been only remotely and

* London and Edinburgh Journal of Science, Nos. 67, 68.
† Philosophical Magazine, Vol. XVIII. 3rd Series, p. 515.

dimly foreseen and hoped for, a practical *Law*—THE LAW OF STORMS; of the first use to the mariner, and of the highest importance to every naval and commercial nation.

14. Thus far may be called the history of the Law of Storms[*] from its early suppositions and proofs to its establishment. I may add, that Dr. A. THOM, of H. M. 86th Regt., has published also, in 1845, a valuable work on the science generally, and more particularly relating to the Cyclones of the Mauritius and Southern Indian Ocean. A list of my own former publications on the subject, which have mostly appeared in the Journal of the Asiatic Society of Bengal, from 1839 to the present day, relating to the China, Arabian and Andaman Seas, Bay of Bengal and Southern Indian Ocean, now amounting, if collected, to a volume of 1700 pp. with 95 Charts and diagrams, will be seen preceding this chapter. Amongst our results here are: The ascertaining of the average tracks of storms in various seas: The *division* of storms: The frequent occurrence of double storms: The *incurving* of winds in a hurricane, and the proofs of the *storm waves* and *storm currents*, and that of the succession of barometrical waves in a Cyclone, with minor matters, to all of which I shall subsequently refer. The invention of the Horn Cards should not perhaps be classed amongst the minor matters; for they have been, I have reason to think, the means of enabling very many persons to comprehend the subject and profit by the science, who would otherwise have been much puzzled by it. It was worth noting, however, before concluding this section, how it seems here and there to have been impressed on the minds of many observant seamen, long ago, that great storms and hurricanes were great whirlwinds. Mr. REDFIELD quotes a letter of Capt. WATERMAN of the "*Illinois*," in his memoir, 1831, p. 24, which concludes thus—"I have only to add that from an experience of 20 or 30 years, during which time I have been constantly navigating the Atlantic, my mind is fully made up that heavy winds, or hurricanes run in the form of whirlwinds."

15. WORKS ON STORMS.—In the former editions of this work I

---

[*] I have not alluded to Mr. ESPY's theory, as not being ours, and as I think being wholly contradicted by facts. It consists of two parts, the causes of hurricanes, and their effects. To this theory I shall allude subsequently.

gave a list of all the publications on the Law of Storms which had come to my knowledge up to the time at which the editions appeared, the object of which was to shew, as part of the history of the science, how it was gradually attracting attention, and also to assist the seaman in his choice of books for his "Storm Library." But these works are now generally so well known, and there are so many of them, that I have thought it better to economise the space which the catalogue would occupy for matter of more interest to the mariner. I should, however, fail in being, so far as my poor abilities may enable me, a safe and trustworthy guide to him, did I not most earnestly warn him against being misled into the absurd notion taken up by some of the Authors of recent works on the science that any positive rules can be given for the management of ships from the mere veering of the wind, leaving out of the calculation all the other elements. One of these books indeed, named below, goes so far as to give showy diagrams and to announce that the rules engraved upon them are to be followed "*whether the Cyclone has recurved or not*," thus coolly throwing the most important element of the sailor's calculation overboard!* The scientific seaman will not require any farther evidence of the danger of trusting to such a guide, and I must refer those who have had less advantages in the way of education, to Part III. where I have endeavoured to set down carefully, and in a classified arrangement, all the considerations which should guide the careful seaman in the vicinity of or within the limits of the vortex of a Cyclone, none of which he can safely neglect.

16. EXPLANATION OF THE WORDS '*Law of Storms*,' AND OF SOME OTHER TERMS. In the foregoing pages I have spoken of the *Law* of Storms as a Theory from which, when confirmed as a Law, Col. REID has deduced the rules which render it of practical utility.†

---

* This wretchedly ignorant work, entitled "The True Principle of the Law of Storms" (SEDGWICK), has been the principal cause of the erroneous ideas which this mistaken notion has given rise to.

† *Theory* and *Law*. The seaman may best understand these two words by his quadrant. As long as people who paid attention to these things *supposed* that light when reflected from a mirror was always so at a certain angle, depending somehow on the direction in which the original light fell upon it, this was a *Theory*. When it was proved by experiment that the angle of reflection was always equal to the

The words LAW OF STORMS, then, signify first, that it has now been proved by the examination, and careful analysis of perhaps more than two thousand logs and of some hundreds of storms by the authors already referred to, and by many other observers in periodical publications, as well as some whose results have not yet been published, that the wind in hurricanes, and frequently in severe storms in the higher latitudes on both sides of the Equator, has two motions. It turns or blows *round* a focus or *centre* in a more or less circular form,* and at the same time has a straight or curved motion forward, so that, like a great whirlwind, it is both turning round and as it were *rolling* forward at the same time.

17. Next it is proved that it turns, when it occurs on the North side of the Equator, from the east, or the right hand, by the north, towards the west, or *contrary* to the hands of a watch, or ◯; and in the Southern hemisphere that its motion is the other way, or *with* the hands of a watch ◯; being thus, as expressed by Professor DOVE of Berlin, S. E. N. W. for the Northern hemisphere, and N. E. S. W. for the Southern hemisphere, if we begin always at the right hand, or east side of the circles.

18. These two principal Laws constitute the rule or LAW OF STORMS. And it has, as before said, been abundantly demonstrated to hold good for several parts of the world, but in others, though our evidence is very deficient and sometimes indeed we have none at all, we must *assume* it only to be true; but we do so on very strong grounds; those of the great analogy usually existing in the laws of nature, and the fact that every new investigation affords us new proofs of the truth of our law in both hemispheres.

19. In its application also to Nautical uses the new Science is called the *Law of Storms*, and here it means that it offers a kind of

angle of incidence this became the *Law* of reflection, and when Hadley applied it to obtain correct altitudes, and to double the angle by the two reflections of the quadrant, he used it for a nautical object of the first importance and of daily practical utility. These are the three great steps of human knowledge and progress. The Theory, or supposition that a thing always occurs according to certain rules, the proof or Law that it does and will always so occur, and the Application of that law to the business of common life.

* Look at the Storm Cards in the pockets of the book.

PART I. § 20.]   *Explanation of terms.*   9

knowledge which, in most cases, will afford the seaman—FIRST, the best chance of avoiding the most violent and dangerous part of a hurricane, which is always near the centre of it; NEXT, the safest way of managing his vessel, if he is involved in one; and THIRDLY, the means of *profiting* by a storm! by sailing on a circular course round it, instead of upon a straight one through it; supposing always in this last case that he has sea-room.

20. STORM.—This term "storm" is not used so much with relation to the *force* of the wind in a storm, as to its *motion*.

A storm or tempest may mean either a gale or hurricane; but it always means in our science a storm *of wind*: and not, as frequently used by landsmen, one of thunder and lightning only.

A GALE means a storm of wind, the direction of which is tolerably steady for a long time; sometimes not only for days, but for weeks, as the common monsoon gales, and the winter gales of the Atlantic, the Channel, and the Bay of Biscay.

A HURRICANE generally means a *turning* storm of wind blowing with great violence, and often shifting more or less suddenly, so as to blow half or entirely round the compass in a few hours. The words *storm* and *hurricane* are both used to indicate a *turning* gale: hurricane when it is of excessive violence.

But these words Storm, Gale, Hurricane, Tempest, &c., are very liable to be used indiscriminately and confounded, and thus produce some perplexity, and even mislead the plain seaman. Our new science having demonstrated a circular or vortex-like (vorticular) motion, requires a new word to distinguish winds of all kinds having greatly curved courses, from those which, like the trades and monsoons, we may assume to be blowing in straight or nearly straight lines. Copying from Mr. Redfield's first memoir " On facts in Meteorology," most of the following names, I have classed them, not as he has done according to their strength, but according to the nature of their motion, noting with an (?) those of which we are as yet doubtful.

CLASS I.
Right-lined (straight) winds.

Trade winds.
Monsoons.
Some gales of high latitudes. (?)
White squalls, (Coast of Africa).
* Tornado, (sometimes ?)
Flaws and Gusts.
Some common squalls.
Harmattan. (?)
Land and Sea breezes.
North Westers (of Bengal).
Willy-waws of the Straits of Magellan.
Sumatras (of the Straits of Malacca).
North-Westers of India.
Helm wind (of Cumberland and mountainous Countries).
Rain-winds; from a fall of rain or of water near the spot, as at cascades, &c.
Scirocco. (?)
Etesian winds. (?)
Pamperos of La Plata.†

CLASS II.
Circular (or highly curved) winds.

Tyfoons and Hurricane Storms.
Some gales of high latitudes ?
Whirlwinds—of wind, rain, dust, &c.—called in Spanish, French, Portuguese, &c. *Turbo, Turbonado, Tourbillon, Tourmente.*
African Tornado, (sometimes ?)
Water-spouts.
Bursting of spouts, water-spouts, &c.
Samiel.
Simoon.

I am not altogether averse to new names, but I well know how sailors, and indeed many landsmen, dislike them; I suggested, however, in the former editions that we might perhaps for all this last class, of circular or highly curved winds, adopt the term "*Cyclone*" from the Greek Κυκλος (which signifies, amongst other things, the coil of a snake,) as neither affirming the circle to be a true one, though

---

* Mr. Redfield distinctly proves one tornado, the New Brunswick one, to have had a whirling motion, and Colonel Reid adduces a log of H. M. S. Tartar, off the coast of Africa, proving that the tornados are, sometimes at least, rotatory. –

† The squalls appear, though originating and sometimes blowing, as straight-lined winds; yet the Pamperos not unfrequently, and the Willy-waws almost constantly, take the whirlwind form. So that thus they may be considered as being of an intermediate class, rather than of either of those which we have distinguished.

the circuit may be complete, yet expressing sufficiently the *tendency* to circular motion in these meteors. We should by the use of it be able to speak without confounding names which may express either straight or circular winds—such as "gale, storm, hurricane," &c.—with those which are more frequently used (as hurricane) to designate merely their strength. This is what leads to confusion, for we say of, and we the authors ourselves write about, ships and places in the same "storm" having "the *storm*" commencing—"the *gale* increasing"—"the *hurricane* passing over"—and the like; merely because the ships or localities of which we speak had the wind of different degrees of strength, though the whole were experiencing parts of the same circular storm. *Cycloidal* is a known word, but it expresses relation to a defined geometrical curve, and one not sufficiently approaching our usual views, which are those of something *nearly*, though not perfectly, circular. Now if we used a single word and said, The "*Cyclone*" commenced, increased, passed over, &c. we should get rid of all this ambiguity, and use the same word to express the same thing in all cases; and this without any relation to the *strength* of the wind, for which we might freely use all the words "breeze," "gale," "storm," "tempest," "hurricane," "tyfoon," &c. as we pleased.* In the first edition of this work I ventured so far to *propose* the new word as to add it in a parenthesis wherever I wished to express a wind blowing in a circuit, whether a circle, or an ellipse, or a wind describing a spiral by its progression while turning, and as the

* I have now before me (October 1846) a newspaper extract giving an account of a meeting of the florists, market gardeners, &c. of the South London district, to consider of means to repair their losses in a severe thunder and hail-storm which occurred in the month of July or August of that year. In this article the words "gale," "storm" and "hurricane" are used to speak of the "meteor" as the French would call it. And in another, giving an account of the storm which visited Edinburgh on the 4th March, the tempest is alternately spoken of as a "gale and a hurricane," and Professor Nichol, of the Glasgow observatory, finally speaks of it as a "storm of translation," or a moving whirlwind, "of large radius but of immense power." The simple word *Cyclone* would express all which Professor Nichol wished to say, and he would then, as well as the editor of the newspaper, have had all his words to express the *violence* of the (hail and thunder) storm; without, by using the term "hurricane" for instance, leading their readers to suppose that there was any thing rotary about the wind if they did not mean to express this.

word has been generally approved of, and in all cases found to be not only unobjectionable, but often highly convenient, I have now, as noted at the heading of this part, adopted it with its natural derivatives, Cyclonic and Cyclonal, and even Cyclonology, when speaking of our new science throughout the present edition. And it has been so generally adopted that I have reason to believe it has now become a recognised English, as well as a professional and scientific word.

There are in tropical latitudes and as far as 50° or 55° N. and S. of the equator two kinds of tempests or storms. The monsoon or trade wind or winter *gales*, in which the barometer remains high and the wind steady, and the hurricanes, or tyfoons (Cyclones), often blowing with irresistible fury, and almost invariably accompanied by a falling barometer. Perhaps they occur even farther from the equator, but we have no good evidence beyond the storms of the Channel, and some off the Cape of Good Hope and Cape Horn.

21. THE STORM WAVE is a mass of water of greater or less diameter according to the storm, raised above the usual level of the ocean by the diminished atmospheric pressure, and perhaps other causes, and driven bodily along with the storm or before it, and when it reaches bays or river mouths, or other confined situations, causing by its further rise when contracting dreadful inundations; but upon open coasts rarely so, or not in so great a degree, as it can there spread out quickly and find its level. The STORM CURRENTS may be briefly described as circular streams on the circumferences of Cyclones, and of these also we have evidence enough for the mariner at all times to admit, and be on his guard against the *possibility* or *great probability* of them.

The deep-sea wave also, (the *flot de fond* of the French writers) no doubt assists the inundation; but as this is not a surface cause, I do not allude to it here.*

We have thus in every Cyclone two sets of forces (currents) independent of that of the wind, acting upon a ship; the one carrying her bodily onward on its track, and the other drifting her round the periphery of that part of the Cyclone circle, in which she may be.

22. STORM CARDS. The horn plates in the pockets of this book

* See EMY "*Du mouvement des Ondes et des Travaux Hydrauliques maritimes.*"

are what are called Col. REID's Hurricane, or Storm Circles, or Cards.

The use of these is to lay down and move upon any part of a chart. They may be supposed to represent a Cyclone of fifty, or five hundred miles in diameter, as we please; and one which would fill up the North part of the Bay of Bengal, would shew the wind in the same Cyclone, South on the coast of Arracan; East on the Sand Heads; North on the coast of Orissa; and West across the middle of the Bay; and if we move it over a chart, the changes of the wind for a ship or an island on its track will be seen. If placed with the centre between Barbadoes, St. Vincent's and St. Lucia on a Chart of the West Indies, it will shew how the wind may be in their hurricanes, for these Islands or for ships in those positions, N.N. Easterly and N. N. Westerly for St. Vincent's and St. Lucia, and S. S. Westerly for Barbadoes at the same hour. This is with the card for the Northern Hemisphere. With that for the Southern Hemisphere everything is the contrary way; and if the Card for that Hemisphere is laid between the Mauritius and Bourbon, it will be seen then that a S. E. Cyclonal wind at Bourbon may, if the circle is large enough to reach so far, be a N. Westerly one at the Mauritius; and so on all over the world, in each hemisphere.

23. The TRACK of a storm is the line or path along which it moves, and we usually speak of the imaginary line or path passed over by the centre as *the* track, though we might call the whole breadth passed over by the storm its track or road, if we pleased. These tracks are different in different parts of the world and in different latitudes, and, together with the ascertaining *certainly* if the usual law of veering prevails every where, are the great objects of the researches now going on, because much of the management of a ship depends upon our knowledge of the track of the storm, as will be subsequently seen.

24. The TURNING and the VEERING of the wind in a Cyclone. The seaman should be careful not to confound these two words, though so nearly alike in meaning and in the ideas they convey. Thus if, as in the figure below, two ships A. and B. in the Northern hemisphere are at 80 miles distance from each other, and a storm is travelling

towards them from the W. b. S. or WSW. to the ENE., as in the Northern Atlantic Ocean, it will be seen by moving the centre of the horn-card along the arrow that the successive changes are—

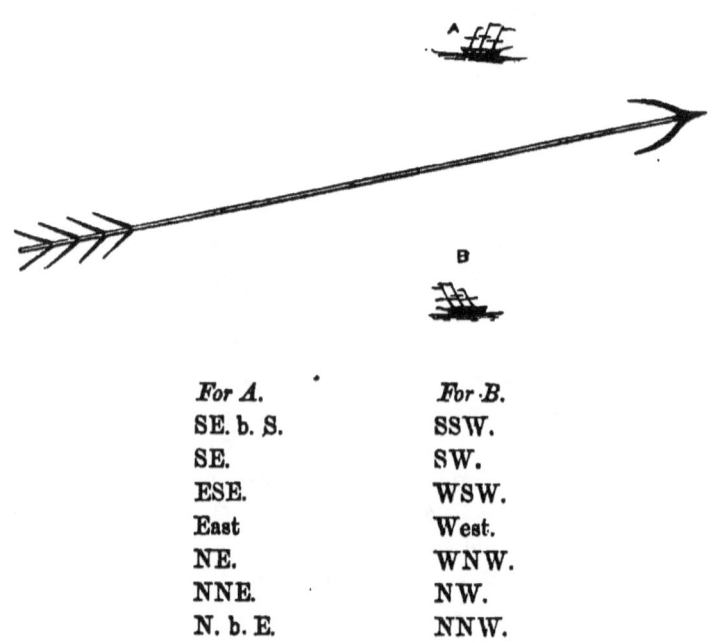

For A.	For B.
SE. b. S.	SSW.
SE.	SW.
ESE.	WSW.
East	West.
NE.	WNW.
NNE.	NW.
N. b. E.	NNW.

This is the *veering* of the wind, and when we say that in either hemisphere the wind *turns* so and so in a rotatory gale (Cyclone) we mean to express thereby that the *whole body* of the storm, whatever be its extent, is whirling round (turning) with or against the hands (hours) of a watch; but when we say the wind *veers*, or will *veer*, in a given direction, we mean then to speak always of its changes or shifts, whether gradual or sudden, in different parts of the circle, so that the wind in a hurricane or Cylone, while it really *turns* but one way, according to the hemisphere in which the storm occurs, may with ships on different sides of its track, be *veering* in apparent contradiction with each other, and with the law given; and one of these veerings is what sailors call "*backing*" round, and this explains, what

PART I. § 24.] *Words explained.* 15

is so often matter of contradiction and confusion, a ship sometimes in one-half of a storm (Cyclone) having the wind '*veering*' and then *backing*—when she gets into the opposite half of the storm circle. In this example we have supposed the ships lying to; if scudding and thus rapidly changing their positions, the changes of wind will of course differ. Mr. REDFIELD says, p. 175 of "American Journal of Science and Arts," Second Series, No. 2, touching this subject:—

"*The paradox of revolving winds.* It is still possible that some persons may not at all be able to understand, clearly, how the wind in a progressive storm which revolves in one constant direction around its axis, can at the same time be found to veer in opposite directions, on the opposite sides of the axis line, as is seen in Tables I. and III. respectively. But this fact, of which an explanation has already been attempted, may be seen to be a necessary result of the law of rotation, as manifested in all revolving bodies, and failing to understand this law, no one can intelligently pursue the enquiry.

"Let a circular disk of stiff paper be written upon in one or more circular lines around its centre, either in concentric, or vorticose form; then put this disk in rotation upon its centre, and pass two fingers across it in parallel directions, one on each side of the axis, and it will be found that one finger passes the circular writing in the order in which the words are written, while on the opposite side of the axis the other finger, though moving in the same direction, will pass over the writing in the opposite or reverse order to that in which the words are written. Of course this will equally follow in case the revolving disk be advanced under the fingers, as when the fingers are advanced over the disk.

"The two opposite orders of succession in which the letters are thus presented on the revolving disk, are equivalent to those of the winds which are presented to separate observers on the two opposite sides of the storm. This then being the law of rotation, it follows, that if the general course pursued by a storm be known, two rough observations of the order of changes in the wind, one on each side of its axis path, may be quite sufficient to determine its revolving character; provided that the early and later winds near the axis path have blown transversely to the course of progression; to determine which even, the same observations may suffice."

Perhaps a still more simple way of understanding this may be given, and it is as follows. If we look at a carriage wheel going fast over a muddy road, we shall see that while the whole wheel (the body of the storm, or *Cyclone*) moves forward on the road or track, the dirt on the upper rim is thrown forward and that on the lower rim backwards. If we suppose arrows (wind arrows) painted on the felly of our wheel and that it was a disk of air a mile or two high* moving *horizontally* over a mass of water, we shall at once see how the different changes take place to ships on opposite sides of the storm circle.

25. A STORM DISK is the thin whirling stratum of air which constitutes the Cyclone, and which though from 50 to 100, and even a thousand miles in diameter, is not probably at any time more than from one to ten miles in perpendicular height, as will be subsequently shewn.

26. THE CENTRE, OR FOCUS, or calm centre, *or focus,* or Lull, or Eye of the Storm, is a space of calm often, but not always, formed about the centre of Cyclones, and which varies like the Cyclones themselves in its diameter, and, according to their rate of motion also in its duration, while passing over an Island or station, or while a ship may be drifting through it or round it.

With these explanations of our *words,* we shall better understand the *things* spoken of, and in the following parts of the work the proofs of these various definitions will be found.

27. VARIOUS THEORIES as to the motions of the wind, and of the body of the Cyclone, and briefly as to the causes of them.

A theory is, as already explained at first, a *supposition.* It obtains proofs by continual observations, and by experiments when these can be made,† and then we set it down as a Law. The principal Theories of Storms—that is of the motion of the wind in them, and not as to what causes this motion—are,

a. That of MR. REDFIELD and Colonel REID already explained; and

28. *b.* That of Mr. T. P. ESPY of Philadelphia, supported by Professor HARE and some other American philosophers. These gentlemen maintain that in Cyclones the winds do not blow *round* a circle, but *inwards from the circumference of a circle to its centre,* or to a space along a line in the direction of the storm's progress; for they agree with REDFIELD and REID in the fact of the progressive motion of Cyclones; so that looking at the Compass card one may suppose 32

---

* See Part V. at the section on the Barometer for the height of storms above the surface of the ocean, and for another simple and practical demonstration of the movement of winds in a Cyclone.

† We *have* had some experiments made in hurricane-Cyclones, and very curious and perfectly convincing ones they are, as will afterwards be seen.

or 64 or 128 winds! all blowing at the same time inwards to the pivot! or to the needle, if it lies in the track of the storm ; being one for every point, one for every half, or one for every quarter point, and they affirm that it is this which produces the calm at the centre.

They suppose, as we do, this centre to be moving onwards, so that if moving North, for instance, it must, as it were, annihilate more or less all the winds from, say NW. to NE. as it advances; and apparently to obviate this difficulty, amongst others, they suppose that at or near the centre, or central line or axis, the wind curves upwards, and that thus the centre is a huge funnel or chimney, like the base of a water spout.*

29. Mr. REDFIELD also admits of or supposes *some degree* of upward (spirally upward) curving of the winds at the centre, but this may easily be admitted without any great violence to our notions. Mr. THOM, while agreeing with REID and REDFIELD as to the circular motion of the wind at the surface, thinks there is an up-current always formed, and in a figure given in his work this is delineated. The difficulty in this respect in Mr. ESPY's and Mr. THOM's theories is to conceive what can become of the enormous volume of air, and why it does not at least carry the rain upwards ? if not the ships, or their masts and yards?

30. VARIOUS THEORIES AS TO THE CAUSES OF CYCLONES. The seaman must be careful not to confound here, as is often done, the CAUSE with the *effects* of Cyclones. We can readily suppose a beginning somehow and somewhere, but when we inquire into the causes of the storms we go back beyond this, which is but *the beginning of the effect*, to look for the cause of that effect, or in other words, we inquire in plain language thus—" By which one, or more, of the known (or any unknown) powers of nature; as the force of winds, electricity, heat, evolved by condensation of the vapour in the

---

* The complete contradiction of this theory is found in the fact that, in numerous well attested cases where the track of a storm has been perfectly ascertained, the shift of wind for ships on its direct path has been *perpendicular* to, or across, and not *right against* the track, which Mr. ESPY's theory requires it to be. See also the convincing instances of the " *Charles Heddle*," and the experiment of the " *Hindostan* " Steamer, subsequently alluded to.

atmosphere, &c. &c. is this Cyclone set a-whirling and moving along?" My present limits and the object of this work do not allow me to go into much detail on this subject, and indeed to do so would but be to discuss what are yet but plausible theories—some of them probable ones it is true—and would suppose my readers also more conversant with the chemistry of the atmosphere and the higher researches of meteorology than most of them probably will be.

31. Mr. REDFIELD who, as already shewn (§ 10, p. 4) is the father of the research in recent times, had no particular theory as to the *causes* of Cyclones. He thought that our knowledge of their effects was not far enough advanced, and that it was unscientific to attempt to account for them, till better informed, by the exclusive action of any one or more causes. In his late publication he inclines to think them produced by the conflicts of prevailing currents in different strata of the atmosphere giving rise to circular movements, which increase and dilate to storms.

32. Col. REID avoids any general speculation as to the causes of Cyclones. He adverts to the possibility of there being some connection between storms and Electricity and Magnetism (Law of Storms, Chap. XII.) but goes no farther than to detail an experiment which appears, he thinks, partially to confirm his views.

33. Mr. T. P. ESPY in America, has published a thick volume entitled the "Philosophy of Storms," in which he lays down as before noted (§ 28 *b*. p. 16) a theory of the *effects* and one of the *causes* of storms. This last is, as briefly as can be explained, this—First, that upon any partial heating of the air at the surface of the globe it rises in columns more or less charged with vapour, which as they rise have this vapour condensed into clouds or rain.

Next, in this changing of state the vapour communicates its *latent caloric** to the surrounding air, which also expands, is cooled itself by that expansion, but also gives heat to that part of the air in which it

---

* The scientific (chemical) term for a quantity of heat which all bodies contain and give out when passing from one state to another which is *less fluid*, i. e. more condensed, as water passing to ice. If the process is reversed, that is, if the body passes from a *more* condensed state to one *less* so,—as water into vapour or steam—then heat is *absorbed* by it to form *latent* caloric, or hidden heat, not shewn by our instruments.

then is; and becoming lighter is carried farther up, so that what Mr. Espy terms an up-moving column (or columns) is always thus formed before rain is produced; and the air, rushing in to supply the partial vacuum at the base of this chimney-like column, forms thus the centripetal (moving towards a centre) streams of wind, which, as just described is, he affirms, the true motion of the wind in all storms, and especially in Cyclones, which are thus, not curved and nearly circular as we suppose, but *straight-lined* winds rushing to a centre according to his theory. To put this in plainer words he conceives the calm centre or lull of a Cyclone to be the base of a huge moving chimney, circular or of any longitudinal shape, the draught of which is occasioned by an extensive condensation of vapour above. He accounts for the production of clouds, the rise and depression of the Barometer, &c. by this cause, inferring that at a certain height the rising air *overflows* the rest of the atmosphere, forming a ring or annulus of cloud and vapour and air, which pressing on that below, occasions the rise of the Barometer found at the edges or approach of storms. To examine critically this and other theories, is not the object of this work.

34. Mr. THOS. HOPKINS, of Manchester, in a work published in 1844, entitled "On the Atmospheric changes which produce Rain, Wind and Storms," and which contains many novel, highly interesting and lucid common-sense views, admits with Mr. ESPY and other Meteorologists who had long preceded him,[*] the ascent and condensation of vapour in the air from various causes, and that all horizontal winds are thus produced. He considers moreover that these ascending winds produce *descending* ones, and that the rain produced in the higher regions brings air and steam[†] (vapour) with it in its descent, and thus constitutes lower atmospheric currents. And finally, that

[*] The original view is that of Daniell, who was the first to announce the important part which the latent caloric (see Note, p. 18) of the vapour, must perform in all atmospheric changes.

[†] Mr. Hopkins constantly and properly uses the word " steam " for the " vapour " of Meteorologists in general, who, if they object that *steam* induces the idea of heat, should recollect that *vapour* is the name of water and other bodies which have *not* changed their chemical state. Mr. Hopkins' Steam is the invisible vapour of a fine day which gives the dew point. The vapour of the Meteorologists should be confined to fogs and other appearances in which the steam, being condensed, though still buoyant in the air, is visible.

storms are produced by the same causes that produce other winds, and that the greatest storms *are descending winds.*

35. The late Dr. ALEX. THOM, H.M. 86th Regt., author of a most important work "On the nature and course of Storms in the Indian Ocean, South of the Equator," &c. is of opinion, with respect to this tract, that the principal cause of the rotatory motion of Cyclones is, at first, opposing currents of air on the borders of the monsoons and trade winds, which differ widely as to temperature, humidity, specific gravity and electricity. These, he thinks, give rise to a revolving action which originates the storm, which subsequently acquires " an intestine and specific action involving the neighbouring currents of the atmosphere, and enabling the storms to advance through the trade wind to its opposite limits;" and he gives a diagram to shew how this may occur. He farther inclines to believe that " as the external motion is imparted to the interior motion of the mass, and centrifugal action begins to withdraw the air from the centre and from an up-current the whole will soon be involved in the same vortical action." The up-current he explains as being formed by the pressure being removed from the centre, when the air there "increases in bulk, diminishes in specific gravity, and its upward tendency follows as a matter of course."

36. There is, however, another point of view in which some writers have considered the formation and continuance of these Cyclones. They suppose them, as Mr. THOM does, formed by opposite currents of air producing whirlpools as in water, but they do not consider with him, that they are produced at the *edges* of the streams as we see in water whirlpools. These authors incline to the belief that the whirls originate between the upper and lower *surfaces* of strata of air of different temperatures, degrees of moisture, &c. and moving in different directions. These whirls they suppose first formed above, and then to descend to the surface of the earth, just as we see a water spout begin at sea with a slight swelling of the lower part of a cloud and then a gradual descent of it. In a word, they look upon Cyclones as *wind spouts.**

---

* The Abbe Rochon describing the hurricanes (Cyclones) of the Mauritius, 1771, says they are "a kind of water spout which seems to threaten the spot over which it hangs with entire subversion." See GRANT's History of the Mauritius, p. 173.

One writer, KAEMTZ,* has seen these whirls of wind formed, though not quite under the circumstances which we require. Still his description is worth copying, as assisting the reader to form an idea of what a first rate Meteorologist considers to be the causes of these phænomena when they occur on the small scale.

First, he says (p. 47), after alluding to the whirls and eddies formed by conflicting currents at their meeting on the edges:—

"In like manner when the N. E. wind prevails below and the S. W. above, violent whirlings are formed at their limits, which descend to the surface of the earth, and are often endowed with prodigious force."

And at p. 116, he describes an instance of the whirlwinds, though not of their descending, as follows:—

"When a moist wind determines an ascending current along the sides of a mountain, it at last reaches atmospheric strata whose temperature is such that the vapour of water is instantly precipitated. This is especially the case when opposite winds meet on the summit. I have often witnessed these phænomena on the Alps—I will content myself with relating in detail the following fact. A very strong south wind was blowing on the summit of the Rigi, and the clouds that were passing at a great height above my head followed the same direction. The north wind was blowing at Zurich, and ascended along the southern flank of the mountain. When it attained the summit, light vapours were formed, which seemed desirous of passing over the ridge; but the south wind drove them back, and they ascended toward the north at an angle of 45°, and disappeared not far from the ridge. The conflict of the two contrary currents lasted several hours. A great many whirls were formed at the point where the two winds met; and travellers, who took little notice of the rest of the meteorological phænomena, were struck with this singular spectacle."

37. The following are the different views of Sir JOHN HERSCHEL on the causes of Cyclones, which I copy as abridged in PURDY's Memoir of the Atlantic Ocean.

"It seems worth inquiry, whether Hurricanes in tropical climates may not arise from portions of the upper currents prematurely diverted downwards before their relative velocity has been sufficiently reduced by friction on, and gradually mixing with the lower strata; and so dashing upon the earth with that tremendous velocity, which gives them their destructive character, and of which hardly any rational account has yet been given. Their course, generally speaking, is in opposition to the regular Trade-wind, as it ought to be, in conformity with this idea.—(Young's

---

* KAEMTZ' Meteorology, translated by Walker, London, 1845.

Lectures, I. 704) but it by no means follows that this must always be the case. In general, a rapid transfer either way, in latitude, of any mass of air which local or temporary causes might carry above the immediate reach of the friction of the earth's surface, would give a fearful exaggeration to its velocity. Wherever such a mass should strike the earth, a hurricane might arise; and should two such masses encounter in mid-air, a tornado of any degree of intensity on record might easily result from their combination."—*Astronomy*, p. 132.

Again: Sir JOHN HERSCHEL, in investigating the observations* made at the different Meteorological Observatories that have been established in various parts of the world, has arrived at the conclusion, that there are, at times, barometric waves, or undulations, in the atmosphere, of immense extent; he has denominated them barometric waves, from their being made evident by the fluctuations of the barometer, which, as was before described, exhibits perfectly the weight, and therefore the quantity of air above the station. One of these waves has been traced as extending from the Cape of Good Hope, through intermediate stations, to the Observatory at Toronto in Canada, under the superintendence of Lieutenant-Colonel SABINE. As an explanation of the origin of the rotary storms under consideration, Sir J. HERSCHEL has proposed the idea, that two or more of these extensive atmospheric undulations, or barometric waves, may, from traversing in different directions, intersect each other, and from their opposing forces cause the phænomena of hurricanes or rotary storms.

88. My own views are, and they will be found in some detail in Part V., that Cyclones are purely electric phænomena formed in the higher regions of the atmosphere, and descending in a flattened, disk-like shape to the surface of the ocean, where they progress more or less rapidly. I think that the whirling tornados, spouts, and dust-storms, are certainly connected with them, i.e. that they are the same meteor in a concentrated form, but we cannot at present say where the law which regulates the motions of the larger kinds, ceases to be an invariable one.

89. Other suggestions have been thrown out, and instances adduced by different writers as to the possibility of volcanoes, and even fires, originating violent circular motions of the atmosphere; and that volcanic eruptions are often accompanied by violent storms and heavy falls of rain there is no doubt. I have myself pointed out—though my published researches have been confined, like those of REDFIELD and REID to the effects, as the sure eventual index to guide us back-

* See Report of the British Association, 1843-4, Vol. XI.

PART I. § 39.] *Theories; Volcanic action.* 23

ward to the *causes* of storms—that in the China Sea and Bay of Bengal* there is much to countenance the idea that Cyclones in some parts of the world *may originate* at great volcanic centres, and I am inclined to believe also that their tracks are partly over the great internal chasms of our globe, by which perhaps the volcanic centres and bands communicate with each other. If we produce at both ends the line of the track of the great Cuba Cyclone of 1844, we shall find that it extends from the great and highly active volcano of Cosseguina on the Pacific shore of Central America, to Hecla in Iceland! And in 1821, the breaking out of the great volcano of Eyafjeld Yokul in Iceland, which had been quiet since 1612, was followed all over Europe by dreadful storms of wind, hail, and rain. In Iceland the Barometer fell from the day before the eruption till the twenty-sixth day after.† Mr. ESPY quotes several other cases, and the authority of Humboldt for South America, to shew that nothing is better established than the fact of the connexion of volcanoes with rains and storms. PURDY (Atlantic Memoir) also alludes to the supposed focus of sub-marine volcanic action on the Equator in that sea, as the spot to which the southern extremes of the West Indian hurricane (Cyclone) tracks would tend, if continued. And the following well authenticated fact has recently been published in the Newspapers. "On Saturday information was received at Lloyd's under date Liverpool, Feb. 4th, 1852, of an extraordinary marine convulsion experienced by the *Mary*, on her passage thence to Caldera. On the morning of the 13th October, the ship being twelve miles from the Equator, in Long. 19° W., a rumbling noise appeared to issue from the ocean, which gradually increased in sound till the uproar became deafening. The Sea rose in mountainous waves; the wind blowing from all quarters, the control over the ship was lost, and she pitched and rose frightfully, all on board expecting each moment to be their last. This continued fifteen minutes; the water then gradually subsided, when several vessels in sight at the commencement of the convulsion were found to have disappeared. Shortly afterwards a quantity of wreck, a part

---

* Sixth Memoir, Storms of the China Sea, Journal of Asiatic Society of Bengal, vol. XI. p. 717.

† Espy, p. 67-68; not correctly printed.

of a screw steamer, was passed, so that some vessels and lives were lost, and it will be observed that this phænomenon is stated to have occurred in October, one of the hurricane months of the West Indies." If I advert to these speculations, it is with the hope of drawing the attention of intelligent mariners to them.

PART II.

1. WHAT IS MEANT BY THE TRACKS OF CYCLONES.—2. AVERAGE TRACKS IN VARIOUS PARTS OF THE WORLD.—3. RATES OF TRAVELLING ON DIFFERENT TRACKS.—4. STATIONARY CYCLONES OR SUCH AS ARE NEARLY SO.—5. SIZES OF CYCLONES.—6. CONTEMPORANEOUS, PARALLEL, AND DIVIDING CYCLONES.

40. In explaining the terms used in our new science, I have already said (§ 23, *p.* 13) that the TRACK OF A STORM is the path or line in which it travels, like the track of a ship, and that of the centre is usually meant; and if I repeat this here, it is only again to caution the novice and seaman against confounding the rotatory motion, or whirling round of the whole Cyclone, with the *progressive* motion or moving forward; the line on which this last is done, whether straight or curved, being the track. If we suppose a Cyclone only to revolve and not to move forward in any way, such a storm would have no track at all. I shall allude at the close of this part to the probability of such Cyclones taking place, and to the certainty that some have such a very slow motion that we may almost call them *stationary* Cyclones. In the practical part (*Part III.*) I shall shew how the attentive mariner may often calculate pretty nearly, by the bearing of the centre, veering of the wind, and his run or drift, what the track of a Cyclone is, and take his precautions accordingly.

41. In speaking of the tracks of Cyclones for the use of the seaman, I mean of course to allude to their routes at sea. Much stress

has been laid, and some part of Mr. Espy's theory rests, upon what he has deduced to be the track of a Cyclone and the various directions of the wind as ascertained on shore. Now, without meaning any discourtesy, if we except some low countries and islands, and on low coasts, it seems to me, speaking as a sailor, sheer nonsense to discuss the question of how *the* wind blows, inland, in reference to any theory which must depend upon the direction of winds for short periods and in storms. For no man can have looked down from an eminence on even a moderately uneven country—to say nothing of a *very* hilly or mountainous one—without allowing, especially if he is a seaman, that almost every village must have a different stream of wind for its weathercocks; and some of these differing from four to eight points! We have only to look at maps on a large scale shewing the ranges of hills in any country, and especially some modern ones in which the elevations are shewn by what are called *Contour lines*,* to be satisfied of the futility of these data, except for the general purpose of shewing the *averages* of winds throughout the year. Mr. REDFIELD in his Memoir on the Cuba and other Cyclones, alludes also to this difficulty. (Am. Jour. Science, No. 1, p. 321. New Series.)

And LUKE HOWARD in a note to the Introduction to his "Climate of London," p. viii. speaking of Eddy winds says, "The site of Geneva remarkably exemplifies the effects of local position in this respect. Here, owing to the direction of the valley, the vanes point almost constantly S. W. or N. E. the cross winds going over above their level."

42. AVERAGE TRACKS OF CYCLONES. We are very far indeed from knowing what the average tracks of Cyclones in *all* parts of the world are, even in some frequented seas. In unfrequented ones we can at present only reason from analogy and from the evidence drawn from single logs. We trust always to the good sense of every rightminded seaman to furnish the labourers in the field with data to trace them out gradually. I proceed to detail what is known in various parts of the world, noting also briefly the tracks of ocean for which we have but little information. If I notice any inland Cyclones it is

* Lines drawn through all the points of the same height above or below any given level.

because some of them may become also ocean storms, and thus interest the mariner, for whom I am writing, as well as the Meteorologist.

43. Before entering on the details of the tracks of Cyclones in various parts of the world, as far as they are known, or for the purpose of indicating where they are not so, the seaman should be acquainted with Col. REID's, Mr. REDFIELD's, and Professor DOVE's general views of what *may* be the average system of Cyclone tracks in both hemispheres; modified of course in a hundred ways by local circumstances, but always *tending* towards the systems they suppose theoretically to exist. The difference between Theory and Law before adverted to should here be carefully kept in mind, as it is the difference between a probable supposition and a positive rule proved by abundant instances, and this is of importance, because, as we shall afterwards find, the questions of scudding or heaving to, and of profiting or not by a Cyclone, or of getting out of the way of it, depend wholly upon our knowledge of the track upon which it is travelling.

44. Colonel REID and Mr. REDFIELD, then, suppose that in those parts of the tropics which are nearest to the Equator, the Cyclones move nearly direct from the Eastward to the Westward; that as they progress they gradually take a more and more Northerly direction in the Northern Hemisphere, and a more Southerly one in the Southern; and then, say about Lat. 20° or 25° North or South, or near to the tropics, they curve more and more rapidly, till beyond them they recurve back to the Eastward again, forming great Parabola-like curves, of which the branches are more or less open according to circumstances. Colonel REID gives the diagram copied in the following page to illustrate his view of the *tendency* of the Cyclone tracks. True Cyclones have been traced to Newfoundland in Lat. 48° N. and found yet to be travelling to the E. N. E. and N. E. and more recently across the Northern Atlantic to Latitude 60° in his Chart of the Sept. Cyclone of 1853.

The most Southerly Cyclones we can yet pronounce to be true rotatory ones are those of the ship *Havering* in Lat. 48° 45' S.; Long. 32° to 50° East in October, 1849. And that of the ship *Marlborough* in Lat. 46° 45' S.; Long. 95° East in October, 1853. H. M. S. *Havannah*, Captain ERSKINE, and the ships *Barham* and

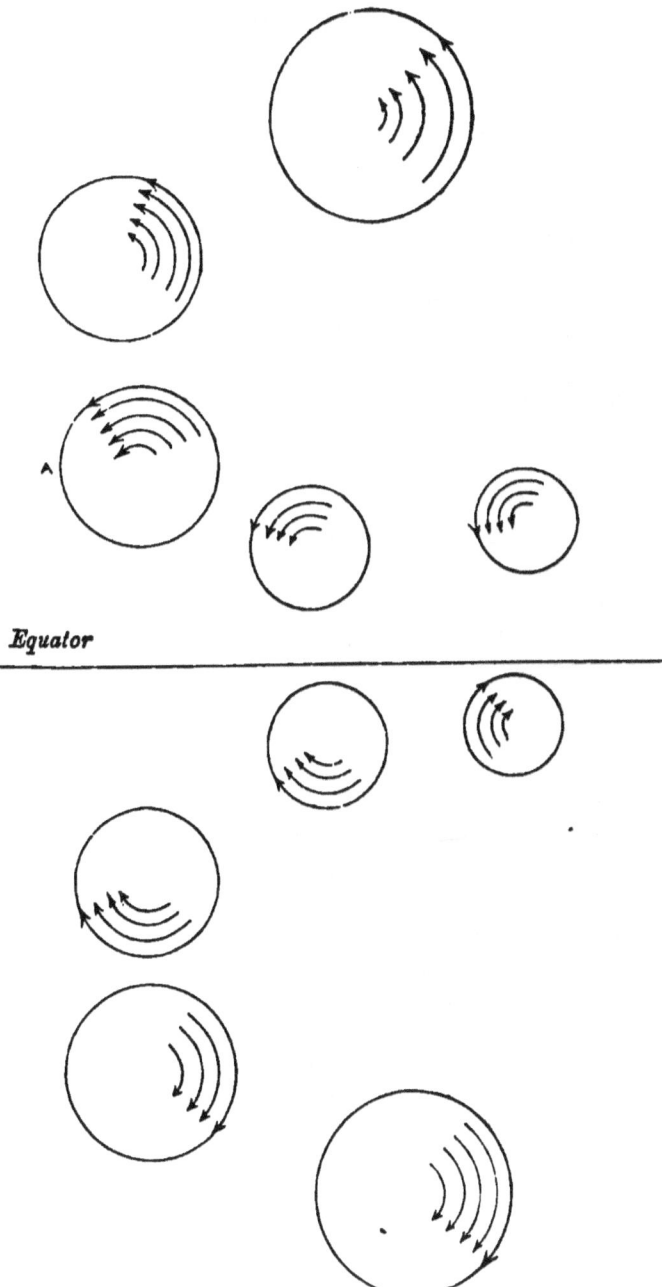

Equator

28   *Tracks—General Theory.*   [PART II. § 45.

Bucephalus had also true Cyclones in 42° and 43° South, between the meridians of the Cape and Amsterdam.*

45. THE WEST INDIES, CARIBBEAN SEA, GULF OF MEXICO, COAST OF NORTH AMERICA AND NORTH ATLANTIC TO THE COASTS OF EUROPE. In this tract we have, thanks to the labours of Mr. REDFIELD and Col. REID, Cyclone of 1853 with addition from Sir WILLIAM REID, Mr. MILNE and my own notes as far as the Bermudas, most careful and extensive researches to guide us.

There appear to be (See Chart I.†) two classes of tracks for both the West Indian and the North American Cyclones. We may conveniently distinguish them as straight-lined and curved tracks. Of the straight-lined tracks of the West Indies one group seems to arise from Lat. 10° to 23° North, and as far as we yet know to the Westward of 55° West (Tracks L and M and F on the Chart). The most Southern of them (F) travel to the W. N. W. past Tobago and Trinidad, along the Northern shores of South America, crossing the Peninsula of Yucatan, where the traces cease for the present. The next, arising farther North (L), pursue also a W. N. W. course along a line drawn from Barbuda and Antigua to the middle of the Gulf of Mexico, and the third, still farther North (M), travels directly to the West from the Atlantic to the shores of Mexico, between Tampico and Matamoros. Between these limits no doubt, other straight-lined tracks will be found to occur. The Tobago Cyclone of Oct. 1847 (O) is also laid down by Col. Reid as a straight-lined track travelling down from the E. N. E. to the Coast of Cumana. These visitations are stated to be very rare at Tobago and Trinidad. At Tobago indeed it is said that it had suffered from them but to a trifling extent since 1780, but this Cyclone was of frightful violence, though lasting only in its full fury for three hours.

* A writer in the Nautical Magazine for 1845, p. 704, concludes, from extracts from the logs of the U. S. Surveying expedition, that they had one or more Cyclones in January and February in Lat. 61° to 64° S.; Long. 137° to 150° East progressing to the South Eastward.

† This Chart is mostly from Mr. REDFIELD's to his Memoir of the Atlantic Cyclone of 1844, extended and augmented by Col. REID for his new work (1849) "On the Progress and Development of the Law of Storms." The *Roman* numerals refer to Mr. REDFIELD's researches and many new tracks are added by myself.

PART II. § 46.] *Tracts: West Indies and North America.* 29

Three more straight (or nearly straight) line tracks, (marked *I J* and *P*) will be seen on our Chart. These, taking their rise in the interior of the great continent of North America, appear to travel directly out to seaward from about W. S. W. to E. N. E., the two Northernmost of them crossing the Lakes and forming there the violent and destructive Lake, and Gulf of St. Lawrence storms of October and November.

It will be remarked that those within the Tropic, or the West Indian straight-lined tracks, *come in* from the Atlantic towards the Continent, and between these limits, Cyclones have been observed and tracked (as the track *E* and the branch of *F*) which first pursue a straight path, and then, about the meridian of 80° West, curve to the Northward, reaching the North Western Coast of the Gulf of Mexico between Matamoras and Mobile.

In a fourth kind of these straight-lined tracks, the Cyclone appears to begin in the Caribbean Sea, or to the Eastward and Westward of Yucatan (Tracks *G* and *H*), pursuing a straight or slightly curved track out *from* the Continent to the N. East and E. N. Eastward.

46. The curved tracks are numerous, and appear at times for a large part of their course to be straight, but end by curving more or less.*

In the former editions of this work it was stated that the curved tracks took their rise in about 10° to the Eastward of the Leeward Islands and sometimes as near the Equator as 10° N. but, thanks to the labours of the lamented REDFIELD, we have them now mapped out as arising sometimes at or near the Cape de Verd Islands (in 14° N.) crossing the Atlantic on a W. b. N. and W. N. Westerly course so as to skirt the outer West India Islands near the Bahamas, or to cross, or pass far within, the Windward Islands, or even over St. Domingo and across the Island of Cuba, to New Orleans. They then

---

* The seaman will not overlook the scale of our Chart ; nor forget that, except where the curves are very sudden, the track is for all practical considerations a straight one, and the object of these general views of the *tending* of the tracks is to enable him by the methods which I shall explain in the next part, to judge pretty correctly of the path of a Cyclone when he is coming into or overtaken by one, and hence to decide upon what may be the best to be done to avoid it altogether ; to keep clear of the dangerous centre ; or to profit by it.

curve gradually to the Northward and North Eastward, forming more or less of a paraboloid, and appearing to be mainly influenced by the Gulf Stream; and thus, following the direction of the coasts of Florida and North America, pass out into the Northern Atlantic by E. N. E., or even more Easterly, course between the Bermudas and Newfoundland; pursue their route to the Bay of Biscay, the Channel or the Northern Coasts of the British Islands, or curving up more Northerly, towards Iceland or Spitzbergen. Tracks Q to R comprise the limits within which these storms have usually occurred, and the seaman from amongst them will have no difficulty in judging of the mean track of those which pass along the Coast at sea, *excluding* those which cross the S. West parts of the United States, which become straight-lined tracks when they pass out to the ocean. Col. REID, p. 447, describes a track of much import to the mariner, being that of September 1839,* (A) which was especially a true Atlantic Cyclone of which the first ship's notice is in Lat. $20\frac{1}{4}$ N. Long. 47° West, and then successive ones, from others, till it passed close to the Bermudas and reached Nova Scotia and the Gulf of St. Lawrence, but was not felt in any of the Windward Islands.

Col. REID has also traced another track which forms a very remarkable anomaly. It is that of the Antigua Hurricane of 2d August, 1837, of which the centre after coming in from the Atlantic in about 17° North, and passing Antigua and the North extremes of Porto Rico like L and other usual tracks there, skirts the Bahama banks and islands on a N. W. course, as if it was about curving to the Northward and N. Eastward as usual, but between Lat. 30° and 31° N. and Long. 79° and 80° West, it suddenly takes a turn to the W. N. W. again, and travels in upon the American continent striking the shore about Doboy, and travelling for some distance inland. I have marked this double curve N on our Chart.

He also in his new work 'On the progress and development of the Law of Storms,' &c. p. 22, says:—

"Nor have gales always an easterly progression in high latitudes. I witnessed one

---

* He refers to a Chart of it, but I have not that Chart with my copy, nor have I found it. It is, however, inserted in the Chart to his new work, and will be found on ours, which is taken from it.

which passed over Bermuda on the 18th August, 1843, moving on a north-west progression and towards the Bay of Fundy. That gale would set in over the Bay of Fundy at the northward of East, and end at the southward of West. Sir James Ross has observed some gales between the Falkland Islands and Cape Horn, beginning south-westerly and ending about north-westerly also indicative of westerly progression."

This track I have marked *S* on the Chart.

47. From Latitude 45° N. and about Longitude 50° West—and the Cyclonic storms are usually here dilated to a very large size—they may possibly progress to the shores of Europe, forming extensive S. Easterly, Easterly and N. Easterly gales on their North sides, and North Westerly, Westerly, and South Westerly ones on their Southern parts; and by placing the transparent Horn Card for the Northern Hemisphere on a Chart of the Atlantic, the seaman will see how this occurs, and by moving it along, how ships which meet with such Cyclones in the Atlantic will have the winds. Other Cyclones, as before said, may pursue a route further to the North, and terminate on the Eastern shores of Greenland or between Iceland and Europe.

On the Chart No. 1, will be seen (from Col. Reid's new work) two large storm circles, the one approaching the coasts of Europe and the other leaving Newfoundland and the Coasts of America. The first of these represents the Cyclones of Nov. 1838, as traced by Mr. Milne, of which the two tracks are marked *V* and *W*, the latter being that of the 28th Nov. which was the severest, and which the circle represents.

The American circle is that of the Cyclone of 15th Dec., 1839, as traced by Mr. Redfield over the Banks of Newfoundland. This Cyclone reversed the trade wind over the Bahama Islands to S. W. and N. W. with a force of 5 on the Admiralty scale.

Mr. Martin, to whose work I refer again at p. 33, says, after investigating eleven tracks in the North Atlantic, "it will be seen that if a line be drawn from the Bermudas to Shetland they (the tracks) will all lie within two other lines drawn parallel to this at ten degrees on either side of it, except those which come down the valley of the St. Lawrence."

48. In Purdy's Atlantic Memoir mention is made of a Cyclone in the beginning of 1828, in which H. M. Sloops *Avon, Contest,* and

Sappho foundered between the Bermudas and Halifax, the *Tyne* in company weathering it. H. M. brig *Beaver* seems, from this account, perhaps to have accompanied the Cyclone or to have run into, or frequently overtaken it on its Southern Quadrants, having severe gales from the Westward, and so heavy a sea that she was near foundering, and obliged to heave her guns overboard. This weather "reached Plymouth Sound, on Sunday, January 13th, when, shortly after midnight, a violent hurricane came on from the S. W. with vivid lightning and thunder, and 13 vessels out of 21 were driven on shore. Plymouth did not suffer greatly, but the *General Palmer*, East Indiaman, met with threatening weather off Portland, when a sudden gust or squall carried away all her three masts." This would appear, as nearly as we can judge, to have really been an instance of a Cyclone travelling the whole distance across the Atlantic. We want, yet, a correct and undoubted tracing of such an instance, and it would be invaluable if we could obtain with it its effects on the currents of the Channel and the shores of the Bay of Biscay. We shall subsequently see what is said of the Cyclones of the Chops of the Channel.

49. At the Azores it would appear from a register supplied by Mr. HUNT, the British Consul,* that the tracks of Cyclones mainly follow that of the Gulf stream, (of which the usual Atlantic current felt there is a continuation,) or to the E. N. E. They are, it would appear, sometimes deflected hereabouts, but always to the Southward and not to the Northward as far as hitherto known. It is possible that these deflected storms may be at times the Madeira, and even the Canary Island storms, which, if not Cyclones, rise at least to the strength of hurricanes, or are not far from it.

50. The Madeira storms are at times very violent, and Col. Reid in his new work has traced one track (*T* on our Chart) of October 1842, which he conceives may "have come from the African continent and was re-curving while passing over the Canary Islands." He also farther supposes it to have reached the coasts of Spain, where several vessels were severely damaged off Cadiz, and H. M. S. *Warspite* was also within its influence. Capt. FITZROY, R. N. (Voyage of the Beagle, Vol. II. p. 54) speaks of the severity of the *gales* in the

* Nautical Magazine for 1842.

vicinity of the Cape de Verd Islands, and says that at Porto Praya a South West *gale* is usually experienced in September, in which—

"From five to ten hours before its commencement a dark bank of clouds is seen in the southern horizon which is a sure forerunner of the gale ; should a vessel be at anchor in the port at such a time she ought to weigh and put to sea."

He gives the case of an American which did so in September 1831 and came back safe, while a slaver which remained at anchor was wrecked, but we could not from this have inferred that these *gales* are Cyclones were we not enabled now to place upon the Chart, certainly, one September Cyclone in the neighbourhood of the Cape de Verds (Track *b*. Chart I.) being that experienced by the ship *Devonshire*, Captain CONSITT,[*] from London to Madras, 29th and 30th of September, 1848. The detailed log was unfortunately not sent, but from the extract there is every probability that the track was nearly as I have marked it, that of a Cyclone of no great extent passing out from the Coast of Africa to the Northward of the Cape de Verds on a W. b. N. or W. N. W. course, at about a degree or so distant, giving to the *Devonshire* as she first stood to the S. S. W. and then (mistakenly) hove to at about 120 miles to the Westward of St. Antonio, a severe gale from N. E. to South by compass, which would have been a Westerly and S. Westerly gale at Porto Praya if it reached so far.

In Mr. MARTIN's work, p. 109,[†] a very imperfect extract from the log of the ship *Sir Edward Parry* is given, describing a severe Cyclone off the Cape de Verds, commencing on the 3rd September 1850 with the Island of St. Antonio, bearing E. S. E. distant from 20 to 25 leagues, and of which Mr. MARTIN states farther details were collected by Captain WAINWRIGHT, R. N. and that it was travelling to the Westward in about Lat. 17° N.

---

[*] It is worth noting here how the very material for these labours must, so to say, be "chased and captured" bit by bit. I heard of the storm of 1848, from my friend, Dr. COLLINS of the *Queen*, East Indiaman, an earnest advocate of the science, but his efforts to obtain the log in England failed. I then watched for the arrival of the *Devonshire* at Madras, and through the zealous aid of Captain BIDEN, to whom the science in India is so much indebted, it reached India but just in time for the second edition. Yet still it is but the extract !

[†] A Memoir on the Equinoctial Storms of March, April, 1850, by F. P. B. Martin, Esq. M. A. 1852.

The most remarkable and distinct evidence, however, which we have of the Cyclones of September, in this tract, will be found in Mr. REDFIELD's splendid Storm Memoir on the Atlantic Cyclone of September, 1853.

51. Those strong but sudden storms often met with between the Meridians of 20° West, and the Coasts of Europe may possibly be sometimes the breaking up of the Greater Cyclones, for, as I shall subsequently shew, there is no doubt that Cyclones do sometimes divide, especially towards their termination. At times, also, as before said, the great Atlantic Cyclones reach the coasts of Europe from the Burlings and Cape Finisterre to the Channel, travelling to the Eastward,* and the changes and directions of the wind are such as prove their rotatory character. Thus a great Cyclone of 500 or 1000 miles in diameter coming in from the Northward of the Azores and travelling E. N. E. to the Channel, might be, at the same time, an Easterly gale at Cape Clear, a Westerly one at Cape Finisterre, South Westerly and Southerly in the Bay of Biscay, and S. Easterly from Ushant across to Cape Clear, and in the chops of the Channel. If the seaman will place the Horn Card thereabouts, with its centre on what he may judge the average track, say one to the E. b. N. or E. N. E. upon a general chart of a moderate scale, he will quickly and clearly perceive how this takes place; and perhaps have some suspicion of how a "Bay of Biscay sea" may get up.

Amongst some valuable notes and criticisms, for which I am indebted to Capt. J. C. Ross of the Keeling and Cocos Islands settlement, is the following remark on this paragraph.

"A peculiar circumstance which occurred at the time makes me recollect, that while Westerly gales were causing the destruction of the Trafalgar prizes off Cadiz, fresh Easterly and North Easterly winds were blowing over the Zetland Isles."

52. THE EASTERN COAST OF SOUTH AMERICA, FROM TRINIDAD TO CAPE HORN.—I have met with little or no information as to whether any true Cyclones occur on this Coast, which is but little frequented to the South of the Rio de la Plata; but to the North of it so much so that if any tempests were frequent we should certainly

* Or say between E. N. E. or N. E., and E. S. E. and S. East; for we do not know yet their tracks nor how the approach to the land affects them.

hear of them. Mr. LUCCOCK in his notes speaks of a *gale* which began in 34° S. and 200 miles from the land, at N. W. and lasted ten days, drifting the vessel to 36¼° South. Capt. FITZROY (Voyage of the *Beagle*, Vol. II. p. 83) describes the gales off Santa Martha, Lat. 28¼° South, as follows :—

"Gales in the Latitude of Santa Martha generally commence with Northwesterly winds, thick cloudy weather, rain and lightning. When at their height the barometer begins to rise (having previously fallen considerably,) soon after which the wind flies round by the west to south-west, and from that quarter usually blows hard for several hours. But these which are the ordinary gales blow off from, or along the land, and do not often raise such a sea as is sometimes found off this coast during a South-East storm."

This description leaves little doubt that these *gales* are Cyclones, of which the tracks of the centre pass usually to the Southward of Santa Martha and are from the West to the Eastward.

The Barque *Trident*, Captain LYALL, from Valparaiso to Liverpool, in August, 1850, experienced between 30° S. and 33° 45′ W. and 22° 30′ S. and 29° 15′ W., on her run from the neighbourhood of the Falkland Islands to pass to the Eastward of the Island of Trinidad a strong gale with violent squalls of rain and hail from S. S. E. to S. West, with a tremendous sea from the Southward ; the Barometer, however, falling only in two days from 30.00 to 29.70 when lowest. If this was a Cyclone, the ship was on its Western verge, and as she was running to the N. N. E. by compass throughout, it was also travelling up in a somewhat parallel track, the ship overrunning it till within a day's sail of Trinidad, where the weather became fine.

The ship *England*, Captain GARNETT, from Liverpool to Calcutta, in the month of September, 1850, between Lats. 28° 38 and 30° 00′ S. and between Longs. 23° 52′ and 23° 20′ West, experienced a severe gale beginning at N. N. E. and veering to N. East and East, with a calm interval of some hours, and a shift, though not so violent, to South and S. East. During the calm a "fearful sea" is described in which the rudder stock was carried away and the wheel disabled. The Barometer fell from 29.76 in two days to 29.55, and in the calm to 29.33. From the winds and ship's run, it would appear that this was a Cyclone travelling down and overtaking the ship from the N. West. The *England* had to put into Rio to repair her damages.

53. THE SOUTHERN ATLANTIC FROM CAPE HORN TO THE CAPE OF GOOD HOPE.—Within the limits of the S. E. trade wind in the Atlantic there seem to be but very rarely any storms, and whether these are rotatory or not we have no record for judging. Without the limits of the trade, that is from 28° to 50° South, it is not improbable that Cyclones may exist, and that the storms in the neighbourhood of the Falkland Islands, and from thence to Tristan d'Acunha and the Cape, may also be at times of that class; though they no doubt are, often, the Westerly gales of the high latitudes. The S. W. gales, veering to the N. W. which are experienced to the Eastward of Cape Horn;* and at the Falkland Islands, the Northerly winds veering to the South by the West,† may be parts of Cyclones or of the circuits of wind supposed by Mr. REDFIELD; but we have here, as in so many quarters, every thing to learn. Captain SULLIVAN, R. N. in a paper on the Falkland Islands (Naut. Mag. for 1841), describes the gales at Berkeley Sound as commencing at N.E. veering to N.W. and ending at S. W. This would indicate a tendency to rotation, according to the law for the Southern hemisphere, and that the track was to the S. S. E. and S. E., as if these islands and the adjacent coast of Patagonia were situated at one of the curving localities. See subsequent remarks on storms in high latitudes.

Colonel REID in his new work states that he received sketches and registers of several of these gales, the result of which is that the tracks come from all quarters between the N. W. and S. W. which, as Colonel REID remarks, and we shall subsequently shew, accords with the observations made in the British Islands in Lat. 51° N., for it is found that the gales there come in also from all quarters between N. W. and S. W. Captain MOODY, R. E. who was Governor of the Falkland Islands, also confirms Captain SULLIVAN's account of the winds in all respects.

The Ship *Janet Wallis*, Captain BAXTER, experienced a small but tolerably severe Cyclone to the Westward of Lat. 37° 45′ S. Long. 13° 33 East; (the only position given in the extract from the Log) of which the centre passed close to her, and was twice remarkably

---

* American Exploring Expedition, p. 160, Vol. II.

† Bougainville, as quoted by Romme, Vol. II. p. 12.

visible as "a dense mass of clouds violently agitated." Her Barometer fell 0.7 in an interval of 13 hours.

The Clipper Ship *Lord of the Isles*, on the 22nd of Nov. 1855, in Lat. 28° 6' S. and Long. 20° 30' W. experienced a heavy gale commencing at N. E. veering to S. E. and finishing off at S. S. E.—washed away the quarter boat, blew the fore top mast stay sail and the close reefed fore top sail out of the bolt ropes—moderated on the 23rd Nov.

The Clipper Ship *Light of the Age* also experienced in April, 1856, in about Lat. 40½° S., Long 29½° West, a severe Cyclone; which, though every precaution had been taken, and the vessel was lying under snug sail awaiting the change, with "a low dark ring all round the horizon with occasional flashes of lightning to the N.W." came upon her from the S. S. West, "somewhat resembling a railway screech," and lasted for about seven hours with terrific violence, raising a frightful sea; the ship's run, for she was scudding, brought the wind to S. W. and W. S. W. From the account published in the Illustrated London News of July 12, 1856, this seems to have been a Cyclone of small diameter and nearly stationary; perhaps even one, as was the general impression on board, "descending from aloft." Great credit appears to be due to her Commander, who, warned by the appearances and the fall of his Barometers and Sympiesometer, began to take in sail early in the day, and thus at least saved his masts, if not his ship.

BETWEEN THE MERIDIANS OF TRINIDAD AND TRISTAN D'ACUNHA AND THE CAPE OF GOOD HOPE there appear to occur at times Cyclonal gales, but not of excessive violence or of long duration, though sometimes sudden and severe, and evidently travelling to the Eastward or S. Eastward, but I have no complete logs, and from what are given as extracts the tracks can only be spoken of in general terms sufficiently to put the mariner on his guard.

54. OFF THE CAPE OF GOOD HOPE, AND EASTWARD TO THE MERIDIAN OF NATAL, there is no doubt that Cyclones certainly occur. They are sometimes, as to extent and duration, though of excessive violence, mere whirlwinds or tornados, but at others of much larger extent and perhaps of slower motion. One of the small

ones, a mere tornado as to extent, but a perfect tyfoon as to violence, and which appears to have been travelling to the S. Eastward, overtook and dismasted the French Ship *Paquebot des Mers du Sud* in August, 1841, while running in a heavy gale from the Northward, which might have been the Eastern side of the storm circle; for the whole Cyclone, as well as the tornado part of it, certainly veered according to the law for the Southern Hemisphere; and the course of the Cyclone by a ship which came up with the *Paquebot* the next day was clearly to the S. East.* That is to say, it followed the direction of the prevailing winds and of the theoretical track, as given at page 26, in those latitudes.

M. BOUSQUET, in his additions to the translation of the first edition of this work, has traced four tracks (A B C and D on our Chart No. II.) of Cyclones between these meridians, the Southernmost of which is in 39° S., and I have added a fifth, E, in 42° S. of which the ships *Barham* and *Bucephalus* experienced the Northern quadrants on their outward bound voyage to India. In 1849, a most severe and destructive Cyclone appears also to have swept the Southern Coast of Africa.

So far the first and second editions of this work; and my surmises as to the probable tracks hereabout will be now found much confirmed by the following valuable details.

Captain FISHBOURNE, of H.M. Steamer *Hermes*, who was for some time on the Cape Station, in a paper read before the Meteorological Society of Mauritius, of which he obligingly sent me in January, 1852, a copy from which I abridge what follows, submits theoretically, and supports, by several special instances and his own extensive experience, that—

1. All the winds of any strength off the Cape and along the Coast, nearly to the limits of the trade winds, are sections of Cyclones; and that there are few or no straight-lined winds.

2. That the course of these Cyclones is first about W. S. W. from where they take their rise (being perhaps branches of the great Indian Cyclones crossing the Mosambique Channel?) and that they continue their course along the Northern Frontier of the Cape Colony and

* See 5th Memoir on the Law of Storms, Jour. As. Soc. Beng. Vol. XI.

then curving round, mostly to the Westward of Table Bay, but sometimes short of it, take a Southerly course about the Cape, and return back along the coast on a course about East or E. S. E. by compass.

3. That these winter tracks differ from those of summer by the semidiameter of a Cyclone in those seas, the winter track being to the Southward, so that the coast line is then mostly in the left hand semicircle, so that a gale will commence at North, is strong at N. W., veers to S. W., and terminates there or at South; but in summer, the coast line being in the right hand semicircle, the gales commence gently at East or E. N. E., becoming strong at N. E., and terminate as the centre passes, at S. E. or South; these directions in both cases varying a little with the paths of the Cyclones. Other anomalies also occurring, which, however, are explained by supposing the Cyclone track sometimes to recurve to the Eastward of the Cape; and this recurring also explaining the apparent anomalies in the indications of the Barometer at Cape Town and Simon's Bay.

These views are supported by various instances from cruises in H.M. Steamer *Hermes*, extending from the Cape along the coast of Southern and S. Eastern Africa, and to the Mauritius; and founded upon this evidence, Captain FISHBOURNE gives clear and careful instructions for ships off the Cape.

55. BETWEEN THE NATAL COAST AND THE MERIDIAN OF THE SOUTHERN EXTREME OF MADAGASCAR.—From various documents in my possession, I am inclined to think that in this tract the storms will at times be found to travel from the E. N. E. to the W. S. Westward, but they may curve much more to the South. The seaman in crossing it should be watchful and observant in time, so as to be able to estimate from his run and the veering of the wind, as will be subsequently shewn, the probable track of the Cyclone, and avoid its dangerous centre. There is no doubt that, with the Madagascar channel open, the tempests are at some seasons very sudden and excessively severe, but apparently either of small extent, or travelling very rapidly, for they rarely last long.

It will be seen on the Chart that there is a blank space between the meridians of 40° and 55° East, and between the parallels of

30° and 40° South; in which we have as yet no data for the tracks, and where they may be highly uncertain, having on the East the curving tracks of the Mauritius Cyclones, and to the West the direct ones traced from the Cape to Cape Recif. I regret that the notices which I have, though clearly relating to Cyclones, are for the most part too vague to enable me to lay down tracks from them.

56. THE SOUTHERN OCEAN FROM THE CAPE TO VAN DIEMEN'S LAND.—I have for this tract also several accounts, from which I entertain no doubt that Cyclones often prevail, and that they travel from the West to the Eastward. It is possible that these may also be at times small Cyclones thrown off from the larger ones; for the dividing of storms, or the generating of new ones by the progress, or at the commencement of the main one, and even the occurrence of smaller tornado-Cyclones within the area of a great Cyclone, is, I think—and I am supported in this view by Mr. REDFIELD and Mr. THOM—indisputable, (for the Northern Hemisphere at least), and we look naturally for the same in the Southern one; and this theory will explain the sudden violence and short duration of some of these Cyclones, and why the regular Westerly gale seems to return again after them.*

I have recently received from Capt. ERSKINE, of H.M.S. *Havannah*, a valuable and a practical corroboration of this opinion that the Cyclones in these latitudes travel from the West to the Eastward, which I take the liberty of giving in nearly that Officer's own words, conveying as they do so much which is suggestive of what a little scientific consideration beforehand may enable the careful seaman to accomplish, even in what are supposed to be well known, because much frequented, tracts of ocean—Capt. Erskine says:—

"In corroboration of your opinion, that the Cyclones in the Southern Indian Ocean, between the Cape and Van Diemen's Land, travel from West to East, I beg to send you a copy of the *Havannah's* log, and the track between the Cape and Sydney, which I think will shew that the winds we experienced were a succession of Cyclones, and that by paying attention to the state of the Barometer and Sympiesometer, and keeping in the left hand semicircle or that of westerly winds, I was enabled to make the passage from Simon's Bay to Port Jackson in comparatively moderate weather, in 34 days, including 3 or 4 days of light winds. I was first led to expect that the

* See also Part III., in which these views are more practically considered.

course of the storms was such as I found it from the recommendation given by Flinders, Vol. I. Chap. 3, p. 45, not to run down the Easting too far to the Southward. "Having made this passage three times before," he says, " I am satisfied of the impropriety of running in a high southern latitude, particularly when the sun is in the other hemisphere, and there is nothing in view but to make a good passage; not only from the winds there being often stronger than desired, but because *they will not blow so steadi'y from the* Westward. In the latitude of 42°, I have experienced *heavy gales from the North* and from the *South, and even* from the *Eastward*, in the month of June and July," &c. He continues, " It may not be improper to anticipate upon the subject so far as to state what was the result of keeping in the parallel of 37° in the month of November. From the Cape of Good Hope to the island of Amsterdam, the winds were never so strong as to reduce the *Investigator* to close-reefed topsails, and on the other hand the calms amounted to no more than seven hours in nineteen days. The average run on the log on direct courses, for we had no foul winds, was 140 miles a day; the *Investigator* was not a frigate, but a collier-built ship and deeply laden."

"In the following twelve days' run from Amsterdam to the S. W. Cape of New Holland, the same luck attended us, and 158 miles a day was the average distance without leeway or calm."

The Australian Directory (I do not know on whose authority) says, Vol. 1, page 1, "Ships from the Cape of G. H., bound to Port Jackson, should run down their longitude on the parallel of 39°, where the wind blows almost constantly from *some Western* quarter, and generally not with so much strength as to prevent sail being carried to it. In a high latitude the weather is frequently more boisterous and stormy, and *sudden changes* of winds with wet squally weather are almost constantly to be expected."

These two authorities particularly led me to believe, that by keeping too far to the Southward, a ship would either get into the right hand semicircle or that of Easterly winds, or that the centre of these revolving storms would be constantly passing near her, whereas if she managed to *hit off* the proper place in the other semicircle, she might perhaps avail herself of the gale for some days together, and I determined therefore not to go to the Southward of 39°. It will be seen by inspecting the log that (after two shorter Cyclones, one from the 9th to the 11th, and another from the night of the 13th to the 15th July), the ship ran in one (which made known its approach by a gradual fall of the barometer), from the 17th to the 21st ; the glasses rising and falling occasionally, as she outstripped or fell short of the velocity of the storm. It will be noticed also that on heaving to for an hour and a half, on the night of the 17th, to allow the centre to pass ahead, the barometer rose immediately, and continued steady for some time afterwards, as we bore up and kept way with the gale. It finally got ahead on the 21st, having apparently run 1185 miles in 5 days, or at the rate of nearly 10 knots an hour.

Although on my arrival here, I could get no precise accounts of the weather which followed this gale along the South coast of Australia, further than it had blown very hard about the last week in July, yet I have no doubt that this Cyclone swept the whole of that coast, as on closing it on the 28th, we found, instead of the usual Easterly current of a mile or more per hour, a set of 14 miles N. 73° W.; on the 29th, one of 27 miles West; and on the 30th, 13 miles N. 41° W., the reflux evidently of the storm wave, whilst on the 4th of August the current had resumed nearly its ordinary course of N. 45° E. 20 miles."

I have placed on Chart No. II. a track (marked G) representing the Cyclone of the 17th to 21st July, at an average distance of from 3° to 4½° to the South of the *Havannah's* position on those days, as also two others (marked E and F) from the logs of the two Indian passenger ships, the *Barham*, Captain GIMBLETT, and *Bucephalus*, Captain BELL, at a day's sail from each other, in September, 1849, which apparently trended somewhat to the Northward of East.

57. SOUTHERN INDIAN OCEAN, FROM LONG. 50° EAST TO THE COASTS OF AUSTRALIA, AND FROM THE EQUATOR TO LAT. 30° SOUTH.

This space, which we can only designate vaguely, and part of the Cyclones of which (those of the Coasts of Australia) will be referred to under another head, will be found comprised in Chart II.

In considering this Chart the seaman should observe, as before said, that its object is not so much to lay down the exact track of any particular Cyclone, as to shew him the *tendency* of the paths of the usual Cyclones, with, at times, those which are irregular, and thus to aid him to form his judgment of that one in which he may be involved, or near to which he may suppose himself to be. Though the greater part of the tracks have been laid down from the logs of single vessels, yet he may be satisfied that they are quite correct enough to serve this purpose; and where the logs have been so vague and ill kept as to prevent any correct judgment being formed, or where the Cyclone itself has appeared to have but a very little progressive motion, small wind-circles have been introduced, to shew that it was either uncertain as to its track, though certainly revolving, or perhaps a *stationary* Cyclone,* which last class will in some parts of the world be found much more common than is yet supposed; and my object in some cases where these wind-circles are marked, has been to put the mariner on his guard against them, and against the small tornado-hurricanes;

* Subsequently described, (p. 91.)

for such some of them may be called from their apparently limited extent. Whether these, afterwards, in any case progress and dilate into true Cyclones, or whether they exhaust themselves and break up, or *rise* up, not far from the spot where they are raging (and they seem to do all three at times), we have yet to learn. What is certain is, that with very brief warning they are violent enough to dismast a vessel; and from dismasting to worse mischief the interval is not great; and such mischief it is our object to avert.

To avoid crowding I have by no means placed on the Charts all the tracks which I could have delineated from the numerous logs sent me. These may be divided into three classes.

1. Those which belong to some of the average tracks.
2. Those which give average tracks, but in new tracts of the ocean where we had none before laid down.
3. Those which are altogether new and anomalous, whether occurring in known tracts of ocean or in parts where we had no previous information. Of the anomalous tracks occurring in known dangerous parts, the Cyclone of H. M. S. *Jumna* (Nineteenth Memoir, Jour. As. Soc. Beng. Vol. XIX.), of which a diagram is given on Chart II. will be found to be one of the most remarkable—and of the average tracks, or Cyclones in parts where only a single log was before available, the coincidences, even to dates, will be found very remarkable.

In some parts it will be seen that the anomalous tracks cross the average one, almost at right angles! and all thése I have been careful to enter, as putting the seaman on his guard against the too common impression which prevails, and through which many I know have lost their masts, that there is any *positive law* for the tracks, though there is certainly a general rule.

The following is the Table of References to the Chart, which are too numerous to be placed upon it, but are also annexed to it on a separate page.

REFERENCES TO CHART No. II.—THE SOUTHERN INDIAN OCEAN.

		Authority.
I.	Track of the Rodriguez Cyclone,	Mr. Thom.
II.	Average Northern limit of Cyclones,	Mr. Thom.
III.	——— Southern limit,	Mr. Thom.

IV.	Track curving to the Southward about Mauritius; Blenheim, 1807, ...	Mr. Thom and Col. Reid.
V.	About Mauritius and Bourbon, . . .	Mr. Thom.
VI.	Curve of the Culloden's Cyclone, March, 1809,	Col. Reid.
VII.	H. M. S. Serpent, Feb. 1846, ...	H. Piddington.
VIII.	Futtle Rozack and other ships, November 1843, ...	H. P. in 11th Memoir, Jour. As. Soc.
IX.	Charles Heddle's Cyclone, March, 1845, .	H.P. 13th Memoir, J.A.S.
X.	Fr. Frigate La Belle Poule, and Corvette Le Berceau, Dec. 1846, (See also O) .	H. P.
XI.	H. M. S. Albion, Nov. 1808, . . .	Mr. Thom.
XII.	H. M. S. Bridgewater, March, 1830, . .	Col. Reid and Mr. Thom.
XIII.	H. C. S. Abercrombie, Jan. 1812, . . .	Mr. Thom.
XIV.	Magnasha, Feb. 1843, ...	Mr. Thom.
XV.	H. C. S. Ceres,—1839, ...	Mr. Thom.
XVI	Timor and Rottee Cyclone, April, 1843, . .	Mr. Thom.
XVII.	Malabar, Jan. 1840, ...	Mr. Thom.
XVIII.	Boyne, Jan. 1835, ...	Col. Reid.
a.	Mauritius, March, 1811, ...	Col. Reid.
b.	————, Feb. 1818, ...	Mr. Thom.
c.	————, Jan. 1819, ...	Mr. Thom.
d.	H. C. S. Dunira, Jan. 1825, . . .	Mr. Piddington.*
e.	H. C. S. Princess Charlotte of Wales, Feb. 1826,	H. P.
f.	H. C. S. Orwell, Feb. 1827, ...	H. P.
g.	Thalia, April, 1827, ...	H. P.
h.	H. M. S. Boadicea, April, 1827, . .	H. P.
i.	H. C. S. Macqueen, Jan. 1827, . .	H. P.
j.	————Buckinghamshire, Feb. 1828, . .	H. P.
k.	————Princess Charlotte of Wales, March, 1823, ...	H. P.
l.	Reliance, Jan. 1831, ...	H. P.
m.	American Ship Panama, Jan. 1832, . .	Mr. Redfield.
n.	Mauritius and Duke of Buccleugh, Jan. 1834,	Col. Reid and Thom.
o.	Neptune, Feb. 1835, ...	Col. Reid.
p.	Thomas Grenville, Jan. 1836. . . .	H. P.
q.	Northumberland, March, 1839, . . .	H. P.
r.	American Whaler, Feb. 1839, . . .	H. P.
s.	Exmouth, Moscow, &c. May, 1840, . . .	Thom.
t.	Kandiana, Sept. 1840, ...	H. P.

* This and several more form a large series of Logs furnished to me by the attention of the Hon'ble Court of Directors of the East India Company.

PART II. § 57.]   *References to Chart No.* II.   45

u.	Mauritius, April, 1840, ...	Thom.
v.	William Nicol, May, 1842, . . .	H. P.
w.	Barque Tropic, March, 1842, . . .	H. P.
x.	———— Elizabeth, April, 1844, and Barque William Gales, Jan. 1848, . . .	} H. P.
y.	Unicorn, October, 1844, ...	H. P.
z.	Dorre Island, Captain Grey and American Ship Russell, Feb. 1844., ...	} Thom, and H. P.
aa.	Swan River, H. M. S. Beagle, May, 1844, .	)
bb.	Houtman's Abrolhos, May, 1844, . . .	} Naut. Magazine.
cc.	H. M. S. Beagle (position uncertain) Nov. 1840,	)
dd.	Candahar, Nov. 1842, ...	H. P.
ee.	Tigris, April, 1840, ...	H. P.
ff.	Earl of Hardwicke, Feb. 1839, and Fleur de Lys, March, 1856, ...	} H. P.
gg.	Windsor, Feb. 1857, ...	H. P.
hh.	French Ship Archibald, March, 1846, . .	H. P.
ii.	Orient, Duncan, and Fr. Ship Grand Duquesne, March, 1846, ...	H. P.
jj.	Maria Somes and American Ship Loo-choo, March, 1846, ...	H. P.
kk.	Manchester, Jan. 1847, ...	H. P.
ll.	H. C. S. London, Feb. 1827, ...	H. P.
mm.	American Ship Howqua, Jan. 1848, . .	H. P.
nn.	Windsor Castle, March, 1817. and Barque West, Nov. 1851, ...	H. P.
oo.	American Ship Pathfinder, Dec. 1847, . .	H. P
pp.	Sophia Fraser, Nov. 1847, ...	H. P.
qq.	John McVicar, Jan. 1849, ...	H. P.
rr.	Dutch Ship Roompot, Dec. 1847, . .	H. P.
ss.	American Ship Hannah Sprague, April, 1845,	H. P.
tt.	H. M. B. Jumna, April, 1848, . . .	H. P.
A.	4th to 6th April, 1848, ...	M. Bousquet.
B.	1st April, 1848, ...	B.
C.	Thomas Blyth, Feb. and March, 1848, . .	B.
D.	Mercury, Feb. and March, 1848, . .	B.
E.	Barham and Bucephalus, Sept. 1849, . .	H. P.
F.	Ditto ditto Sept. 1849, . .	H. P.
G.	H. M. S. Havannah, July, 1848, . .	H. P.
H.	Jan. 1848, ...	B.
J.	June, 1848, ...	B.
K.	March, 1848, ...	B.

L.	Jan. 1848, ...	B.
M.	Cocos Islands, Capt. J. C. Ross, . . .	H. P.
N.	November, Dec. 1847, ...	B.
O.	Berceau and Belle Poule, Dec. 1846,* . . .	B.

From the foregoing, the researches of Colonel REID, Mr. THOM, M. BOUSQUET, and my own, with much material yet unpublished, besides numerous corroborations and coincidences, even as to the month in which the Cyclone has again occurred in the same tract of ocean, and from the anomalies or varying tracks being at times most remarkable and instructive, as shewing how fatal may be the notion that there is any positive law for the tracks,—which I repeat are only laid down as *averages* and *warnings*, and not as rules—we can now speak with great confidence of the usual tracks of Cyclones over this great extent of Ocean, as being generally from the E. N. E. to the W. S. W., but from Long. 80° East, westward, and particularly when approaching the Mauritius group† having a tendency to curve to the Southward, and frequently, according to Colonel REID's results, *back* to the E. S. E.‡ when they reach the Southern Tropic.

Whilst the first edition of this work was printing I found in a newspaper, the *Mauritien*, copied from the *Hebdomadaire* of Bourbon, an account of a Cyclone between the Island of Bourbon and the Island of Sainte Marie, on the East Coast of Madagascar, which gives us another track for that locality. The French Frigate *La Belle Poule* was overtaken by a Cyclone, in the night between the 15th and 16th December, 1846, or rather she seems, as far as one can judge from a newspaper notice, to have partly run into it, being bound *from* Bour-

---

* According to M. Bousquet.   † Rodriguez, Mauritius and Bourbon.

‡ Mr. THOM differs from Colonel REID in this view. From a close examination of the documents, I agree entirely with Colonel REID; and while preparing the first edition of this work for the press, I received a report from Commander NEVILLE, of H. M. S. *Serpent*, detailing a storm in 27° South, 64° E., from which he escaped by heaving to in due time, of which the track (No. VII.) was certainly from about N. 16° West, to S. 16° East, and it is within 8° of the spot where the curving of the *Culloden*'s storm is marked on No. VIII. of Colonel REID's Charts. The *Serpent's* was one of the small tornado-like storms, and moving with great rapidity. Other tracks shewing both the *tendency* and the actual curving will be seen on the Chart, and no doubt they sometimes become the Easterly tracks of the higher latitudes.

bon to St. Marie, where her fair course was about W. N. W., the wind gradually increasing from the S. E. with a falling barometer, till it became of hurricane violence, blowing everything to pieces, then ceasing with a calm of an hour or two, and recommencing again with terrible fury from the N. West, at which point it seems to have abated after completely disabling the Frigate. The Corvette *Le Berceau*, in company with her, foundered. This Cyclone occurred about 50 leagues from the Island of Cape St. Marie, and thus nearly midway between that port and Bourbon, and the shift described would give a track from about N. E. to S. W. for this locality. It will also be seen by reference to the chart, Tracks IX. and X. that the storm of the *Charles Heddle* in the same locality, which will be subsequently referred to, had about the same tendency to the S. W. The one, Track IX. is laid down as traced in my 13th Memoir, the other is made a little more Easterly and Westerly to allow room for the figures, and the seaman will understand that both these, as before explained, shew that the *average* tracks, as far as yet we know, are as here delineated. M. BOUSQUET, p. 26, cites several logs and reports which lead him to conclude, I think correctly, that this Cyclone travelled in the direction of the curve O on our chart. (See below, p. 108).

At the Keeling (or Cocos) group in 12° 5′ S., 96° 53′ East, it is said by Captain FITZROY, R.N. (Voyage of the *Beagle*, Vol. II. p. 637, Note) that in November and April, 1835, severe gales, amounting almost to hurricanes, were felt blowing from S. E. and South, and veering and shifting to the Westward, which would indicate tracks from the Northward to the South. One of the gales lasted upwards of 24 hours, the other only about two. I have marked a track M to indicate these Cyclones. In my Sixteenth Memoir, Journal As. Soc. of Beng. Vol. XVII. I have shewn that in March, 1846, there were *three separate Cyclones raging at the same time!* in the space between 14° and 19° S., and 75° and 78° East, of which two were certainly severe, and one of terrific violence, dismasting three stout merchantmen, and nearly destroying one of them. These Cyclones were moving on an average track from the N. E. to the S. W. ¼ S. and their mean rate of travelling, which varied from 1.8 to 5.5 miles, was about 3.9 miles per hour. They are the tracks ii and jj on the Chart.

The tracks of the Cyclones or Cyclonal gales at these Islands, and near them, appear to be very variable. In Captain Ross' notes already alluded to (p. 34), memoranda are given of a gale in November, 1834, in which the wind veered from S. b. E. to East, indicating if the gale was part of a Cyclone, a track from the S. E. ¼ E. to the N. W. ¼ W. and passing to the N. East of the Islands; and again, in April, 1835, of a decided Cyclone, of sufficient violence to break cocoa-nut trees, which veered from East to S. S. E., South and S. S. West,* indicating a track from the N. W. b. N. to the S. E. by S. and also passing to the N. East of the Islands. The Western quadrants of the H. C. Ship *Macqueen's* Cyclone in January, 1827, appear also from Captain Ross' journal to have reached these islands, so that we have in their vicinity tracks in two opposite directions as marked. These gales, or Cyclones, however, Captain Ross says, are not by any means frequent, and with proper ground tackle, vessels are quite secure within the lagoon anchorages.

58. Between the Latitudes of 5° and 25° South, and the Longitudes of 75° and 105° East, is a space where, from some as yet unknown cause, frequent Cyclones appear to rise, and to progress from thence, at first slowly and then more rapidly; their little progression, or uncertain tracks, entitling them almost to be called *stationary* Cyclones. Colonel REID thinks the *Albion's* storm in from 5° to 15° South, and 80° to 90° East, one which may have been, in its first commencement at least, "floating with an irregular motion as waterspouts do in calm weather." The *Futtle Rozack's* Cyclone, track VIII. in this tract, in November, 1843, investigated by myself† from the logs of numerous vessels, moved up first from the S. E. to the N. W. for two days very slowly, and then curved abruptly away to the S. W., its mean rate not exceeding 2¾ miles per hour, and for the first 2 days only 2 miles.

Mr. THOM inclines to think that many more of the Mauritius Cyclones take their rise hereabouts, and travel the whole distance. We

---

* This note from Captain FITZROY's voyage of the *Beagle* is said by Captain Ross to be erroneous, but I have allowed it to stand to keep the Mariner on his guard in this most treacherous part of the ocean.

† Eleventh Memoir, Journal As. Soc. Beng. Vol. XIV. p. 10.

PART II. § 59.] *Storm Tract and Mozambique Channel.* 49

have not, I think, sufficient evidence for this inference, though there is nothing impossible in it, and some of Mr. THOM's storms are traced a long way: see also preceding page 48.

Within this storm tract, and I suspect also within other Cyclone tracts, there appear at times also to occur, as already noticed, and as will be subsequently quoted for the Bay of Bengal, violent little tornado-like Cyclones which, for the time they last, are true hurricanes as to strength, and if taking an unprepared ship, and at night, might be fatal to her. The H. C. S. *Rose*, Captain PALMER, on the 25th August, 1815, in Lat. 9° 55' South, Long. 87° 35' E. from a fresh gale at N. E. at noon had a calm of half an hour, and 11 P.M. was taken aback by a hurricane at South, which lasted till 3 A.M., carrying away main topmast, main yard, &c. &c., and blowing her boats to pieces against the mizen rigging. The H. C. S. *Macqueen* and *Orwell* (tracks *i* and *f* on the Chart) both experienced severe Cyclones hereabouts which were travelling from the North in a Southerly direction.

59. CYCLONES OF THE MOZAMBIQUE CHANNEL AND TO THE NORTH AND SOUTH OF IT.—The tracks of two Cyclones investigated by Colonel REID and Mr. THOM within the northern part of the Channel are all we have, but these seem, like the track of a very remarkable one in the Andaman sea, to follow the usual law of the storms in the neighbouring great ocean, and to move from the E. N. E. to the W. S. Westward.

We have no records of any Cyclone to the North of the Comoro Islands, nor to the West of Cape St. Marie, North of its parallel. The Cyclones of the Northern part of the Channel, however, for such they are as to violence, though perhaps mere tornados as to size, reach Mozambique, and I think even as far north as to Zanzibar. At the Seychelles and in the Chagos Archipelago, gales and hurricanes are said by the French Creoles of the Mauritius and Bourbon to be unknown; but it would seem that this is not correct, for in September, 1851, the ship *Seringapatam*, Captain FURNELL, experienced a severe Cyclone there, which was apparently travelling to the W. b. S. or W. S. W. Captain FURNELL, warned by his barometer and the sea, very properly hove to in 7° S., Long. 58° East, till the centre

E

had passed him; his barometer falling from 30.50 to 29.50. Hence ships should be on their guard even in this low latitude.

60. THE WESTERN COAST OF AUSTRALIA. From the N. W. point of Australia, Southward to Cape Leuwin, it would appear that storms of great violence, and these often Cyclones, occur. At Dorre Island in Shark's Bay, we have a good account, in Capt. GREY's voyage, of one which certainly came from the N.E. and was travelling to the S.W., and this is also the first instance we have on record, in the Eastern Hemisphere, of a Cyclone travelling out to sea from a great continent, which this Cyclone certainly did; for the American Whaler *Russell*, Capt. LONG, which conveyed Capt. GREY to Shark's Bay, having steered off to the Westward for her cruising ground, encountered apparently the same Cyclone, lost her three topmasts, boats, &c. and put into the Cocos to refit.* In the Western Hemisphere we have two instances, Mr. REDFIFLD's Cyclones of November, 1835, and December, 1839, in which this has occurred, and these Cyclones have been traced from the interior of North America to the banks of Newfoundland.† To return to the coast of Australia: Lieut. WICKHAM, of H. M. S. *Beagle*, says, Nautical Magazine for 1841, p. 725 :—

"The West Coast of New Holland is at times visited by sudden squalls, resembling hurricanes, as I was told by the master of an American whaler that in March 1839, when in company with several whalers off Shark's Bay, he experienced some very bad weather without any previous warning, but it was not of long continuance. The gusts of wind were very violent, shifting to all points of the compass, some of the ships lost topmasts, sails, &c.; I think the first squall was from N.E., off the land."

The analogy of these with the Ceylon and other tornado-like Cyclones, which will be subsequently described, will strike every one. The Cyclone marked *dd* on the Chart on the South Coast is one of a like kind, and is described by Capt. RIDLEY of the *Candahar*, as forming with a densely black sky to the W. N. W. in an incredibly short space of time, and within an hour, "bursting upon us like a clap of thunder, and blew a perfect hurricane," which lasted for about 15 hours, and must have dismasted the vessel if not prepared, as she

* Captain Ross' notes.
† I have little doubt that in India we shall also find this to occur with some of the Arabian Sea Storms, near the Coast of Malabar and the Gulf of Cutch.

PART II. § 62.]   *Western Coast of Australia.*   51

was laid with her gunwale in the water. The wind does not appear to have veered more than from W. N. W. to W. S. W., but I have marked this as a Cyclone to put the seaman on his guard hereabouts, as off the Cape of Good Hope.

61. Off Rottnest Island, the French discovery ship *Le Geographe*, appears by M. PERON's report, as translated by PINKERTON, Vol. XI., to have experienced in June, 1801, a Cyclone of which the track appears to have been from the S. b. W. to the N. b. E., for the wind " veered in a short time from W. S. W. to S. E.," and as it is tolerably clear that the Cyclones about that part of the coast, and as far North as Shark's Bay, travel both *from* the coast, and *in* towards it from the ocean, this anomaly is by no means an improbable one; though we have but a scanty note and the very imperfect authority of a naturalist who was not, probably, a sailor, for the original account, and the chances of errors of the press and translators. We have however now for these nearly *meridional* tracks, as we may call them, several instances. Track XXVI. on Chart No. IV. of the *Golconda's* Cyclone being the first, the Ship *Buckinghamshire's* and *Cleopatra* Steamer's which is described in the next paragraph as travelling at most N. N. W., and this one, with one or two in the Bay of Bengal.

At Swan River, Lat. 32° South, very severe gales, of full hurricane violence are also felt. The anchorage in Gage's Roads is considered as insecure from the N. W. *gales* as they are called, during the winter. Lieut. WICKHAM of H. M. S. *Beagle*, (Nautical Magazine, 1841, p. 724,) describes one of these storms at Swan River in the year 1838, which evidently was a Cyclone, and *came in* from the W. N. W. to the E. S. E. (Track *aa*). He describes also another at Houtman's Abrolhos, Lat. 28° (Track *bb*) in May, 1840, which also *came in* from the Ocean towards the land, or from the Westward to the Eastward. We shall subsequently see that from about 20° to 40° S. on the Coast of South America, the storms also appear to *come in* from the Westward.

62. STORMS OF THE ARABIAN SEA.—Including in this title the whole space North of the Equator, to the Coast of Arabia, Persia and Beloochistan, and from the Western Coasts of India and Ceylon to the Eastern Coasts of Africa.

We have hitherto but scanty data for the tract which extends West from Ceylon to Longitude 50°, and what we possess is mostly confined in that part between Longitude 67° and the coast of India. It will be seen on Chart No. III. that at present, and with some exceptions to be presently noticed, there appear to be two classes of Cyclone tracks in this sea. The first (that of the Cyclones investigated by me in my Eighth and Fourteenth Memoirs, (Jour. As. Soc. Beng. Vols. XII. and XIV.) some of which originating in the Bay of Bengal (Tracks VIII. and XIV.) force their way across the peninsula or the Island of Ceylon, while others are either connected with the former or take their rise about the Laccadives; but both these kinds, which I distinguish as the first class, travel it would seem (Tracks VIII. G. and E.) out to the W. N. W. and N. W. towards the coasts of Persia and Arabia, and are felt there, and even as far as Aden, as stormy weather.

By the attention of Commodore LUSHINGTON, commanding the Indian Navy, who has sent me a report and a fragment of the ship's log, I am enabled to state that the H. C. Cruizer *Ternate*, Lieut. POOLE, on the 8th and 9th of June, 1836, ran down into a Cyclone nearly on the tropic, or about one hundred miles from the Coast of Beloochistan, between Long. 63° and 65° East. She reached the centre, broached to, and was obliged to cut away her main and mizen masts; and by the wind shifting to the Southward, she was drifted in towards the Coast, where she anchored and rigged jurymasts. The track of this Cyclone by the ship from N. N. E. to S. S. E. appears to have been one from East to West, but as she appears to have run down to the S. E. the wind mostly at N. N. E., it may have come up a little from the Southward of East, as E. b. S. or even E. S. E.; and have curved somewhat towards the Gulf of Persia when its Northern quadrants touched the land.

From an article in the *Bombay Times* of 20th November, 1855, it appears that the French ship *Bangalore* suffered very severely in a Cyclonic storm in the Arabian Sea in October, 1855, in Lat. 16° 30′ North, Long. 58° East, which subsequently reached the English ship *Chevalier* of Glasgow, in 14° 24′ North, Long. 55° 28′ East, or about midway between the island of Socotra and Cape Ras al Morebat. The centre of the Cyclone passed close to the *Chevalier*, which ship lost both her fore and main masts from their being struck with lightning in the height of the Cyclone. Its track appears to have been from about East to West, and that it was travelling at from 7 to 10 miles per hour.

63. From a passage in Horsburgh, Vol. I. p. 444 of 6th edition, we may infer that Cyclones sometimes occur in April and May in Surat Roads. The appearances are said to be dark cloudy weather, gloomy and black to the S. Eastward, with lightning and faint variable breezes, and several ships have been lost by remaining too long at their anchors when the wind veered round to the Westward. The storm of the 20th April, 1782, is cited as one of especial severity, and one ship is said to have rolled away her masts, and foundered when the wind had veered, and was blowing hard *from* the land, by her labouring between the wind, tide, and high cross sea from the Southward and Westward. There is every probability that this was a true Cyclone, but from the last paragraph, it would seem that the track was from the Northward and Westward.

Those of the second class which are investigated in my Fifteenth Memoir (Jour. As. Soc. Beng. Vol. XVII.) seem to originate in the space between the Maldives and Laccadives to the West, and Ceylon and the coast of Malabar to the East; and then to travel up in a line more or less parallel to the Coast to the N. N. Westward (Tracks *B*, *D* and *E*); some of these Cyclones are of terrific violence and apparently equal to the tyfoons of the China Sea.

The exception to these two classes of tracks is the track marked *A* on our Chart, which is a distinct and well confirmed instance of a Cyclone, though of no great extent or violence, travelling *in* towards the coast or to the N. Eastward. In this instance the diameter of it did not exceed 100 miles and was perhaps nearer to 50; nor did it reach the coast, as it was not felt at Trevandrum in Lat. 8° 30' N. where there is an observatory. It is possibly, like track *C* at Trincomalee, an instance of a tornado-Cyclone developing itself through a limited space only; though as to time, from its slow motion, it may, as in the case of the *East London*, last at least 24 hours. The seaman requires, however, to be advised of these as well of the larger and more usual ones, for they may dismast him, which is no trifling mischief, and when occurring near coasts, or in narrow or intricate passages, may add to his danger and inconvenience, since disabled vessels may be easily driven on shore. We have seen in the preceding paragraph that there are also Cyclones on the Western Coast of Australia, and on the Coast of South America, which like this one *come in* from the Westward. All these anomalies are matters of future research for us, till we find by multiplied and careful observation, whether they are governed by any fixed law.

64. CYCLONES OF THE BAY OF BENGAL.—I have investigated a considerable number of these, and have published part of my investigations in twenty Memoirs in the Jour. of the As. Soc. of Bengal, and I have much yet unpublished. The result however of these pretty extensive labours, as shewn on Chart III. is, that in the Northern and N. Western parts of the Bay, the Cyclones travel from between E. N. E. and East round to S. E. and even S. S. E. and S. b. East towards the opposite points of the compass. The rhumbs between E. and S. E. however seem by far the most usual; those to the Northward of East, or nearer to the meridian than S. E. b. S. seem to be comparatively rare.

65. Between the Andamans and Madras the tracks appear usually to be nearly from S. East and East to the Westward and Northwestward; but when Cyclones occur in about 6° or 8° of Latitude North, and to the Eastward as far as 90° East, they seem first to travel up from the S. S. E. or S. East, and then to curve away to the Eastward towards the Coast of Ceylon or the Southern part of the Coromandel Coast; though sometimes such Cyclones travel almost due East* or E. b. N. from Long. 90° to the Coast of Ceylon. As yet we do not know if the tracks are influenced by the monsoons, and are thus more inclined to particular routes according to the season of the year; and this is one of many subjects for inquiry.

66. Off the Coast of Ceylon there appears at times to form a sort of very violent but confined storm, though a true Cyclone as to rotation, but which rather deserves the title of a tornado-Cyclone from its size; these *may* come in from the middle of the Bay, but I

* See 14th Memoir, Journal Asiatic Society, Beng. Vol. XIV.

PART II. § 66.]   *References; Chart No.* III.   55

suspect that they are sometimes formed very suddenly and not far from the shore towards which they travel. The tracks marked X. 2 *b.* and *c.* are instances of these ; *c.* is that of a tornado in which H. M. S. *Sheerness* was blown from her anchors, and bilged on the rocks in Trincomalee, a land-locked harbour!, in January, 1805. (Asiatic Annual Register, Vol. VII.) About Cape Negrais also these little tornado-Cyclones are met with, and are excessively violent; one traced by me, that of the *Cashmere Merchant* (marked on the Chart II. 2, 5th Nov. 1839), travelled from the S. S. E. or S. b. E. to the Northward.

The following is the table of References to Chart No. III. which comprises also the Malabar Coast and the Eastern part of the Arabian Sea.

References to Chart No. III. *Cyclones of the Bay of Bengal and Arabian Sea.*

BAY OF BENGAL.

I.—Sand Heads,	June 1839.
II.—Coringa Cyclone,	Nov. 1839.
II. 2.—Cashmere Merchant's Tornado,	Nov. 1839.
III.—Cuttack,	April, May 1840.
V.—Madras,	May 1841.
VII.—Calcutta,	June 1842.
VIII.—Madras and Arabian Seas, (and Eliza Leishman, 1849,)	Oct. Nov. 1842.
IX.—Pooree, Cuttack and Gya,	Oct. 1842.
X.—Madras and Masulipatam,	May 1843.
X.—Coringa Packet,	May 1843.
XI.—Bay of Bengal and Southern Indian Ocean,	Nov. Dec. 1842.
XII.—Briton and Runnimede,	Nov. 1844.
XIV.—Bay of Bengal and Arabian Sea Cyclone,	Nov. Dec. 1845.
XVIII.—October Cyclone,	Oct. 1849.

The above with Roman numerals refer to my Memoirs with the same numbers published in the Journal of the Asiatic Society of Bengal.

a.	Ongole Cyclone,	Oct. 1800. H. Piddington.
b.	H. M. S. Centurion,	Dec. 1803. H. P.
c.	H. M. S. Sheerness,	Jan. 1805. H. P.
d.	Dover Frigate's,	May 1811. H. P.
e.	Kistnapatam ; Palmer's,	Mar. 1820. H. P.

f.	Burrisal and Backergunge,	June 1822.	H. P.
g.	Liverpool and Oracabessa,	May 1823.	H. P.
h.	Balasore and Cuttack,	Oct. 1831.	H. P.
i.	London's,	Oct. 1832.	H. P.
j.	Duke of York's,	May 1833.	H. P.
k.	Calcutta,	Aug. 1835.	H. P.
l.	Madras,	Oct. 1836.	H. P.
m.	Protector's	Oct. 1838.	H. P.
n.	Kyook Phyoo	May 1834.	H. P.
o.	Kyook Phyoo; Syren and Java's,	Nov. 1838.	H. P.
p.	Madras,	Oct. 1818.	H. P.
q.	H. M. S. Cornwallis and Point Pedro,	Nov. 1815.	H. P.
r.	H. C. S. Minerva,	Nov. 1797.	H. P.
s.	Coromandel and Malabar Coasts,	May 1820.	H. P.
t.	William Wilson,	Oct. 1836.	H. P.

ARABIAN SEA.

A	East London's Cyclone, XVI. Memoir,	April 1847.
B	Buckinghamshire's, XVI. Memoir,	April 1847.
C	Higginson and Lucy Wright, VIII. Memoir,	Oct. 1842.
D	H. C. S. Essex, XVI. Memoir,	June 1811.
E	Bombay, XVI. Memoir,	June 1837.
F	Rajasthan, } XIV. Memoir,	Dec. 1845.
G	Monarch,	Dec. 1845.

67. THE ANDAMAN SEA.—This tract, confined as it is, is yet subject to Cyclones of terrific violence, though they seem to be of rare occurrence, for I had met with no record, nor had I heard of any one suspecting even their existence, till the wrecks of the *Runnimede* and *Briton* transports in November, 1844, which I was fortunately enabled fully to investigate,* and two others subsequently noticed, have shewn that, at times, Cyclones fully equal to those of the China Sea or Bay of Bengal may arise there. This proves that a wide extent of sea is by no means necessary for the development of these meteors in their most violent form, though of very small extent.

The track of the first, which apparently took its rise in about 11° North and 96° East, was from the E. S. E. to the W. N. W., but that of the second, the Brig *Erin's* Cyclone, in Nov. 1850 (Twenty-second Memoir, Jour. As. Soc. Beng. Vol. XXIII., Track XXI.) proceeds

[*See XIIth Memoir, Jour. As. Soc. Beng. Vol. XIV.]

up from the S. E. b. S. to the N. W. b. N. passing between the two Volcanic Islands of Narcondam and Barren Island, and out into the Bay of Bengal by the Preparis Passage.

We should thus have said, formerly, that the average tracks were from the East and S. East to the N. West in this Sea; but in April, 1854, the H. C. Steamer *Pluto*, Captain BOON, proceeding with a relief of troops on board, from Moulmein to Bassein River, was assailed by a furious Cyclone in the Gulf of Martaban, in which but for Captain BOON's admirable management in boldly steaming out for the middle of the Gulf, so as to avoid the dangerous surf-sea of the small water near Point Baragui, she would certainly have foundered. This track, as Captain BOON well knew, took him directly into the centre, but of two evils he was driven to choose the least; and his log, with that of other vessels, while furnishing me with materials for my twenty-fifth Memoir, and with some curious facts in reply to queries sent to him, has also indubitably established the anomaly of a track from S. West to N. East in the locality; commencing, as will be seen, about Narcondam and travelling perhaps towards some other volcanic focus in the Shan or Siamese ranges of mountains in the unknown interior.

68. THE CHINA SEA: SEE CHART IV.—In the China Sea, from the result of the analysis of all the records of storms which I could obtain with the assistance of the Honourable the Court of Directors of the E. I. C., from 1780 to 1841, which will be found detailed at length in my Sixth Memoir,* and from additional information collected by me and laid down on the Chart to December, 1846, and subsequently in my Seventeenth Memoir (Journal As. Soc. Beng. Vol. XVIII) to 1848, and now to December, 1857, it appears, that the *mean* tracks of the Tyfoons, for the seven months from June to November, in which they chiefly occur,† are as follows:—

In June, the tracks are from between S. 30 East and East to West.‡

* Journal Asiatic Society, Beng. Vol. XI.
† Journal of the Asiatic Society, Beng. Vol. XI.
‡ On the 27th and 28th June, 1846, the H. C. Steamer *Pluto* ran into a terrific tyfoon, in which she nearly foundered, the track of the Cyclone was about from the S. 30° E. to N. 30° West.

In July, the tracks are from between N. E. and S. E. by E. to the N. Westward.

In August, the tracks are from between East and S. 40° E. to the Westward and N. Westward.

In September, the tracks are from between N. 60° E. to S. 10° E. to the S. b. Westward and N. b. Westward.

In October, the tracks are from between N. 12° E. and S. 45° E. to the S. b. Westward and N. Westward.

In November, the tracks are from between N. E. and S. E. to the S. Westward and N. Westward, evidently varying with the opposing strength of the monsoon and trade wind, and probably also influenced by the vicinity of the land.*

Three remarkable curved tracks XXV., III., and *n*, will also be seen on the Chart. These have been laid down on the clearest evidence, and may be taken as nearly correct. Tracks XXVI. and XXVII. are, though differing so widely, Cyclones which occurred at the same time; and at their junction it is probable that the *Golconda* troop ship, with 300 Native troops on board, foundered.†

There will also be seen on the Chart a very remarkable track marked in the Formosa Channel, off Amoy (Track *u*), which is one coming off *from* the Coast of China, and travelling the N. E. b. E. and two others near to and within the Bashee Passage. The first of these is (Track *ff*) one travelling to the N. N. E.; and the last (*x*), which is perfectly well ascertained by a good log and notes from the commander of the vessel, Captain SHIRE, of the Barque *Easurain*, is one first travelling out from the Coast of Luzon and then curving to the N. b. E., but when arriving at about 50 miles from the South point of Formosa, suddenly *re*-curving again, so as to travel out to the Eastward! making thus, if we suppose the Cyclone to have originally come up from the E. S. Eastward from the Pacific Ocean, in about Lat. 15° and Long. 130° a double curve, analogous to the

* Horsburgh (p. 280) speaks of "a storm coming out of the Gulf of Tonkin." Now this, to a ship crossing from the Coast of Cochin China to Hainam, is really the Southern half of a great whirlwind, travelling from the Eastward along the South Coast of China, as will be seen by placing the hurricane card on our Chart, and following any of the tracks from X. to V. with the centre of it.

† See 4th Memoir, Jour. As. Soc. Beng. Vol. IX.

PART II. § 68.] *The Palawan Passage.* 59

curves of the Cyclones of the Gulf of Mexico, as shewn on Chart No. 1, and occurring not far from the tropic, or with us in Lat. 20°, and in the West Indies from 25° to 36° N.

Track *w* is also one to be noted as shewing that the Cyclones from the N. Eastern part of the China Sea sometimes travel in to the land on the Coast of Cochin China. This track is one of a Cyclone carefully observed by Lieutenant CRAWFORD PASCO, of H. M. ¡Steamer *Vulture*, at anchor in Turon Harbour. It brought with it a sufficient rise of the Storm (and tidal ?) wave to overflow all the low land, and destroy great numbers of native houses.

THE PALAWAN PASSAGE.—It would appear that in this dangerous passage, Cyclones may possibly occur, and this in the low latitude of 9° North. I have marked on the Chart No. IV. (Track XXVIII.) the *Magicienne** and *St. Paul's* Cyclones in Nov. 1840; and I am now indebted to Captain MCLEOD, of the Ship *John MacVicar*, for a note of one in 9° 19′ N., Long 117° East, in which the wind, which fortunately veered only from West to S. S. W., while the ship was scudding under a close-reefed main-topsail, enabled them to keep in the channel, though " the sky was overcast with masses of dark electric clouds, with sudden gusts of *heated* wind! blowing with violence enough to carry the masts by the board, the sea raging in pyramids with great impetuosity and sweeping all before it." The Barometer was unfortunately broken. The Barque *Moulmain* was totally wrecked about 100 miles to the N. N. W. of the *John MacVicar* on the same night.

The following is a list of references to the tracks on this Chart:—
REFERENCES TO CHART No. IV.—THE CHINA AND LOOCHOO SEAS AND ADJACENT PACIFIC OCEAN.

No. of Track.	Dates.	Names.	Authority.
I.	17th July, 1780	H.C.S. London	H. Piddington
II.	19th June, 1797	H.C.S. Buccleugh	H. P.†

* I applied officially through the " Ministre de la Marine" for the Log and other details of the loss of this ship, a fine French frigate, but was informed that it could not be obtained ! I have, however, been assured from another quarter that the track is quite correct.

† A large proportion of the data of these tracks I owe to the attention of the Hon'ble the Court of Directors of the East India Company from their records, and several to the kindness of my friend Captain Biden, Master Attendant at Madras, who has been indefatigable for many years in providing me with materials.

REFERENCES TO CHART IV.—*continued.*

No. of Track.	Dates.	Names.	Authority.
III.	20th to 22nd Sept. 1803	H.C.S. Coutts, Camden, &c.	H. Piddington.
IV.	20th September, 1803	H.C.S. Royal George, Warley, &c.	H. P.
V.	28th and 29th Sept. 1809	H.C.S. True Briton and Fleet	H. P.
VI.	29th and 30th Sept. 1810	H.C.S. Elphinstone, Fleet	H. P.
VI. 2,	1837	H.C.S. Vansittart	H. P.
VII.	8th and 9th Sept. 1812	H.M.S. Theban and Fleet	H. P.
VIII.	28th and 29th Oct. 1819	H.C.S. Minerva	H. P.
IX.	29th November, 1820	H.C.S. Lord Castlereagh	H. P.
X.	18th and 19th Oct. 1821	H.C.S. General Kyd and General Harris	H. P.
XI.	14th and 15th Sept. 1822	H.C.S. Macqueen	H. P.
XII.	25th to 27th Sept. 1826	H.C.S. Castle Huntley	H. P.
XIII.	9th August, 1829	H.C.S. Bridgewater	H. P.
XIV.	8th and 9th August, 1829	H.C.S. Chas. Grant, Lady Melville, &c.	H. P.
XV.	23rd September, 1831	At Canton	H. P.
XVI.	23rd October, 1831	At Manilla and the Panama's first Storm	Mr. Redfield.
XVII.	24th and 25th Oct. 1831	Panama's second Storm	Mr. Redfield.
XVIII.	25th and 26th Oct. 1831	Fort William	H. Piddington.
XIX.	3rd August, 1832	At Canton and Macao	H. P.
XX.	22nd to 25th Oct. 1832	Moffatt	H. P.
XXI.	28th August, 1833	Brigs Virginia and Bee	H. P.
XXII.	12th October, 1833	H.C.S. Lowther Castle	H. P.
XXIII.	3rd July, 1835	Barque Troughton	H. P.
XXIV.	4th August, 1835	H.M.S. Raleigh	Mr. Redfield.
XXV.	16th to 22nd Nov. 1837	Ariel and Vansittart	H. Piddington.
XXVI. XXVII.	22nd to 24th Sept. 1839	Cal. Thetis, London Thetis, and Golconda	H. P.
XXVIII.	29th and 30th Nov. 1840	French Frigate Magicienne and St. Paul	H. P.
XXIX.	21st July, 1841	Hong Kong Fleet	H. P.
XXX.	26th July, 1841	Hong Kong Fleet	H. P.

The foregoing are tracks taken from the Chart to my Sixth Memoir, "THE STORMS OF THE CHINA SEA." *(Journal As. Soc. Beng. Vol. XI.) The following are from my Seventeenth Memoir (Storms of the China and Loo-Choo Seas and Pacific Ocean) Journal As. Soc. Beng. Vol. XVIII.*

PART II. § 68.] *Tracks; References to Chart.* 61

d.	November, 1816,	Spanish Brig *Vigilante*,	H. Piddington.
e.	31st October to 2nd November, 1843,	*Atist Rohoman* and *Shaw Allum*	H. P.
f.	31st October, 1843,	American ship *Unicorn*, Manilla	H. P.
g.	16th & 17th Nov. 1844,	Ship *Edmonstone*	H. P.
h.	10th & 11th Oct. 1845,	H.M.S. *Espiegle*, Ship *Ann*, and others	H. P.
i.	6th to 8th Oct. 1845,	H. M. Steamer *Driver*,	H. P.
j.	28th & 29th June, 1846,	H. C. Steamer *Pluto*	H. P.
k.	21st and 22d July, 1846,	Ship *Hyderee*	H. P.
l.	14th & 15th Sept. 1846,	H.M.S. *Agincourt*	Mr. J. E. Elliott, Master H. M. S. *Agincourt*.
m.	15th & 16th Sept. 1846,	H.M.S. *Ringdove*	H. Piddington.
n.	24th to 26th Sept. 1846,	Clipper *Mischief*	H. P.
o.	Manilla, October, 1843,		H. P.
p.	Manilla, October, 1831,		H. P.
q.	Manilla, November, 1845,		H. P.
r.	Chusan, Sept. 1843,		Capt. Vyner, R.N.
s.	Loo Choo Sea, *Casique*,		H. P.
t.t.	Pacific Ocean and China Sea H.M.S. Str. *Driver*, Ships *Ann, John o' Gaunt, &c.* Oct. 1845,		H. P
u.	*Don Juan*, Straits of Formosa, Sept. 1846,		H. P.
v.	Brig *Guess*, July, 1847,		H. P.
w.	H. M. Str. *Vulture*, Nov. 1847,		H. P.
x.	*Rob Roy* and *Swallow*, Nov. 1847,		H. P.
y.	*Rob Roy*, Manilla, Nov. 1847,		H. P.
z.	*Easurain*, Coast of Luconia and Bashee Passage, Nov. 1847,		H. P.
aa.	H. M. S. *Childers*, Shanghai, July, 1848		A. R. Elliot, R. N. Master H. M. S. *Childers*.

PACIFIC OCEAN.

bb.	*Duke of Buccleugh*, and Fleet, July, 1797,		H. P.
cc.	Ditto ditto	ditto ditto,	H. P.
dd.	Ditto ditto	ditto ditto,	H. P.
ee.ff.	*King William IV.* Pacific Ocean and China Sea, July, 1845,		H. P.

Reviewing the (apparent) labyrinth of tracks on this Chart No. IV. the seaman will not fail to remark that while some tracks absolutely contradict each other, some again proceeding from S. East and South, cross the other and more numerous class almost at right angles, but when bad weather is coming on, while he will not perhaps regret to find a track close to his ship's position and agreeing with his own estimate of that of the Cyclone he expects, yet he will bear this in mind and above all things be on his guard against the mischievous and ignorant notion that there is any fixed law for the tracks of these terrific meteors, especially in narrow seas with volcanic islands or continents within, or near to, or limiting them.

THE JAVA AND ANAMBA SEAS.—In the China Sea, I have found no record of any Tyfoon (Cyclone) South of Lat. 9° North, nor from thence through the extensive tract which I have entitled the Anamba Sea, lying between Borneo and the Malay Peninsula, Banca and Sumatra, to 3° South, where the Java Sea begins. In the Java Sea itself, during the N. W. monsoon, strong *gales* are often experienced, but whether these are at any time rotatory we cannot yet say.

A notice of the loss of the Brig *Minerva*, from Moulmein, bound to Hong Kong, in the *Singapore Free Press of* 15*th Feb.* 1849, states— That she sailed from Singapore on the 3rd Nov. 1848, and on the 13th in Lat. 6° 49' N.; Long. 111° 28' East; she experienced a severe "gale of wind" which lasted three days, *the wind veering round the Compass*, causing a tremendous sea, from which the vessel suffered so much that in endeavouring to make some place of shelter, she was lost on a reef in South Bay on Balambangan. From this statement we may suppose that this was a true Cyclone in the Anamba Sea.

§ 69. THE BANDA, CELEBES, AND SOOLOO SEAS.—One of these Seas has an open space of upwards of forty square degrees, though as just marked (§ 66) a large space is by no means indispensable to the production of the most violent hurricane-Cyclones, and in reference to these and the Anamba Sea, and as we shall subsequently shew, for the Pacific Ocean at least, rotatory storms may occur close to the Equator! We do not yet know of any true Cyclones occurring in them, though strong *gales* are certainly of more frequent occurrence, than from their vicinity to the equator and land-locked position would be supposed. The numerous volcanoes (§ 39, p. 23) with which the Moluccas abound, *may* have some influence on the meteorological phænomena. The following is from a Singapore paper—Sarangani Island is off the South end of Mindanao, in 5° 23' North.

PART II. § 70.]   Tracks; Timor Sea.   63

The American whaling ship *Octavia*, Captain Pell, arrived here on the evening of the 25th inst. to refit, having on the 30th March last, while off Hummock or Sarangani Island, encountered a most severe hurricane, in which she lost all her masts, and received extensive damage otherwise. They afterwards proceeded to Leron Harbour in the Island of Salibabo or Lirog, one of the Talaut group of islands, where they made temporary repairs and afterwards proceeded to Monado, and from thence came on to Singapore.—*Singapore Free Press, July* 27, 1849.

Capt. Ross in his notes from the Cocos Island says that Banda is visited at times by severe Cyclones, one of which, in 1778, destroyed 85 per cent. of all the nutmeg trees. And again that in 1811 another was scarcely less destructive, and that those of lesser intensity are rather frequent. As this sea is wholly volcanic, Burning Island and Banda as well as the islands to the South being in full activity, farther information would be of much importance.

70. THE TIMOR SEA.—This great, and yet imperfectly known extent of Ocean, extending from Java and the N. W. point of New Holland to Torres Straits, is subject to severe hurricanes, which are true Cyclones, revolving according to the law for those in the Southern Hemisphere, and travelling from the E. N. E. to the W. S. W. or S. Westward; at least those from the meridian of Timor to the westward appear to do so. East of this meridian we know only of one Cyclone, which occurred at Port Essington, Lat. 11° 22′ S. Long. 132° 10′ E. in December, 1839, in which H. M. S. *Pelorus* was driven on shore, and the new settlement nearly destroyed. It was of small extent though its violence was excessive, and the Barometer sunk to 28° 52′ Its track may have been from the East to the West, or not far from those points.

On the Chart No. II. which comprises the tracks of this sea, will be noted a very remarkable one marked *m m*, which is that of the American ship *Howqua* dismasted in January 1848, in a Cyclone of terrific violence in which her Barometer fell 2.20 and she was near foundering. She first ran up with a S. Westerly gale till she had a shift to South which broached her to, and she shortly fell into the calm centre with the wind at S. East and East. She then had it N. Easterly and North, and unless there were two Cyclones, which may have been the case, we can only account for these shifts and lulls by a track, such as I have marked, travelling up from the South and then curving away to the North West. The log of this vessel has not reached me, and this estimate is only from the newspaper and a private account. I have thus marked it both with a track and a wind circle of which the centre in 15° S. and 115° East is at the spot where the ship was dismasted.

As the second edition was going to press, I received from Captain S. VAN DELDEN, commanding the Dutch Ship *Roompot*, from Sourabay to Middleburg in Dec. 1847, some valuable logs, among which was one of a Cyclone which he met with on quitting the straits of Bally, but fully aware of his danger Captain VAN DELDEN kept away to the North with the wind at W. b. S (though homeward bound) to give the Cyclone sea-room, and then cautiously followed in its wake to the S. W. b. S. on its Eastern edge, keeping the wind at about N. N. W. from the 13th to the 19th. Other ships which were not so scientifically managed, suffered in this Cyclone, of which the track is marked r. r. on the Chart, but from Captain VAN DELDEN's Logs only, as he was not able to procure me the others.

71. The Western Coast of Australia I have noticed at p. 50. Along the South Coast of Australia to Bass' Straits, though the usually Westerly winds and gales of the higher latitudes in both hemispheres prevail during the greater part of the year, I have no doubt that Cyclones also often occur. The *gales* are described as beginning at N. W. with a low Barometer,* and thick rainy weather, increasing at West and S. W. and gradually veering and moderating to the S. and even S. S. E. with the barometer rising to above 30 inches. Sometimes also they return to West or more Northerly, with a fall in the barometer, but the gale is not then over, though it may diminish or die away for a day or two. *Sometimes the wind flies round suddenly* from N. W. to S. W., and the rainy thick weather then continues a longer time. This account is a little abridged from Horsburgh, Vol. I. p. 121,† and if the Northern half of the Horn Card for the Southern Hemisphere be moved slowly from West to East over a dot marked on a Chart, in about the parallel of 40° S., the usual track of ships proceeding towards Bass' Straits, it will be seen at once how closely the first part of this description answers to the case of a Cyclone overtaking and striking a ship with its N. Eastern quadrants, and

---

* 29.5, or lower.

† The original is by Flinders, but the reference has not been given in Horsburgh.

travelling faster than the ship, or say on a parallel course. The centre is nearest to her, when the wind is at West and therefore the gale strongest, and it leaves her with the Southerly winds as it proceeds ahead of her. The case of the renewal of the gale may be that of the ship's overtaking it again, or of another vortex following closely on the first, and the "flying round suddenly" occurs, unquestionably where the central portion of the Cyclone reaches the ship. The long duration of the storms is doubtless owing to ships usually scudding with them, being a fair wind. All this is however but "*great probability,*" and to make it a full certainty, we should have the logs of Whalers or other vessels in higher latitudes with Easterly gales at the same time. In the Chapter on which I speak of the Storm Wave and Storm Currents, the seaman will find a caution of importance to him when approaching Bass' Straits.

We have no good records for the storms off the coast of Van Diemen's Land, nor any for those to the North between Bass' Straits and Port Jackson. Within Bass' Straits I find in a review of the voyage of H. M. S. *Beagle,* that Commander STOKES says:

"The gales in Bass' Straits begin at N. N. W., and draw gradually by W. to S. W., when they subside. If it backs it will continue, but the barometer will indicate its duration; it is seldom fine with a pressure of 29.95 in. and always bad if it falls to 29.70 in." This exactly represents the Northern half of a Cyclone travelling through the Strait from the Westward. The "backing" would be the case of a Cyclone passing up from the S. W. to the N. E. and lasting longer, either because its diameter was greater, or because its motion was slower, having been checked in its passage, over the N. W. part of Van Diemen's Land; for the passage of a Cyclone over high land certainly checks, for a time, its rate of travelling.

72. THE NORTHERN PACIFIC OCEAN AND LOO CHOO SEA.—Distinguishing by this last name the space of ocean included by the chain of islands, which extends from the N. E. point of Formosa to the Meacsima Islands at the south extremity of the Japanese Group, and of which the Loo Choos are the centre:* off the South Coast of

---

* Called by the Chinese the TUNG HAI or Eastern Sea.

Japan we have one account of a Cyclone by Admiral KRUSENSTERN in 1804,* which appears to have been travelling from the Southward due North, as deduced from the shift of wind which he experienced. It was in Lat. 31° to 32° North, that this Tyfoon was experienced and in sight of the South Coast of Japan.

This track lying between the Bonin Islands and the S. Eastern Coast of Japan may probably be one of the curving localities for the Cyclones of the Northern Pacific, which would account for the apparent meridional tendency of the tracks, and it is also sometimes visited by sudden and severe Cyclones in which there is no warning from the sea, so that we may suppose them to *descend*. The ship *Akbar*, of London, Captain D. MILNE, bound to San Francisco with emigrants, experienced hereabouts, in Lat. $32\frac{1}{4}$° North, Long. 136° to 137° East, on the 27th and 28th May, 1852, a very severe and sudden Cyclone, in which the ship lost everything but her lower masts, the Barometer falling in about ten hours from 29.95 to 29.05. It is worthy of attention that this Cyclone, though blowing with terrific fury for a time, was not only not preceded by any sea, but Captain MILNE, on the contrary, notes particularly the smoothness of the water until the gale began, when a "fierce short sea getting up" is noticed. He estimates the diameter of this Cyclone to have been about 200 miles, and that it travelled at the rate of 15 miles per hour. After the Cyclone the current was found to have been running to the E. S. E. 45 miles in 24 hours.

I am indebted to Lieutenant M. T. MAURY, U. S. S. *Saratoga*, for a note of a true Cyclone experienced by the barque *Louisiana*, Captain Crosby, on the 25th, 26th, and 27th of May, 1852, between Lat. 23° 23′ N. Long. 126° 46′ East, and 25° 09′ North, Long. 126° 44′ East, on her passage from Hong Kong to San Francisco, in which her Barometer fell from 29.65 to 28.30, and she was obliged to put back to Hong Kong to repair the heavy damages she had suffered. From the best estimate I can make from the notes and the veering of the wind with the ship's drift during the Cyclone, I am inclined to think that it was one coming up from the S. East or S. E. b. S. and travelling to the N. N. W. or N. W. b. N.; and being to the S. East of

---

* Chinese Repository, September 1839.

Typinshan, it affords us for the month of May a track in that quarter. The *Louisiana's* Cyclone was preceded by a heavy sea from the S. E.

The Ship *Lady Amherst*, Captain DANDO, on her return voyage from California to China *appears* to have followed a Cyclone travelling to the W. N. W. from midnight of 20th, 21st September, 1852, to the 27th of that month, and from about Lat. 19° N. Long. 142° East to Lat. 23° N. Long. 127° East, keeping carefully on its N. Eastern and Eastern quadrants, and heaving to when warned by the Barometer and the increasing severity of the squalls that he was approaching it too closely, Capt. DANDO says:—

"The appearance was very remarkable, *i.e.* a dense body of clouds to the South Westward giving off squalls from the S. E. and Southward, the violence of which was very great. It put me in mind of an immense coach-wheel revolving horizontally and throwing dirt off its circumference. The main body must have been at some distance off, from the fact of the dense clouds not being at any time higher than 25° or 30° above the horizon."

At Shanghae in Lat. 31¼ N.; a severe Cyclone (Track *aa* on the Chart) was experienced in July, 1848, of which I have given the records from A. R. ELLIOTT, Esq. Master, H. M. S. *Childers*, at Shanghae, and Commander HAY, H. M. S. *Columbine*, in my Seventeenth Memoir (Jour. As. Soc. of Bengal, Vol. XVIII.) This Cyclone raised the waters of the great Yang-Tse-Kiang river 20 feet above low water mark against an ebb tide! At Chusan, Lat. 36° N. in September, 1843, a severe Cyclone was experienced, which undoubtedly came from the S. E. They are said to be uncommon there. In Lat. 27° N. Long. 122° E., in the Loo Choo Sea, I have also a newspaper record of a Cyclone experienced by the Ship *Cacique*, in September, 1843 (Track *s* on the Chart), in which the shift, with a short calm, was from N. E. to S. W. shewing also a track from the S. Eastward to the N. Westward at that season of the year. Our track *m* also, within the Straits of Formosa, which is that of H. M. S. *Ringdove*, in the month of Sept. 1846, is from the S. Eastward, and it will not escape the reflecting seaman that here, as in the corresponding latitude in the West Indies, we have a wide extent of Ocean to the Eastward, on the limits of the trade winds, then a sort of barrier of islands formed by the Japan, Loo Choo, Typinshan, and

Formosa Archipelagos, and within these a confined sea of seven or eight degrees in width, and eight to ten in length, bounded by the great Continent of Asia. When we collect more data, we shall be better able to pronounce if the effects of these similar circumstances are also similar; at present we can only notice the probability and regret our want of materials. Kotzebue, in the *Rurick's* voyage of discovery, Vol. II. p. 160 of the 8vo. edition, describes a storm of hurricane *violence* on the 18th of April in Lat. 44° 30′ N, Long. 181° West, but he gives no account of the veering of the wind. Again in the voyage of the *Rurick*, Vol. I. p. 264, after a smart gale in the neighbourhood of the St. Lawrence Islands, he was informed by the Tchukutskoi of St. Lawrence Bay, on the Asiatic Coast of Behring's Straits in 65° 40′ North, "that the time of violent storms was at hand, and that the last had been only a faint wind. He gave us to understand that in a *real* storm nobody was able to stand on their legs, but that they were obliged to lay themselves flat on the ground." This is exactly, as to violence, the description which a Carib of the West Indian Islands might have given to Columbus, or which a Mauritius or Jamaica negro would give of their hurricanes in the present day. I have heard it often said in descriptions of hurricanes there, by persons of all classes, that fearful that the dwelling house might be blown down, the family crept on the ground on all fours (lying flat down when the gusts were most furious) to reach the nearest negro hut or other low sheltered spot, or a "hurricane house" built of stones for such occasions. Are these Behring's Straits' storms analogous to those which arise in the interior of the continent of North America (Track *J* on Chart No. 1) and pass out to sea over Newfoundland? and do the Asiatic storms arise in the plains of Eastern Siberia and travel *out* towards Behring's Straits? When the Government of India in 1847, determined to send a mission to Western Tartary, which was expected to winter at Yarkund, I drew up a detailed set of instructions for observing the violent storms of the Tartarian steppes, to ascertain if they were circular or straight-lined winds, with queries for the natives, but the expedition, unfortunately, did not proceed. We have evidence enough from HUMBOLDT, CHAPPE, EHRMANN and others, of the violence of the Siberian *gales*, but no detail of the veering of the wind that I have met with, or recollect.

These queries, 37 in number, were subsequently published in the Journal of the Asiatic Society of Bengal, and elicited from Dr. A. CAMPBELL, Resident of Darjeeling, a valuable paper of replies from information collected by him in Eastern Thibet. From this it appears that in the Chang country, of which the Southern boundary is in Lat. 31° N. extensive and dangerous *whirlwinds* at least, and in all probability true Cyclones, of which the Southern quadrants sweep in westerly gales over that district, are well known. One of the peculiarities mentioned by Dr. CAMPBELL's informant, of the most violent kind of storms, called in the language of the country *Babiuk*, is that they are generally preceded by a clattering noise, as of galloping horses, which is intermitting, and in reference to what will be said in Part IV. of the noises of Cyclones and Hail Storms this is of much interest, fanciful as it may at first sight appear to those who do not consider by what singular, and even childish comparisons and metaphors unusual meteorological phenomena are always described by half civilized people and savages, and even by the illiterate of civilized countries, as witness the term "merry dancers" to describe the Aurora Borealis. M. PELTIER (see Kaemtz, p. 383, note), describes the noise of a hail-storm as resembling the gallop of a regiment of cavalry.

At the Bonin Islands, Lat 27° N. Long 142° East, tempests of hurricane violence are said to be frequent, and to come from the Eastward.

In the space comprehended between the Philippines, the Bashee Islands, Formosa, and the Loo Choo Group to the West and North, and the Bonin and Mariana Islands to the East and South, Tyfoons (Cyclones) are excessively sudden, and certainly come from the Eastward. The ship *Montreal*, of Boston, from the Sandwich Islands to Hong Kong, was overtaken, November 7th, 1845, in Lat. 19° N. Long. 142° East, by a severe Tornado-Cyclone which, by the log of another ship which kept way with her the whole time, and was during the Cyclone only from 30 to 65 miles distant from her, was of very small extent; but it was so violent that Captain LOVETT, of the *Montreal*, to whom I am indebted for this and several other communications, was on the point of cutting away her masts, when hove on her beam-ends by the shift of wind at the centre. The track of this Cyclone appears to have been about from S. E. to N. W. and the *Montreal's* Barometer, fell, after fluctuating remarkably on the approach of the tempest, from 29.60 to 29.30; and what is very remarkable it rose to 29.40 just as she first reached the edge of the central space. On the Eastern Coast of the Philippines, severe gales are known to prevail, which become true Cyclones.*

It should be borne in mind also when navigating in this great tract, that is, from the latitude of the Bashees and Formosa to 45° North, and perhaps to the Aleutskoi Islands and for an unknown distance to the Eastward, that there seems to exist a current like that of the Gulf Stream. Mr. REDFIELD, in his Essay "On the Tides and Prevailing Currents of the Ocean and Atmosphere,"† says of it:

"Hence on the coast of China and Japan we find a current which fully represents the Gulf Stream of the Atlantic. This current I find was fully noticed, incidentally, by the officers of Cook's last exploring expedition, and its velocity stated in some instances at five miles an hour. Other observations to which I have had access have confirmed the existence of this current and have shewn the elevated temperature which this stream carries from the lower latitudes; so that near one thousand miles east of the coast of Japan in Lat. 41° North the temperature of the water has been found at 79½ of Fahrenheit."‡

Now when we take into account that this warm ocean stream is found exactly on the lines which connect the greatest volcanic chain of our globe (that extending from the Bay of Bengal through Sumatra and Java to the Philippines, Japan and Kamschatka, and to the Aleutskoi Islands and Mount St. Elias) we are struck with the analogy which this presents with the Florida Gulf Stream; and we call necessarily to mind the prevalence of Cyclones within its limits and

---

* The Natives of the Philippines, who like all nations of Malay origin are good fishermen and daring seamen after their own fashion, perfectly distinguish between the tyfoons or turning gales (Cyclones) and the straight-lined Monsoon gales. They call the first *Bagyo* and the latter *Sigua*.

† American Journal of Science and Arts, Vol. XLV. No. 2.

‡ Voyage of Captain Dupetit Thouars; other and earlier observations had attracted my attention, particularly in the cruising voyages of our American whalers, but I now refer to this as a more recent and convenient authority.

their tendency to follow its track. See Part I. § 39, for what is there noticed as to the possible connection of Cyclones and Volcanoes.

It will be seen on inspection of the chart, and its Table of References, that I have been able to trace, partly from documents now half a century old, but highly authentic, being the logs of the East India Company's Fleet of 1797, of which the *Buoclough* formed one, and in which H. M. S. *Swift* foundered while convoying it,* a few tracks, which are again corroborated by that of the Cyclone experienced by H. M. S. *Driver*, in October, 1845 (*i*). If we follow these tracks with the eye to the shores of the great island of Luzon, with its extensive chains of lofty mountains, we shall easily conceive how in forcing their way over amongst these obstacles, as well as in passing through the Bashee Passage, many of the Storms may be deflected from their course to the North West to one towards the West and S. W. since there is no sort of doubt that curved tracks occur in the open ocean, in the China Sea, Bay of Bengal, Southern Indian Ocean and West Indies.

Hence we may perhaps say for this track, *i. e.* from 7° or 8° to 30° North, and from the meridian of the Marianas to the Coast of China and the Philippines, that the tracks are generally from between the S. E. and N. E. to the N. W. and S. W., perhaps with some a little nearer to the meridians, according to the varying state of the Monsoons and Trades, and other causes of which we are as yet ignorant, but that towards and beyond the tropic of Capricorn, the seaman must be on his guard lest he should meet with a re-curving storm like track *z* on our Chart (See p. 58 for a brief notice of this), or like tracks *w* and *ff* if they also continued their course to the N. Eastward.

Amongst the Marianas and Carolinas storms appear to occur, and in some seasons frequently, but we have no authority except that of analogy, and what I have given above, to pronounce them Cyclones or to deduce their probable tracks.

I find in the *Revue des Deux Mondes* for January, 1852, p. 208, in an article written by M. Jurieu de la GRAVIERE, commanding the French Corvette *La Bayonnaise* in the cruise in the China Seas and

---

* See Part III. for details of the lesson afforded us now by this dismal event.

on the coast of China, an instance of a severe Tyfoon Cyclone occurring at the end of June, 1848, in the port of San Luis de Apra in Guam, which, beginning at East and veering to South East, shifted *(sauta)* to the North West. This shift, and there is no doubt of the gale being a true Cyclone, for it is so described (as a "*circular storm*") by M. de la GRAVIERE, would give a track to the North East, which is a very probable one for that position.

Captain T. B. SIMPSON, commanding the Brig *Freak*, of Sydney, has obliged me with a letter of great interest (published also in the Nautical Magazine for 1851, p. 272), describing a Cyclone encountered by him, 1st to 3rd of May, 1850, to the Westward of the Marianas, between Lat. 19° 28′ and 19° 56′ N. and Long. 138° 44′ and 135° 50′ East, which was at first coming up towards the vessel on a North Westerly course. The Brig scudded to the Westward with the Easterly gales till it became dangerous to run on, and then hove to. After some anxious hours he found that the Cyclone had evidently, and fortunately, curved away to the North Eastward where it was last seen in a dense bank of clouds. The range of the *Freak's* Barometer in this Cyclone was from 29.75 to 28.87. The Nautical Magazine for February, 1855, gives another instance in the log of the Brig *Gifford*, Captain B. Briard, which in 27° 40′ N. Long. 134° 10′ East was overtaken by a Cyclone coming from the Eastward, and which curved off to the Northward and North Eastward in front of her, the Brig being very properly hove to, though not quite soon enough. Her Barometer fell from 29.60 to 28.70, and as the centre seems to have been close upon her, she lost, by the wind and heavy sea, her bowsprit, foremast and main topmast, and had her ballast partially shifted. In the CONVERSATIONS ABOUT HURRICANES, p. 50, there is a supposed case which exactly illustrates these two vessels' positions and the Cyclone curving up in front of them.

Of the great extent of Ocean in the Northern Pacific, from the Marianas and Carolinas to the Sandwich Islands and West Coast of North America, we have also but very little notice, and none from which we can deduce the nature of the storms, or their tracks if they are rotatory; but from what will be presently said of the storms of the Southern Pacific, there seems no reason to doubt that as the same causes exist so the same effects are produced, and that Cyclones are found in the Eastern part of the Northern Pacific turning according to the law of their revolutions for the Northern Hemisphere ◯ or against the hands of a watch, and usually travelling on tracks tending gradually to the Northward, until influenced by the West Coast of North America, for which I have also been unable to obtain any data.

KOTZEBUE, in the voyage of the *Eurick*, says that at the Radack Islands in 10° N. 170° East, the winds are usually from the E. N. E. but that in the months of September and October it sometimes "blows from the S. W. and not seldom rises to a furious hurricane rooting up the cocoa and bread-fruit trees, and desolating the islands on the Western point of the group *which he* (the native informant) *assured me were sometimes swallowed up by the waves:*" probably a storm wave like that which so recently swept over some of the Laccadives, as will be subsequently described.

73. IN THE TROPICAL REGIONS OF THE SOUTHERN PACIFIC, from the Barrier Reefs of New Holland, through the numerous groups of islands to the Low Archipelago, and perhaps even nearer to the Coast of S. America, and from the Equator to Lat. 25° S. there is no doubt that true Cyclones occur, of as great violence at least as those in the Northern Pacific just alluded to; but from the scattered accounts from single ships, or missionary residents on the various islands we cannot with confidence say any thing positive as to their tracks, though they appear to come from the Eastward amongst the islands, and sometimes to curve to the Southward. The following are a few notes. The seasons at which they prevail seem also to be the same as those of Mauritius and Bourbon.

At Vitileva, in the Feejee Group,* in February, 1841, a well defined Cyclone tolerably observed, seems to have moved to the Southward, and though it lasted four days was not felt at Tonga, 8 or 10 degrees to the S. E. of it.

At Apia Harbour in the Samoan Group of the Navigator's Islands, Lat. 14° S. on the 10th December, 1840, a true Cyclone of great violence, with a fall of 4 inches of the mercury, (by a damaged barometer) was observed, moving from the North to the Southward, and

* U. S. Exploring Expedition. Vol. iii. p. 321.

four years previous, another, also well defined, moving from the N. W. to the S. Eastward, the change of wind being from S. E. to N. W. The space between the Samoan (Navigator's) Islands and Friendly Islands is said, expressly, to be subject to violent *hurricanes*, and that scarcely a year passes without some of the Friendly Islands suffering from them. Their violence is such, that many of the American Whalers have been made complete wrecks of by them—two were lost about 1842* at the Navigator's Islands.

The *Samoan Reporter*, No. 11, of July, 1850, forwarded to me by the Rev. W. MILLS, gives a brief account of a hurricane of great violence but not of any great extent, from which I extract what is essential to our research. The account is evidently not written by a sailor, and hence is, for us, very imperfect, but the general remarks on the seasons for their gales and Cyclones (taking their hurricanes to be such) are of much interest in relation to those of the Southern Indian Ocean, where, as will be seen in our Table in Part V., they occur at the same seasons. The Reporter says:

"Between the months of November and March, we have generally a gale of wind for a few days on these islands. In the month of January last, we had some high winds and bad weather, and, as March passed away, we were thinking the stormy season was over. On the 5th of April, however, we had a hurricane not only unusually late in the season but also very sudden and severe. It was scarcely felt on the island of Savaii; one side of Manono escaped, but all along both sides of Upolu it raged with fearful violence, gathering strength, apparently, as it advanced in its course from the West towards the East end of the Island. It extended to Tutuila, but was not felt there to the same extent as on Upolu. It commenced to blow in the morning from the South, shifted to the West about midday, and by 2 P. M. bread-fruit and other trees innumerable were laid low: upwards of two thousand native houses, chapels, and other buildings were in ruins, and three vessels at anchor in Apia harbour were driven on shore among the rocks."

After describing the mischief on shore, it concludes: "The vessels wrecked at Apia were the whaling ships, *Favourite*, of London, and

* Year uncertain.

the *Hercules*, of New Bedford, and also the French R. C. Mission schooner *Clara*. All on board the vessels were saved, with one exception."

"Captain COURTNEY, of the *Two Friends*, reported to us, that he was about sixty miles to the south of Tutuila during the gale; that they had it strongest twenty-four hours later than us; and that not one on board expected to be saved."

If this shift was distinctly one from S. to West, this would give a track from the N. E. to the S. W. for the Cyclone. A summary of the Log of the *Two Friends*, is also given from a letter by her Captain to Captain ASHMORE, at p. 667 of the Nautical Magazine for 1850, from which it would appear that her Barometer fell in the Cyclone from 29.80 to 28.10! the total fall being thus 1.70! and that as it "ranged high," *i.e.* above a standard Barometer, it really fell to about 27.80. It is not said what it was at before and after the Cyclone in the comparatively fine weather. The Brig was hove to for the whole time and had the wind veering from S. E. to S.W. and S. S. W. when most violent, *i.e.* when the centre was nearest to her as shewn by her Barometer. This with her drift will give also a track from about the N. East to the S. West.

It is also said in this letter that the stranded ship at Upolu had the Barometer down to 27.15, and the shift "from N. W. to S. W." which would give an opposite track or one from West to East, but there are one or two evident errors of the press in the compass points, so that this leaves us in some uncertainty, though the track is in all probability what we have deduced from the Missionaries account (who had full leisure both to observe and to correct their printed article) and from the log of the *Two Friends*.

At the Kingsmill Group *on the Equator!* violent storms, which seem to be Tyfoon-like, are experienced.

At Vavaoo in the Friendly Islands, Lat. 19° S. Long. 173° W. in 1837, the American Whaler *Independence* was driven on shore by "a hurricane," and taken off by the shift of wind.

The account of the storm at Raratonga, in the Harvey Islands, in 19° S. 160° W., described by Mr. WILLIAMS, and quoted by Colonel REID, gives us unfortunately nothing farther than the certainty that Cyclones prevail there at times.

Mr. THOM says, 'p. 846,—" In December, 1842, H. M. S. the *Favorite*, on the way from Tahiti to the island of Mangeea, met with a storm of a rotatory kind, and so severe, that the vessel was hove to under a main topsail. Captain SULLIVAN was warned of a hurricane before his departure, which shews that storms of this kind are familiar to the natives."

At New Caledonia and at the Loyalty Islands between it and the New Hebrides, it would appear that *hurricanes* (by which term either Cyclones or the trade winds amounting to the force of a hurricane may be meant) are common and well known to the natives; for in the Sydney Herald of 30th of March, 1848, is an account of the losses of the *Sarah* and *Castlereagh* in a hurricane at Lefoo one of the Loyalty group in Lat. 20° 45′ S., Long. 167° 45′ East, on the 12th and 13th Feb. 1848, with notes of several other craft, then trading at the New Caledonia group, which experienced the same storm. In the accounts of both vessels it is stated that, from the statements of the natives of New Caledonia and Lefoo, such a terrific *hurricane* had not occurred "for the last eighteen years," and that the Barometer fell to 28.46,* the storm being felt off the South end of New Caledonia and on the West Coast after it had passed Lefoo, and it therefore crossed the large island of New Caledonia. The only note of the direction of the wind is unfortunately that of the hurricane at Lefoo, which veered only from S. E. to E. S. E., so that this being from the quarter of the trade wind, we are not, on such a meagre notice, yet fairly entitled to consider this as a Cyclone, except from the fact of *hurricanes* being so well known there and the great fall of the Barometer.

Since the second edition of this work I am indebted to Mr. SWANEY, chief officer of the ship *Futtay Salam*, for a valuable note on this hurricane, which was doubtless a true Cyclone. He says, writing from memory only, that he was hove to for four days, 12th to 15th February, off the N. W. end of New Caledonia, the gale commencing at S. W. to W. S. W. veering gradually to the N. Westward where it was heaviest and ending at N. E. The *Sarah* and *Castlereagh* were lost on the 13th with the wind at S. East at 4 A. M., Mr. SWANEY

---

* A fall of about an inch; for the lowest average we can suppose in this Latitude must be 29.75 at the most.

having the heaviest of the gale on the same day at W. N. W. or N. W., his position at noon by good observation being 14° 5' S. Long. 163° 55' East. No observation for four days previous. Torrents of rain on the 11th and 12th. He adds, that he learned from subsequent inquiries amongst friends, that the centre of the Cyclone had passed over New Caledonia a little to the Southward of Balade and was felt on the East side some hours before it reached the West side of the Island. Trees of the largest size were torn up by the roots "and blown about like straws," so it was expressed. He learned also that both the *Sarah* and *Castlereagh* might have been saved *if they had shifted their berths in time !* The *Isabella Anna*, an old ship, lying in a secure harbour on the S. E. side of New Caledonia, suffered so severely that she could not go to sea again, and was broken up.

The sea to the Westward of New Caledonia (perhaps the whole extent of it to the Barrier Reefs ?) is certainly subject to Cyclones, and they moreover appear at times to pursue curved tracks, so that this may not improbably, like the corresponding latitudes of the Indian Ocean between Mauritius and the South end of Madagascar, be one of the usually curving tracks for the Cyclones of the Southern Pacific. From a log of the Barque *Rifleman* sent me by Captain HAMMACK, it appears that in running up to pass between the Western Coast of New Caledonia and the Bampton Shoals, she encountered on the 2nd March, 1847, (the hurricane season also at the Mauritius,) a Cyclone in Lat. 21° 22' S. ; Long. 157° 10' East. The wind veered with her from E. b. N. by S. E. and S. W. to West and N. N. W. in 3¼ days, while she was hove to ; her Barometer falling from 29.40 to 28.80 ; and being then obliged to bear up from about Lat. 19° 30' S. to the S. E., to run round the South end of New Caledonia, for passing to the Eastward of that island, as she could not weather the shoals to the Northward ; she again fell in with the Northern quadrant of the Cyclone which left her only in Lat. 24° S. and Long. 164° East. From apparently some error in copying, I am unable to say with perfect certainty that this was a curved track, as there is a remote possibility that there may have been two Cyclones, but I incline strongly to take it as an instance of a re-curving track in the Southern Pacific. Be this as it may, the seaman will now be upon his guard in this passage which is much frequented by vessels from Sydney to China and India, and being embarrassed with shoals and lee shores the Cyclone becomes here doubly formidable.

74. AT NEW ZEALAND there is no doubt that true Cyclones sometimes occur, and these of considerable violence. In the U. S. Exploring Expedition, Vol. II. p. 381, is a very good account of one which occurred on 29th February, 1840, at the Bay of Islands; said to have been the severest which the Missionaries had experienced there. It was felt at other stations, with all the veerings, calm centre, &c., of a true tropical Cyclone, its course being to the S. Westward.* In July, 1840, H. M. S. *Buffalo* was wrecked, on the 28th, in a heavy gale which lasted three days, at Mercury Bay, New Zealand. About that time also three American Whalers were wrecked at Port Leschenault in one of the strongest hurricanes ever experienced by their commanders.

75. In the Nautical Magazine for September, 1847, are some remarks on New Zealand, by Commander C. O. HAYES, of H. M. Steamer *Driver*, and at p. 465, he says, speaking of Cook's Straits—

"The gales always blow between South and East, and North and West. They blow with great violence from both these quarters, but from N. W. they generally come on gradually; they shift suddenly from that quarter to the S. E. and blow with great violence, with exceeding heavy squalls. When these gales come on, it is advisable to seek an anchorage (of which there are plenty) without delay."

* It is said to have passed between the Bay of Islands and the River Thames, which lie about S. S. E. ¼ E. and N. N. W. ¼ W. of each other, hence the track may have been either to the Northward or Southward of S. West. Its track to the S. Westward or perhaps S. S. W. after crossing the Island, I am enabled *perhaps* to corroborate from a log in my possession, of the Ship *Adelaide*, which vessel between the 1st and 2nd March, about 3½ degrees due West of Cape Egmont, experienced a smart " gale," commencing at about E. b. S. or E. S. E. and veering to S. b. E., reducing her to close reefs with a heavy cross sea. This roughly calculated gives 340 miles for 36 hours, or about ten miles per hour. Commodore WILKES suggests that this New Zealand Cyclone may possibly have been the same as that which occurred at the Feejee Group, which is very probable. Mr. SWANEY in his note already referred to (p. 76) says that he was in this gale, which, to the N. W. of Cape Maria Van Diemen, with him was a N. W. gale, and that its centre passed to the S. W over the Northern part of New Zealand, about Doubtful Bay. The *Harriet Falcon*, and many other vessels were lost in the Bay of Islands, Bay of Plenty, and other parts of the coast.

PART II. § 76.]   *Tracks; Van Diemen's Land, Cape Horn.*   79

This shift, which is of course the average one only, would indicate a track *from* the S. W. towards the N. E., or one exactly opposite to that described above.

While this page is passing through the press, I receive from Capt. ERSKINE of H. M. S. *Havannah* the log of that vessel from the 3rd to 5th July, 1849, which while it shews again as before (page 40) the advantages which the scientific or even the attentive seaman may reap from the study of our science, gives us also the anomaly of a track *to* the N. E. in the Southern Pacific, between the parallels of 19° and 27° S. The *Havannah* was bound from New Zealand to Savage Island, Lat. 18° 58′ S.; Long. 169° 51′ East; and when standing to the Eastward as usual in Lat. 27° South at noon on the 3rd July, with Savage Island bearing N. 21° East, distant 530 miles, she fell in with a Cyclone (the Barometer falling from 29.78 to 29.58 in four hours) which Captain ERSKINE judged to be travelling to the N. E. and forthwith altered his course to the N. b. E. to take advantage of the Westerly and W. N. Westerly winds, on its Northerly and N. Easterly quadrants for a run across the trade; so as to make a direct course instead of an angular one as usual; and in this he perfectly succeeded as she carried the Cyclone to within sight of Savage Island at daylight on the 6th, and, with a few hours of calm, had the trade again on the 7th. This track fully confirms the inference above from Commander HAYES' note, and it warns us too that, in a sea so extensively studded with islands, variations from the usual laws are to be looked for.

76. In the great space lying between Van Diemen's Land and Cape Horn, we have scarcely any observations, but in a capitally well kept log of the ship *Lord Lyndoch*, by the late Captain CLAPPERTON, formerly Master Attendant at Calcutta, I find that, in the month of December, 1820, in Lat. 45° South, Long. 117° West, a "gale" was experienced which veered from N. W. to S. W. in 14 hours, or about half a point an hour, in which time the ship, standing to the North 60° East, made about 83 miles on that course. If this was a rotatory storm, (and the Bar. fell from 29.70, to 29.07) it passed her to the Southward, on a track a little to the Northward of West, and travelled at the rate of about 15 miles per hour. Judging from the fall

and subsequent rise of the Barometer, as well as the veering of the wind, there seems no reason to question that it was so.

77. In a newspaper article copied from the *Sydney Herald*, I find the following extracts and notes, evidently written by some person who understood and was endeavouring to teach the science,* but I regret that of a series of papers only a part of one has fallen into my hands. "So completely does the law of rotation appear to be from left to right, in gales of wind off the coasts of Australia and on the neighbouring ocean, that it is scarcely possible to escape the observation, in perusing the log books of any extended cruize. One farther example, to shew this, shall now be quoted.

"The whaler *Merope*, left Sydney 22nd March, 1840, with wind at the South, steering for Lord Howe's Island. On 27th, she was in Lat. 35° 4' S. Long. 158° 35' E. The order of the wind's changes was as follows: 23rd, at S. E. veering N. E.—24th, N. N. E. and N. E.—25th, increasing from N. E., N. N. E., E. N. E. with a tempest.—26th, N. E. to N. with confused sea; N. W. and drawing to W.—27th, S. W., S., S. S. E., and back to S. The wind completed a revolution in five days, on a direct course, from left to right.†

---

* So far the first edition. I have since learned that the author of this and of numerous other meteorological notices in the Sydney papers is the Rev. W. B. CLARKE, St. Leonard's Parsonage, Sydney, who has obliged me by some of his papers. He is most strenuously labouring in the good cause of science, in his endeavours to throw light on the Meteorology of that most interesting quarter of the globe.

† A reference having been made to me from London stating that this passage was thought something obscure, I subjoin my explanation of it.

1. The ship's course *made good*, was about an E. b. S, one from Port Jackson to her position on the 27th, though bound to the E. N. E., *i. e.* she was first driven to the S. E. and S. S. E. and South by the E. N. E. to N. N. E. gales, and then no doubt ran back to the North East and N. N. East, when the wind shifted or veered. How did the changes occur?

2. If we take the wind of the 24th to be the commencement of the Cyclone, which it probably was, its average was N. E. b. N., the centre of the Cyclone then bore N. W. b. W. from the ship, at some unknown distance, of which we can form no judgment, as no Barometer observations are given. The ship probably made nothing better than a S. E. or S. E. b. E. course of three or four knots per hour, or say at best, 100 miles S. E. ¼ E. in this 24 hours.

3. On the 25th, it became a tempest; from N. E. to N. N. E. and E. N. E., or say the average was N. East, so that the centre now bore N. W. of the ship, and was so

"Between Australia and America, a similar course is pursued by the winds to that which is followed between the Cape of Good Hope and Cape Leuwin, and more than one instance has come before us of vessels having been driven *all round the compass before the wind* during a gale, not far from Cape Horn.

"The following examples from Captain STOKES' Journal (given in Vol. III., of the *Beagle's* Voyage) shew the general character of the gales on the West coast of America. About the Lat. of 50° S. on 5th April, 1828, a gale came on from N. off Cape Tres Puntas, blowing on the 6th, 7th, and 8th, from North, N. W. and S. W. with squalls, thick weather and rain. It abated on the 9th, veering to Southward and then to S. E. when it ceased. This was *from left to right*.

"On the 10th April, another gale came on from N. W., which as suddenly subsided in the western quarter. This, Captain STOKES says, 'was singular, for those we have experienced *generally* commenced at North, thence drew round to the Westward, from which point to S.W. they blew with the greatest fury, and hauling to the Southward, usually abated to the Eastward of South,' (p. 192.) These gales therefore rotate *from left to right*."

78. It would seem that, as to violence at least, hurricanes are felt as far South as Patagonia, where (Nautical Magazine for May, 1846,) "a severe S. S. E. hurricane" is said to have "swept the coast" from

much nearer being now much more violent. It had, in short, nearly travelled down in the wake of the ship whatever her drift (or course made good) might have been; and from noon this day to the 26th, we can only allow about 40' or 50' South.

4. On the 26th, it is N. E. to North, with a confused sea. Then N. W. drawing to West. The N. E. to North (if this be not a newspaper misprint) is probably an effect of the incurving of the wind as the centre passed the vessel; and this would give an average wind at N. N. E. or centre bearing (close on the vessel) W. N. W. It then veered or shifted (for the account is too condensed to say which) to N.W. drawing to West, or in other words, the centre passed close to the vessel on about the same track, (N. N. E. to N. W. gives a track to the S. S. E.) and she profiting by the veering to the Westward probably stood up on the other track to the Northward, while the Cyclone was passing away to the S. Eastward of her gradually giving her Westerly, and on the 27th, S. Westerly and Southerly winds as it passed over.

This is all we can deduce from so meagre a notice, but it is quite reconcileable with the hypothesis of its having been a Cyclone, and quoting from the author's own words, I did not like to add to them.

the Bay of Camaras to the island of Desejada, occasioning the loss of twelve English and American vessels. We do not know if it was rotatory, or what was its track.

79. In the Voyage of DON JUAN de ULLOA in 1743, speaking of the weather on the Coast of South America,* we find some account of storms which resemble rotatory ones, and the latter part is in fact a description of one. I have abridged the extract a little, in parts unessential to our present subject.

"Tempestuous weather is equally common in the latitudes of 20° and 23° in the South Sea as in the Oceans of Europe. Along the coasts and adjacent seas, the winter begins in the month of June and lasts till October or November, its greatest violence being past in August or September. Storms which arise with great rapidity are very frequent during the whole winter; Northerly winds are very prevailable, and often of extreme violence, raising a tremendous sea. It often happens that these violent North winds, without the least sign of an approaching change, shift round instantly to the West, which change is called the *travesia*,† but continue to blow with the same force. Sometimes indeed this sudden change is indicated by the horizon clearing up a little in that quarter: but in seven or eight minutes after the appearance of this small gleam of light a second storm comes on; so that when a ship is labouring against the violence of a storm from the N. the greatest care must be taken, on the least appearance to prepare for the *travesia*; indeed its rapidity is often such as not to allow time sufficient for making the necessary preparations, and the danger is sufficiently violent if the ship has her sails set, or is lying to.

"In the month of April, 1743, in the latitude of 40°, I had the misfortune of experiencing the fury of a storm at N., which lasted in its full violence from the 29th March till the 4th of April. Twice the wind shifted to the *travesia*, and veering round to the Southward, returned in a few hours to the North. The first time it shifted to the West, the ship by the vortices formed in the sea by this sudden opposition to the course of its waves, was so covered with water from head to stern, that the officers concluded she had foundered, but fortunately we had our larboard tacks on board, and by a small motion of the helm, the ship followed the change of the wind, and brought to without receiving any damage; whereas we should otherwise, in all probability, have been lost. Another circumstance in our favour was, that the wind was some points to the Westward of N., for though these winds are here called *Nortes* (Northers) they are generally between the North and N. W., and during their season veering in some squalls to the North, in others to the N. W.; sudden calms

---

* I am quoting from a written extract from the English translation, in which the page and volume are not marked.

† A Spanish word meaning Oblique or transverse, used also for a side or contrary wind. The change we see *is* one of eight points as here described.

also intervene ; but if these happen before the wind has passed the *travesia*, it returns in about half, or at least an hour with redoubled fury. These dangerous variations are however indicated by the thickness of the atmosphere, and the dense clouds in the horizon. The duration of these storms is far from being fixed or regular ; though I well know some pilots here will have it that the North wind blows twenty four hours, and then passes the *travesia* ; that it continues there with equal violence three or four hours, accompanied with showers, which abate its first violence ; and that it then veers round till it comes to the S. W., when fair weather succeeds. I own indeed that I have in several voyages found this to be true ; but at other times I experienced that the successive changes of the wind are very different.* The storm at North I before mentioned, began March the 29th, at one in the afternoon, and lasted till the 31st, at ten at night, which made 57 hours ; then the wind shifted to the *travesia*, where it continued to the 1st of April without any abatement, that is, during the space of twenty-two hours. From the West the wind veered round to the W. S. W. and S. W., still blowing with its former violence. Hence a short calm succeeded, after which it a second time shifted to the North, where it continued blowing with its former fury fifteen or twenty hours ; then came on a second *travesia*, and soon after its violence abated, and the next night shifted from S. W. to S. E. Thus the whole continuance of the storm was four natural days and nine hours ; and I have since met with others of the same violence and duration, as I shall mention in their proper place. What I would infer from my own experience, confirmed by the information of several pilots, is, that the duration of these storms is proportional to the latitudes, being between twenty and thirty degrees, neither so violent nor lasting as between thirty and thirty-six, and still increasing in proportion as the latitude is greater.

"These winds have likewise no regular or settled period, the interval betwixt them being sometimes not above eight days, at others much longer ; nor do they always blow with the same violence, but are most uncertain in the winter, rising suddenly when least expected, though not always blowing with the same force.

"In this sea a change of the wind from N. to N.E. is a sure sign of stormy weather, for the wind is never fixed in the N. E., nor does it ever change from thence to the E., its constant variation being to the W. or S. W. *contrary to what is seen in the Northern hemisphere*. Indeed in both the change of the wind usually corresponds with the course of the sun, and hence it is, that as in one hemisphere it changes from E. to S., and thence to the W., conformable to the course of the luminary, so in the other it changes, for the same reason, from the E. to N. and afterwards to the West.

"It is observed, that within thirty or forty leagues of the Coast of Chili, while one part is agitated with storms at N. the S. winds freshen in another. This, however singular it may appear, is no more than what was experienced by the three ships *Esperanza, Belen,* and *Rosa,* which being at the mouth of the bay of Conception, the latter took her leave of them, and bore away with a fresh gale at S. to Valparaiso,

---

* When in a different part of the storm circle.

whilst the others who steered for the islands of Juan Fernandez, were overtaken in their passage by a storm at N."

The foregoing extract is long, but it is the best authority I have for the tracks. The change from due north to due West, with or without an interval of calm, would indicate a storm coming in from the Pacific and from the N. Westward, supposing always the law of rotation to be the same here as it has been shewn to be in other parts of the Southern hemisphere. Don J. de ULLOA's own account in the next paragraph, describes accurately enough either two Cyclones following close upon each other, or a vessel drifted round and round the same vortex, and in the absence of any more detailed account it is impossible to say which it was. The changes are evidently those of a Cyclone of very slow progression, coming in from the N. Westward. And this is farther confirmed by the account of the three ships in the last paragraph. We may therefore venture to say for the present that the storm tracks between 40° S. and 20° S. between the meridian of Juan Fernandez and the coast of South America, appear to be from the Westward (and *probably* from the *North* Westward) towards the Coasts. See p. 50 for what is said there of the tracks on the West Coast of Australia.

80. From 20° South to the Equator and thence to the Gulf of California, our information is still very deficient. Mr. REDFIELD and Colonel REID incline to think that the storms on the coasts of Nicaragua, Guatimala and Mexico are connected with those of the Gulf of Mexico, or perhaps originate there. In a recent memoir,* Mr. REDFIELD says:

"According to Humboldt, both the Eastern and Pacific coasts of Mexico are rendered inaccessible for several months by severe tempests, the Norths prevailing in the Gulf of Mexico, while the Navigation of the western (Pacific) coasts is very dangerous in July and August, when terrible hurricanes blow from the S. W. At that time, and even in September and October, the ports of San Blas and Acapulco are of very difficult access. Even in the fine season, from October to May, this coast is visited by impetuous winds from N. E. and N. N. E., known by the names of Papagallos and Tehuantepec.

"It appears in like manner that the coast of Nicaragua and Guatimala in the Pacific, is visited by violent Southwest gales in the months of August and September,

* American Journal of Science and Arts, March, 1846. New Series, No. 1, p. 164.

known by the name of Tapayaguas, which are accompanied with thunder and excessive rains, while the *Tehuantepec* and *Papagallos* exert their violence during a clear sky.

"This seems to shew that the so-called Papagallos, Tehuantepec, and Norther of Vera Cruz, severally, are but the clear weather side of a revolving gale, like the North-wester of the coast of the United States, each in its turn being but part of a great vortical storm, which in certain other portions of its area, or route, often exhibits an abundance of rain.

"Humboldt suggested that these Northerly winds may blow from the Atlantic and Gulf of Mexico to the Pacific, and that the Tehuantepec and Papagallos may be merely the effect, or rather the continuation of, the North wind of the Mexican Gulf and the Brizottes of Sta. Martha. But the vortical character and determinate progression of violent gales was then unknown, and I cannot doubt that the Northers which visit the Pacific coast and the Gulf of Tehuantepec precede, in point of time, the same storms in the Gulf of Mexico, and are identical with them, having, commonly in this region, a Northerly progression."

Colonel REID says,

"It is possible that the Spaniards may apply the term Nortes, or Northers, to more than one phenomenon; but the violent North winds in the neighbourhood of Vera Cruz are frequently no other than the left hand side of rotatory storms in their Northerly progress across the Gulf of Mexico."* In this remark I fully concur.

In Chap. IX. of a work entitled "Two Years before the Mast," by R. H. DANA, junior, which is not a very authentic authority to quote it is true, were it not that the author has deservedly a high reputation for the fidelity of his descriptions, which every seaman will recognize and admire, it is said, in a description of a S. Easter, obliging his vessel to put to sea from Santa Barbara (Lat. about 24° N.?) on the coast of California, that they seldom last above two days and are often over in twelve hours. In the one described they put to sea and lay to in the S. Easter, till the calm came, and the gale came on again from the N. W. This would indicate a track in towards the shore from the S. W. to the N. East, but the N. W. part (or S. W. quadrant of the storm) seems to have been of very short duration, as we so often find it in all parts of the world.

81. Of the tracks of storms, and even if any are rotatory in the European Seas, we have but little information, and in truth this may remind some to be thankful for what is already collected in distant

* Second edition of "Attempt to develop the Law of Storms," &c. London, 1841.

oceans, when they see how little is yet accomplished at their own doors. Beginning then with the Mediterranean, it would appear from the account published by Mr. THOM, p. 262, that tornados or tornado-storms are certainly experienced there. The N. Westerly gales in the Gulf of Lyons, those which are at some seasons of the year common in the Ægean, and especially those of the Black Sea, may all be either rotatory or straight-lined winds,* or those which appear to be straight-lined, may in reality be parts of great circles or Cyclones. Nelson, whose wonderful blockade of Toulon with an inferior force, and a fleet not over well supplied, is justly accounted a master-piece of seamanship, and is still inexplicable to our neighbours the French, says (Dispatches and Letters, Vol. VI. p. 156) when writing to the Duke of Clarence from off Toulon, and speaking of these gales:

"I have always made it a rule never to contend with these gales, and either to run to the Southward to escape their violence, or furl all the sails and make the ship as easy as possible."

This would lead us to suppose that those in which he had found it advantageous to run to the Southward, might be Cyclones of which the N. West gales were the S. Westerly quadrants, and a short run took the fleet out of the violent part of them, and gave it variable or Southerly winds afterwards to get quickly back to the cruising ground; and that those in which he had to heave to under bare poles were the westerly gales of the Atlantic. He had no doubt his " Barometer of Signs," a term which I shall explain when speaking of the signs of approaching Cyclones, as well as his Mercurial one, and we shall subsequently own that he made excellent use of both. In the Miscellanea, Part VI, (p. 368) will be found some grounds for supposing that this gale of January, 1805, alluded to particularly in the Barometrical

---

* And as not very remote we may add to these the Caspian, which from the days of Horace has been noted for its sudden and violent tempests. In Dr. E. D. CLARKE's Travels, Vol. II. of the octavo edition, is an account of a violent storm which he experienced in the Venetian Brig *Moderato*, at the Mouth of the Straits of Marmora, (Constantinople,) which appears to have been a violent little Tornado storm moving up to the E. b. N. or E. N. E. with some peculiar appearances of " whirling clouds," of which it is to be regretted we have not more details.

Section, (p. 230) in which Villeneuve escaped from Toulon (while Nelson was watering at the Maddalena Islands) but had to put back with most of his fleet disabled, was a Cyclone travelling from about Genoa to the W. S. W. or S. W.

In the Nautical Standard for the 20th January, 1849, is an account of a severe gale at Malta, in which several vessels were seriously damaged and one lost; and this was in the safe creeks of the harbour of La Valette. It is stated that,

"On the 26th December, 1848, the wind was squally from the S. W. with rain,* but nothing indicated the coming storm. The violence of the wind caused many vessels to take shelter in port, and thinking the same wind would hold they anchored, some in Port Calcara and some in Bighi Bay. At 2 A. M. on the 27th the wind suddenly chopped round to N. N. E. and blew so violently that the vessels thus anchored had not time to secure themselves from disasters."

The vessels putting back, and disasters of various kinds are then noticed, and it is remarked that on the evening of the 27th the sea calmed a little, but on the morning of the 28th it began to blow again with greater violence, so that the vessels most exposed to the violence of the sea were all abandoned by their crews until the gale subsided, and all more or less suffered damage. The English and French steamers in the ports kept their steam up constantly for two days. To judge from this imperfect account we may suppose that the "gale" was possibly, and probably, a Cyclone, and if we take the shift to have been from W. S. W. to N. b. E. the mean between two accounts,* this will give with 1½ point of Westerly variation a track from the N. W. b. W. to the S. E. b. E. The "gale," as it is termed, did not finally subside there till the night of the 28th; I have marked this as a track (a) upon Chart No. I. This storm *may* have been the same which five days afterwards reached Constantinople, and its slow progress and curving we may suppose owing to the intervening land. The following is from Bell's Messenger of 27th January, 1849.

DESTRUCTIVE HURRICANE AT CONSTANTINOPLE.—In a letter dated Constantinople, Jan. 7th, we find the following:—"On the 3rd of this month we had here one of those dreadful hurricanes which are only to be found in the annals of the West Indies. At half-past ten at night a heavy shower of rain fell, which was

* The Malta Times says, the wind was blowing steady from West and W. S. W. and shifted to North and N. b. E.

rapidly succeeded by a snow storm; at about midnight it blew a thorough hurricane, which lasted till day-break. Never in the memory of man had such a storm burst over Constantinople. Trees were uprooted, houses unroofed, chimneys hurled to the ground. Several minarets have lost their kullafs, or extinguishers, and the steeple of the Galata Tower was blown clear off. The old bridge was broken to pieces, and, floating about the port, caused considerable damage to the shipping. The losses have been calculated at upwards of 10,000,000 of piastres. Among other losses an English brig, name unknown, was lost at the mouth of the Black Sea."

On the Coast of Syria from Acre to Beirout, and in the Eastern parts of the Mediterranean, the British Fleet on service in that quarter, experienced in December, 1840, a most severe gale, of which in the first edition of this work, I could give only an imperfect notice from the Nautical Magazine. Col. REID in his new work, p. 279, has given the result of the examination of the logs of the fleet, and printed in detail those of H. M. S. *Vanguard, Rodney, Bellerophon,* and *Magicienne,* and on his Chart has laid down the track (Track U on Chart No. I.) as one from about W. $\frac{1}{4}$ S. to the East $\frac{1}{4}$ N. and there is no doubt that it was a true Cyclone.

82. For the British Channel, the Chops of the Channel, the British Islands and the neighbouring seas, our information is as yet very imperfect,* a few Cyclones only having as yet been investigated by Mr. MILNE, Col. REID and others. I have marked most of these *approximately* on the Chart in addition to the two (and V and W) which were already set down by Col. REID as follows:

V and W, 26th and 28th Nov. . . . 1839.
X, 11th Oct. . , , . . 1838.
Y, 28th Oct. . , . . . 1838.
Z, 7th Jan. . . , . . 1839.

From a consideration of these it will appear that the tracks may lie between those from the S. West to the N. East, and from the N.

* And until the Science is made, as it assuredly ought to be made, as much a matter of National research as the Trigonometrical Survey or other establishment of the kind, it is to be feared it will for a long time remain so. If we were even to suppose a sailor of independent fortune, and with all the required talents, devoting himself to the research, he would still want the influence to persuade, and oftener the power to *demand* reports and returns ; and would, in all probability, when writing from London to Oporto, Archangel, Aleppo or South America, for data of great importance, find that his requests were useless, unless he had an agent on the spot !

West to the S. East, the Cyclones being of various sizes and strength, from gales to downright hurricanes.

83. The AVERAGE RATES AT WHICH CYCLONES TRAVEL ON DIFFERENT TRACKS is also an essential part of the seaman's knowledge, to enable him to form the best judgment of how he may avoid or profit by a Cyclone. These rates vary greatly, not only in various parts of the world, but even in the same localities, and at the same season; nor does the size of a Cyclone afford any rule whereby to estimate its rate of travelling, as both large and small ones are known to move with great rapidity, or at moderate or slow rates, without, apparently, any sort of law. It has been conjectured that the vortex below was carried forward by currents of wind above, as we see waterspouts move with the clouds from which they proceed; but it is objected to this, that the upper strata of clouds are often seen, through breaks in the storm, to move across or against the track of the Cyclone.* We are thus as yet as entirely in the dark as to what causes their progressive motion, as we really are as to what gives them their violent gyrations; and we are still more so, if we consider that like sand-storms or dust whirlwinds, some of them appear to remain comparatively stationary for hours, or a day or more (see p. 42) moving at a rate of 1·5 or 2 miles an hour, and then to start off, as it were, on a track upon which their size and velocity gradually increase! We find them also usually much diminished in their velocity when they pass over land, especially if it be high, and it would appear too that storms sometimes *decrease* in size and in rate in travelling, but augment fearfully, in so doing, their violence,† and this when passing over the open ocean.

It is then quite impossible to say to what their progression is owing, but the seaman will not fail, I trust, to recollect that all his

* Which itself has a track against the prevailing winds, as in the Atlantic hurricanes moving to the N. E. *against* the trade; or oblique to them, as the hurricanes of the Southern Indian Ocean, moving to the W. S. W. *across* the S. E. Trade, and those which occur in both monsoons in the Bay of Bengal and in the China Sea also moving obliquely, or at right angles to it.

† A well marked case of this occurred in the Bay of Bengal in Nov. 1839, when Coringa was devastated by a furious Cyclone, which was traced from the Andaman Islands across the Bay. Jour. As. Soc. of Beng. vol. ix. p. 438.

observations may be useful to those who are ready to turn them to the best account, if he will but register and communicate them; and that what may appear to him at the time a very insignificant occurrence may, by chance, aid greatly in working out some mysterious branch of the research. In giving the rates, then, at which Cyclones travel, as far as is yet known, I shall follow the order in which I have described their tracks, passing over localities in which we have no good data for their rates of travelling.

84. In the West Indian and North American Cyclones the highest rate given by Mr. REDFIELD is 43 miles per hour,* and the lowest 9.5 miles. The mean rate is thus 26 miles, which last or perhaps from 16 to 20 miles per hour, should always be taken as the lowest rate at which a storm in this part of the world *may* be moving. The Atlantic Cyclone noted at p. 32 appears to have travelled at about 11.4 miles per hour.

85. In the Southern Indian Ocean Mr. THOM considers that the rates of travelling may be from 9' and 10' to little more than two miles per hour, and that the diminished rate occurs about where he supposes the storms to terminate near the southern tropic, and on the meridians between Madagascar and the Mauritius. Col. REID has laid down from 7 to $12\frac{1}{2}$ miles per day, in his Chart of the Cyclone of 1809, No. VIII.† My own researches certainly prove, that about the part I have called the "Storm tract," (see p. 48) the Cyclones are of very slow progression, being from $2\frac{3}{4}$ to $1\frac{1}{2}$ miles per hour only, and upon a singularly curved track. I shall advert in the following section more fully to these nearly stationary storms.

86. In the Mosambique Channel the rate of the *Boyne's* Cyclone in 1838, is laid down as about 10 miles per hour, though from the log of a single ship all these calculations are very uncertain.

87. In the Arabian Sea, from the scanty data we have, we may estimate from 4 to 16 miles per hour, as the rate of progression. Our information here, however, is yet very deficient.

88. In the Bay of Bengal, my researches, both published and un-

---

* The Cuba Hurricane of Oct. 1844.

† Deduced, however, from imperfect data, the fleet being much together, so that the extent of the storm circles is partly conjectured.

published, enable me to say that the Cyclones travel at rates of from little more than 2 to 39 miles per hour, but this last very high rate has occurred only in one instance, and from 3 to 15 miles may be taken as the usual rates.

The low rate of " little more than two miles an hour" (53 miles in 24 hours) is that of the tremendous Cyclone, and inundation of Burisal and Backergunge at the mouth of the Burrampooter and Ganges, in June 1822; in which upwards of 50,000 souls and vast property in houses, cattle, &c. perished. I have investigated this Cyclone, and its track is that marked f on Chart III.

The great rise of the waters was probably owing in part to its long action over one point, and in part to its being there a S. E. storm all day at Burisal, which is exactly the wind required to dam up the stream of the great estuary of the Burrampooter and Ganges.

89. In the Andaman Sea, where as before noted Cyclones of terrific violence do occur, the one which has been traced travelled at the rate of about four miles an hour only.

90. Off the Coast of Ceylon the small tornado-like storms (true Cyclones) of excessive violence, alluded to at p. 54, travel perhaps at from five to ten miles an hour or more.

91. In the China Sea, from the result of all my investigations, we may safely set down the rate of progression at from seven to 24 miles per hour.

92. The scattered notices and single logs which we have for the Pacific ocean, though they enable us to announce with certainty that the storms there *are* often Cyclones and of great violence, yet do not enable us to give any rate for their progression. The seaman, however, who is threatened with one will do well never to allow less than ten or twelve miles an hour, and always to risk as little as possible upon the chance of the Cyclone's not overtaking him.*

93. STATIONARY CYCLONES. As previously remarked in this chapter, there is no doubt that some Cyclones are so slow in their progress that they may be almost considered as Stationary. Every sea-

---

* Scudding parallel to a Cyclone may be, and behind it, *is* safe enough, but to attempt to scud "in front" of one is often highly dangerous, on these very accounts, namely, because it is impossible to say what is its rate of travelling and its size.

man has remarked of waterspouts, that they often appear stationary for a time and then move forward; and in tropical countries the same certainly occurs with the dust whirlwinds so common there, and with the sand-storms of the desert, according to Bruce's picturesque account. The same on a large scale appears certainly to take place at the commencement of some Cyclones in the Indian Seas. And it is not impossible that it may sometimes also occur towards their close. Those of which we have any distinct notice are the *Albion's* Cyclone of 1808, which, says Colonel REID, "seems more to have resembled the commencement of a whirlwind floating with an irregular motion, as waterspouts do in calm weather." This was in the Storm tract, as I have called it, between 5° and 15° S. and 75° and 90° East; and in the Cyclone of November and December, 1843, from Lat. 5° to 12° S. and between 82° and 90° East, investigated by me* from the logs of several ships, the actual rates of progression were found to be 60, 82, 135, 47, and 57, miles in each twenty-four hours, for five days, during the whole of which it was raging most severely; and the number of the logs and position of the ships enable us to speak of this instance with much confidence.

94. In the Burisal and Backergunge Cyclone; Track *f* on Chart No. III. to which I have just alluded, it appears that although it moved, as I have stated, about 53 miles in the twenty-four hours, yet I have reason to believe that for at least twelve of these hours it was really stationary, or nearly so, over the town of Burisal. In giving these average rates of motion, we can only state what is shewn from one noon to another, and in this case the 53 miles probably were accomplished mostly in the last 12 hours of the day.

95. Of the Cyclone of September, 1838, at the Bahamas, and along the coast of America, in which H. M. S. *Thunder* and *Lark*, were in great peril, as noted by Colonel REID, p. 433, and in the Nautical Magazine for 1839, Colonel REID is of opinion that, over the Bahamas, and for a time at least, this was a hurricane resembling that of the *Albion* in 1809, and floating about in no fixed course. He remarks also that this occurred about the recurving point of the West India Cyclones.

* Eleventh Memoir, Jour. As. Soc. Beng. Vol. XIV.

96. In PURCHAS' Pilgrims, Vol. IV., Book VIII., Cap. 14, p. 1679, is an account of the Azores, taken from LINSCHOTEN, which, after relating the gallant feat of Admiral Sir RICHARD GREENFIELD in the *Revenge*, in engaging the whole Spanish fleet, in September, 1591, goes on to give an account of a dreadful Cyclone following it, in which upwards of 100 of the Spanish fleet, of 140 sail, were lost at Tercera; and it is remarkable that this must either have been a stationary Cyclone, or two Cyclones at least, following so close upon each other, as to admit of no intermittence in the bad weather; for the writer says—

"This Storme continued not onely a day or two with one winde but seven or eight days continually, *the winde turninge round about in all places of the compasse at the least twice or thrice during that time*,* and all alike with a continuall storme and tempest most terrible to behold, even to vs that were on shore, much more then to such as were at sea."

Captain SILVER amongst other notes gives the following which is worth printing as corroborating, in the Western Hemisphere, the beautiful lesson of the *Charles Heddle* in the Indian Ocean. It shows also that there are Cyclones in those seas almost stationary.

"I was once caught in a Norther in the Caribbean Sea between New Grenada and St. Domingo. The Barometer fell very low; ship under close-reefed topsails; I ran round and round the compass for 36 hours in the focus of the storm and lost both top-gallant masts and jib-boom, and on the second day when it moderated and we got sights for chronometer, and Lat. at noon, I found that we were only 40 miles from where I first started."

97. THE SIZES OF CYCLONES. We may suppose that there might be a complete chain, as to size, of Cyclones; from the waterspout, which becomes a whirlwind when it reaches the land, through the Tornado of a few tens or hundreds of yards in diameter, up to the great hurricanes of the Atlantic or Indian Ocean; and so far it is certain that we cannot on the one hand say *how small* true Cyclones may be, as we have traced them to probably less than 100 miles in diameter† and possibly as small as 50‡ in the Indian seas. On the

---

* Italics are mine.

† See Fourteenth Memoir, Journal Asiatic Soc. Beng. Vol. XIV.

‡ Cyclone of the *Cashmere Merchant*, Nov. 1839, Second Memoir, Jour. As. Soc. Beng. Vol. IX.

other hand, when we come to the smaller, Tornado-like, Cyclones below say, fifty miles in diameter, there is no good evidence, as yet, for their revolving invariably in the same direction when occurring in the same hemisphere, as the larger storms; and thus we cannot distinctly affirm that the very small Cyclones are subjected to the same laws, or arise from the same causes. This is one of those questions which require frequent and extensive observation to settle. For the sailor, however, it may suffice to say that Cyclones revolving according to the usual law, may be looked for of all sizes from 50 to 500 or even a thousand miles in diameter; the very large and very small ones being comparatively rare, and the small ones often sudden and severe. There is no doubt also that at times they both dilate and contract, in their progress. We cannot certainly say if, when they become larger, they become more or less severe, but I am inclined to think that when they contract they sometimes augment fearfully their violence, approaching apparently the concentrated power of the Tornado or the Whirlwind. The following are instances of Cyclones dilating and contracting, and of their usual sizes in known localities.

98. In the West Indies the researches of Mr. REDFIELD and Colonel REID seem to shew that though while approaching to, or within the islands, they are sometimes as small as 100 or 150 miles, in diameter; they may, and it would seem most frequently do, after reaching the Atlantic Ocean, dilate considerably, and there often attain to 600 or even 1000 miles in diameter, the wind blowing an excessively severe gale over all this area, and towards its centre becoming of true hurricane violence, and the whole vortex, so whirling, travels over thousands of miles of track. I have inserted on the Atlantic chart, from Colonel REID's new work, his two large dotted circles, one on the American and the other on the European shores of the Atlantic, shewing the vast extent to which the winter Cyclones sometimes extend, the American one being a Cyclone of December, 1839, traced by Mr. REDFIELD, the other one of November, 1838, investigated by Mr. MILNE, of Milne Graden, in the Trans. R. S. Edinburgh, for 1839. It will be seen that the outer limits of these Cyclones extend to 20° of radius from the centre, and the limits of the true Cyclone circle to 12° or from 1400 to 1500 miles of diameter.

PART II. § 101.]   *Sizes of Cyclones varying.*   95

99. In the Southern Indian Ocean, Mr. THOM thinks that "hurricanes when first discovered are from 400 to 600 miles in diameter." My own researches shew that they may be as small as 150, and the researches of Colonel REID and myself both carry them as far as to 600 miles in diameter.

100. In the Arabian sea our researches *as yet* do not warrant our supposing that Cyclones exceed 240 miles in diameter, and indeed they are perhaps often below that size.

101. In the Bay of Bengal, the usual size of the Cyclones is from 300 to 350 miles, but it would appear that they sometimes much exceed that extent for, so far as can be deduced from the imperfect newspaper accounts of the day, the Cyclone marked *s* on our Chart No. III. being the Coromandel and Malabar Coast storm of May, 1820, appears to have extended from about 8° to 18° North if (which we cannot now ascertain) it was not a double storm at one period of its progress. As before said however,—and of this we have a very remarkable and distinct proof in the Coringa Hurricane of November, 1839, Jour. As. Soc. Beng. for 1840—they sometimes contract from 300 or 350, to about 150 miles; augmenting, however, in violence when they do so. The remarkable Cyclone XII. in the Andaman Sea was not at most more than 100 to 150 miles in diameter.

The Madras Cyclone of December 12, 1807, which had a track from the E. N. E. to the W. S. W. (shift from about N. W. to South), and which was of terrific fury, did not much exceed 60 miles in diameter, for it "scarcely reached to the Northward farther than Pulicat," which is 21 miles North of Madras, and was not felt at Pondicherry, 68 miles to the Southward. In this Cyclone the timbers of a large vessel which had been burnt in the roads, in 1799, were thrown on shore! Asiatic Annual Register, Vol. X. p. 129.

Some of the smaller sized Cyclones in the Bay also seem to move at a rapid rate, and are excessively violent, resembling the Tornados on shore, which in tropical climates, and in Bengal especially, destroy literally everything in their progress, though their tracks are but a few hundred or a thousand yards in breadth. These appear also to be common in North America from the St. Lawrence to the West Indies.

In a newspaper account copied into the *Bengal Hurkaru* of July,

1814, it is stated that on the 19th Nov. 1813, "a terrific tornado from the S. Eastern board" was felt at Halifax, which "rushed up the harbour with a violence unequalled since the tornado of Sept. 1798."* More than 190 vessels suffered from its effects, amongst which are named seven sail of men-of-war on shore, many lives lost, and merchantmen sunk.

102. The Typhoons (Cyclones) of the China Sea appear to vary from 60 or 80 miles to three or four degrees in diameter.

We have no sort of details which would enable us to speak of the size of the Cyclones in any part of the Pacific Ocean.

We shall subsequently see how this question of the average size of Cyclones is of great import to the mariner.

103. CONTEMPORANEOUS, PARALLEL, AND DIVIDING CYCLONES. It is important to the seaman also to know that there are at times Cyclones which are contemporaneous (occurring at the same time) or nearly so, and these sometimes so near to each other as to travel along in parallel tracks, leaving but a very short distance between them when parallel; or perhaps, as in the case of hail and thunder storms to be afterwards quoted, falling into each other when the lines of converging tracks approach closely, or follow each other upon nearly the same track. At other times they move forward on tracks which form such angles that the two storms must meet, and at others again we have instances of Cyclones starting from points nearly on the same meridian, *but in opposite hemispheres*, at 5 or 6 degrees of distance each from the Equator, and pursuing each their usual tracks, but revolving in opposite directions, according to the law by which they are affected! In some cases, also, Cyclones seem to be generated about the same time and in the same part of the world, but at very considerable distances from each other, and lastly there is no doubt that violent Cyclones sometimes divide into two or more; each following a track somewhat diverging from the other. The first case of travelling on parallel tracks, or nearly so, has certainly occurred both at the head of the Bay of Bengal,† and at its mouth in the latitude of Ceylon.‡

---

* I can trace no account of this tornado.—H. P.
† See Ninth Memoir, Jour. As. Soc. Beng. Vol. XII.
‡ See Fourteenth Memoir, Jour. As. Soc. Beng. Vol. XIV.

In both these cases they were of small extent, but in the first of great violence.

104. The second case of Cyclones following close upon each other and upon parallel tracks has also occurred in the Bay of Bengal* and in the Southern Indian Ocean as shewn by Mr. THOM. Mr. REDFIELD found that the Cuba Cyclone of 1844, already alluded to, was preceded by another, though smaller and less violent, and he speaks of them as "the two associated storms in the Atlantic."† The third case of contemporaneous Cyclones travelling on converging lines occurred in the China Sea, when in October, 1840, the *Thetis* of Calcutta had a Tyfoon-Cyclone which must have travelled up from the S. S. E. to the N. N. W., at the same time that the *Thetis* of London, only 200 miles distant from her, had a Tyfoon-Cyclone coming up from the W. S. W., these two tracks together making an angle of about 47°. I have no doubt that the ship *Golconda* of Madras, with 300 native troops on board, must have run up to near the meeting of the two, where she foundered and all on board perished.‡

105. In my Sixteenth Memoir, already quoted at p. 47, the case of *three* separate and nearly parallel Cyclones occurs, the two outer ones travelling from the N. N. E. and the centre one making a small angle with them and apparently falling into the Southernmost of the two, with a terrific whirlwind or waterspout occurring about the time and place of their junction. (See Part V, whirlwind of the *Duncan*.)

106. As to what may occur at the meeting of Cyclones we are quite in the dark, but I have found in the course of my researches one instance of the meeting of two tornados—which seem to have been miniature Cyclones—and which was observed at leisure by the inhabitants of a whole city. It is the following, which is slightly abridged from the Annual Register for 1761, being an account of a tornado on the 2nd May of that year at ¼ past 2 P.M. at Charleston in South Carolina, (called a typhon in the letter) which is said to have—

* Ninth Memoir, Jour. As. Soc. Beng. Vol. XII.
† American Jour. of Science, p. 366.
‡ Fourth Memoir, Jour. As.Soc. Beng.Vol. IX.: and Seventeenth Memoir, Journal As. Soc. Beng. Vol. XVIII. where the former deductions from the logs of two ships only are shewn to be perfectly correct by those of three others obtained some years afterwards.

Meeting of Cyclones; Charleston Tornados. [PART II. § 107.

"Passed down Ashley river and have fallen upon the shipping in the Rebellion roads, with such fury as to threaten the destruction of the whole fleet. It was first seen from the town coming swiftly down Wappo creek, resembling a column of smoke and vapour, of which the motion was very irregular and tumultuous. The quantity of vapour which composed this column and its prodigious swiftness gave it such a surprising momentum as to plough Ashley river to the bottom, and leave the channel bare, *floating small craft and boats to a great distance by the flux and reflux. When coming down Ashley river it made a noise like constant thunder;* its diameter at that time was estimated at 300 fathoms and its height at 35 degrees (seen from Charleston). It was met at White point by another gust which came down Cooper's river, which was not however equal to the other, *but upon their meeting together the tumultuous agitation of the air was much greater, insomuch that the froth and vapour seemed to be thrown up to the height of 40 degrees, while the clouds that were driven in all directions towards this place seemed to be precipitated and whirled round at the same time with incredible velocity.* It then fell upon the shipping in the roads, and was scarce three minutes in its passage, though the distance was near 2 leagues. Of 45 sail 5 were sunk outright, and H. M. S. *Dolphin* with eleven others lost their masts. The damage which is valued at £200,000* was done instantaneously, and some which were sunk were buried in the water so suddenly as scarce to give time to those below to get upon deck, yet but four lives were lost in them. The gust from Cooper's river checked the progress of the tornado from Wappo creek, which had it kept on would have driven the town of Charleston before it like chaff. This tremendous column was first seen about noon upwards of 50 miles W. b. S. from Charleston, and destroyed everything on its road, making a complete avenue where it passed amongst trees.

The sinking of the five ships was so sudden that it was a doubt whether it was done by the 'weight of the column,' or by the water being forced from under them and thereby letting them sink so low as to be covered and ingulphed by the mass of water!"

The passages which I have marked in italics are worthy of attention, as shewing the close relation of tornados, of this class at least, to our Cyclones, in their travelling up at 25 miles per hour, their wave, their noise "like thunder," and their whirling; but the most interesting part, to us, is the fact that at the meeting of the two,—for the Cooper's river gust was evidently a smaller tornado—the phænomena were much augmented in violence. This may be some faint index to what occurs at the meeting of two Cyclones, and is certainly a warning to every careful seaman.

107. We have no sort of *proof* that hail storms are meteors analogous to Cyclones, but it is worth remarking here that the Count de Tristan, in his paper "On the progress of Storms in the Department

* £20,000 in original, evidently a mistake if the five ships are included.

PART II. § 109.]   *Hail Storms;· Diverging Cyclones.*   99

of the Loiret,"* lays down after much investigation as amongst the aphorisms (brief phrases or precepts embodying results of the research) to which he reduces our present knowledge of storms (hail and thunder-storms): 1. That storms may be attracted by forests, and more than one may be so attracted at the same time, and that if they unite they may appear stronger afterwards. 2. That one stormy cloud attracts another and makes it deviate from its route, the strongest of course being the least deranged. 3. That one cloud attracted by a stronger one accelerates its motion as it approaches the principal one, and that if it was in action, before being attracted, it may sometimes cease its ravages when approaching the principal storm, and that after their junction the mischief usually increases.

108. The instance in which Cyclones have occurred on the same day, and about on the same meridian, in opposite hemispheres, has occurred also in the Indian Ocean, in the Cyclones investigated by me in my Eleventh Memoir, Jour. As. Society, Beng. Vol. XIV. In these remarkable Cyclones, which are traced by numerous logs from the 28th Nov. to the 3rd of Dec. 1844, two Cyclones were raging from Lat. 6° North and Lat. 7° South,† and in Longitude 87° to 89° East, and each, while revolving in opposite directions, was advancing on tracks diverging from each other and from the equator, along which, and on both sides of it, heavy Westerly gales only were blowing.

109. The case of a Cyclone of great violence dividing as it were or "throwing off" another, or several others, from itself, has been also satisfactorily shewn in our Indian researches. Col. REID had already in his work admitted the probability of this, and Mr. THOM, who had not then seen the Indian Memoirs, thinks, p. 150, that at the formation of a Cyclone there may be "vast revolving discs, which in their early stages are so extensive as sometimes to include *within their central space* ‡ a number of lesser vortices." The case in which

---

* Annales de la Societé Royale des Sciences d'Orleans. See Quarterly Journal of Science for 1829, p. 214.

† It is probable that both were formed on the 26th, but our authority for that of the Northern Hemisphere does not begin till the 28th.

‡ Italics are mine : "within their area" is, I presume, meant?—H. P.

the dividing of Cyclones has been most clearly and unequivocally traced, is that of the Calcutta Cyclone of June, 1842, which is fully investigated in my Seventh Memoir (Jour. As. Soc. Beng. Vol. XI.) In this case the Cyclones marked VII. on our Chart, No. III. came up from the S. S. E., appearing to have formed not very far out at sea, or at most not farther than from Lat. 20° North. It was a severe rotatory gale with a pilot vessel at sea, and a furious hurricane-Cyclone at Calcutta, driving and tearing half the ships from their river moorings, and swamping and sinking those which got adrift, with a dead calm and the shift in the middle of it. It was traced some hundreds of miles inland by a long series of reports; and indubitably, after passing Calcutta, divided, as I have delineated it, into perfect Cyclones of smaller diameter, but of less violence. The rate of travelling of the entire Cyclone from sea to Calcutta was about 5.3 miles an hour, and when it had divided, the branch seems to have travelled much slower, or about at one-half of the rate of the main body.

PART III.

1. THE GENERAL PRACTICAL APPLICATION OF THE LAW OF STORMS, AS TO AVOIDING CYCLONES, PRESERVING VESSELS FROM DAMAGE WHEN INVOLVED IN THEM, AND PROFITING BY THEM WHEN THIS CAN BE SAFELY DONE.—2. BEARING OF THE CENTRE OF A CYCLONE.—3. WIND OR CYCLONE POINTS AND COMPASS POINTS.—4. PROBABILITY OF THE INCURVING OF WINDS AND THE FLATTENING IN OF CYCLONES ON APPROACHING THE LAND.—5. ASCERTAINING THE TRACK OF A CYCLONE.—6. COLONEL REID'S RULE FOR LYING TO.—7. SCUDDING OR HEAVING TO.—8. USES OF THE TRANSPARENT HORN CARDS.—9. SPECIAL EXAMPLES FOR THE WEST INDIES, ATLANTIC OCEAN, COASTS AND SEAS OF EUROPE, SOUTHERN INDIAN OCEAN, ARABIAN SEA, BAY OF BENGAL, CHINA SEA AND PACIFIC OCEANS.—10. RECAPITULATION OF CASES OF ERROR AND OF GOOD MANAGEMENT.—11. CASES OF SHIPS PREVENTED BY THEIR SITUATION ON THE STORM CIRCLE FROM RUNNING INTO A CYCLONE.—12. PROOF OF THE ACCURACY OF THE RULES AND DEDUCTIONS BY INFORMATION OBTAINED AFTER PUBLICATION. 13. USES OF THE LAW, AND RULES, AND APPLICATION OF THE CARDS IN RIVERS, HARBOURS, AND OPEN ANCHORAGES.

110. THE GENERAL PRACTICAL APPLICATION OF THE LAW OF STORMS. Every sailor can understand when he looks at a waterspout or a heavy dust-whirlwind, that a canoe or a Thames wherry caught in it,* however cleverly managed and well prepared, even were she fitted as a life-boat, would run awful risks; that it would be the extreme of folly to attempt to sail *through* one, and for the boat or canoe-man not to do all in his power to keep out of its way; and

* Not only in the *visible* part, but within the invisible whirl of wind which OERSTED and others suppose to exist around the part which we can discern, which last is probably only the centre. See PART V.

finally, that if the waterspouts and whirlwinds had regular veerings of winds and progressions on given tracks, which the boatman perfectly knew, he might generally keep out of the way of harm, and sometimes even profit by them. This is the seaman's own case. His finest merchantman is but "a cock boat," and the proudest line of battle ship "a great war canoe," which in a waterspout or whirlwind, in the shape of a hurricane or Tyfoon, may be destroyed or damaged, or get through it safely, or profit by some parts of the wind if well managed; or the Commander of which may in ignorance run headlong into the mischief he dreads. This is owing to the *scale* on which the operations of nature in Cyclones are performed, which is so large, that to our imperfect faculties the winds always appear—and have hitherto been considered—as blowing in straight lines. The moment they are considered as curves, and probably parts of a circle or ellipse, and that this circle or ellipse is both whirling round and the whole moving bodily forward according to fixed laws, the matter is clear; but there is with many plain seamen great difficulty in understanding how to apply this knowledge, which I shall now try to remove: addressing myself to such as if we were sitting at the cabin table together.

111. It is clear that there are three kinds of cases for the management of your ship, in or at the approach of a Cyclone. You may avoid it altogether with sea-room and management; you may get through it safely by avoiding the dangerous centre; by heaving to on the right tack; or by running off in the proper direction; or you may even *profit* by it!

112. Let us take the first supposition—that of avoiding a Cyclone. Every seaman almost has shortened sail or altered his course, and many have hove to, to avoid a dangerous looking waterspout. Well; this is the most usual method by which circular storms are avoided, for they are neither more nor less (for your purpose) than *windspouts** travelling along given tracks at certain rates, of the neighbourhood of which the weather and your Barometer inform you; and of the course, (*i. e.* track,) of which I shall shortly shew you how to form a

---

* I do not mean here to agree with any of the theories mentioned p. 16 to p. 23, but am simply giving the plain explanation in the commonest words that I can find.

tolerable estimate, by the winds, your transparent Storm Cards, and your log. If that course crosses your track it is plain that you may heave to and let it pass on first. If you are on that side of it that by standing on, or standing back, for a time you increase your distance from it, or get out of its way, of course you will do so; and in both these cases you will have *avoided* a Cyclone by your knowledge of the Law of Storms.

113. But next; from your position, or for want of sea-room, or from the excessive suddenness of the approach of a Cyclone,* you may be involved in one. Your business now is to get out of it as soon, and with as little damage as you can.

The dangers to a vessel *in* a Cyclone are three. The *veering* of the wind; the excessive violence of it near the centre; and the sudden calms and shifts and awful sea *at* the centre. All these involve, as you well know, damage and loss by dismasting, straining, leaks, and distress of various kinds up to foundering; and you will find in the present Chapter ample rules to shew you that if involved in a Cyclone, *even with a fair wind*, you should often heave to, not to approach the centre, or with a foul one that you should, if the weather possibly admits of it, stand back; that if you have to heave to, you will, if you take the right tack, have both the veerings of the wind, and the shift of it such, that you will always "come up" to it, and run no risk of being taken aback; while if hove to on the wrong one, you will certainly have the wind heading you off, and in all probability be taken aback if it shifts. Of the dangers in all these cases every seaman is aware, as for instance that of getting stern-way† in a hurricane; and it is one, and not the least of the practical applications of our

---

* See what is said at p. 89 to p. 92, of the rates of travelling of Cyclones.

† Getting stern-way.—This taking aback in a tempest we all know to be most dangerous, not only on account of the getting stern-way, here mentioned, being pooped, dismasted, and the like; but from another danger, which is not sufficiently adverted to I think; and this is, that a vessel, may, in one of the terrific gusts which accompany these sudden shifts of wind, be thrown on her broadside in the trough of the sea, with her deck *towards* the sea! In such a case, and I have many instances of the kind on record, she is in the position of a vessel which has fallen over to seaward on a reef; and there is every chance that her hatches would be beaten in, which might swamp her, or that if her bulwarks are too high and solid she may be kept down by the weight of water on deck. Hatches are not usually made strong enough.

Science, that there is scarcely yet a case on record, in all the logs that have been examined, where ships which have suffered in a Cyclone might not have escaped with far *less* loss, if they could not have avoided the Cyclone, or perhaps without any, by due attention to our rules; and we can pretty nearly demonstrate also, that many which have foundered must have done so from utter ignorance of the danger they were running into!

114. The last case of which I have spoken is that in which you may often *profit* by a Cyclone. This is strange and startling enough, compared with the feeling of utter helplessness with which we all used, in former days, to regard these phænomena, but it is all comprised in these few words. *If circumstances allow you, sail along with it or run round it, instead of sailing through it.* Go back to the consideration of the wherry or canoe with the waterspout or whirlwind, at the commencement of this chapter, and you will at once see how this *may* be true. I shall now proceed to shew you how it *is* true, and how you may *make* it true for yourself, at least in ordinary cases.

115. BEARING OF THE CENTRE OF A CYCLONE.—Considering then every Cyclone as a great whirlwind, of which the outer part, as to strength, is a common close-reefed topsail gale, such as no seaman cares for, and no seaworthy ship is hurt by, but of which the violence increases with great rapidity as the centre is approached, till close to, or *at* it, it becomes of destructive fury. And considering also the centre, though it may be a space from one to perhaps fifty miles in diameter, as a mere point, round which the whole storm is revolving, our first care must be to find how this point or centre bears from us; for this is what guides us in our future consideration and manœuvres.

116. This is simple enough when the Cyclone has fairly and unequivocally commenced; I mean both as regards the state of the weather and the fall of the Barometer. The seaman should then mark off his place on his Chart, and placing that Horn Card which belongs to the hemisphere he is in, upon it, *with the wind's place, as marked on the outer circle, over the ship's place,* he will see at a glance how the centre bears from him; thus in the Northern hemisphere with the wind at E. N. E., the centre of the card, which is that of the Cyclone, will be seen to bear S. S. E. from the ship, if at East due South, and at E. S. E., S. S. W. of her, and so on.

PART III. § 117.]   *Table of Bearing of Centre.*   105

In the Southern hemisphere again, a N. W. wind would give the centre bearing S. W., and a S. S. E. wind would give it bearing E. N. E., and so on successively. In a word, every wind will be a minute tangent to a circle of greater or less diameter, and for practical purposes we consider the centre bearing about at right angles to it.

117. As seamen however are not accustomed to consider the winds as tangent lines to a circle, and the bearing of the centre perpendicular to them, the consideration of "how the centre bears," even with the aid of the Storm Card, may hence sometimes be found puzzling, I add, as an assistance, the following table, which some may find it useful to glance at. The middle column relates to those at each side.

In the NORTHERN HEMISPHERE when the Wind is	The centre bears about	In the SOUTHERN HEMISPHERE when the Wind is
NORTH.	EAST.	SOUTH.
N. b. E.	E. b. S.	S. b. W.
N. N. E.	E. S. E.	S. S. W.
N. E. b. N.	S. E. b. E.	S. W. b. W.
N. E.	S. E.	S. W.
N. E. b. E.	S. E. b. S.	S. W. b. S.
E. N. E.	S. S. E.	W. S. W.
E. b. N.	S. b. E.	W. b. S.
EAST.	SOUTH.	WEST.
E. b. S.	S. b. W.	W. b. N.
E. S. E.	S. S. W.	W. N. W.
S. E. b. E.	S. W. b. S.	N. W. b. W.
S. E.	S. W.	N. W.
S. E. b. S.	S. W. b. W.	N. W. b. N.
S. S. E.	W. S. W.	N. N. W.
S. b. E.	W. b. S.	N. b. W.
SOUTH.	WEST.	NORTH.
S. b. W.	W. b. N.	N. b. E.
S. S. W.	W. N. W.	N. N. E.
S. W. b. S.	N. W. b. W.	N. E. b. N.
S. W.	N. W.	N. E.
S. W. b. W.	N. W. b. N.	N. E. b. E.
W. S. W.	N. N. W.	E. N. E.
W. b. S.	N. b. W.	E. b. N.
WEST.	NORTH.	EAST.
W. b. N.	N. b. E.	E. b. S.
W. N. W.	N. N. E.	E. S. E.
N. W. b. W.	N. E. b. N.	S. E. b. E.
N. W.	N. E.	S. E.
N. W. b. N.	N. E. b. E.	S. E. b. S.
N. N. W.	E. N. E.	S. S. E.
N. b. W.	E. b. N.	S. b. E.
NORTH.	EAST.	SOUTH.

118. The following is from the Nautical Magazine of December, 1846, p. 651, and as we cannot have too many ways of considering the matter, to facilitate its apprehension to those who find it difficult to understand this all-important point of the bearing of the centre, I have given it here:

"We left you with the first fair wind of a hurricane, from which you were to ascertain at once the direction of the focus from you. The method we gave you was good and simple, but perhaps the following may be more so. Turn your back to the wind, then if you are in North latitude it will be on your left hand, but if you are in South latitude it will be on your right, in both cases at a right angle from the direction in which you are looking. This rule holds good, you will perceive, from the very nature of a whirlwind, in all parts of it clear of the focus. Having determined the direction of the focus from you, the next step to be taken to avoid it depends on the part of the hurricane circle on which you find yourself to be, along with the direction in which it is travelling; your principal object is to avoid the focus as you would a waterspout, and to do this you must give up the idea of keeping your course, even should the hurricane wind which you have be fair—make a fair wind of it if you like, but not to run into the focus. The *Pluto* and the *Maria Soames* did this, and both suffered for it.* Let us suppose you were on board the *Pluto* (in North latitude) when she had her first hurricane, wind E. and E. S. E. Turning your back on the wind, and facing yourself W. N. W., would have given you the focus of the hurricane on your left hand, at a right angle S. S. W. of you. It would then be clear to you that standing down to the Southward would be the worst course you could adopt, as it would take you into the middle of the storm, and knowing that the hurricane was travelling to the westward, as the hurricanes of the China Sea do, you would have immediately put your ship's head to the N. E. and have sacrificed a day to get out of the way of it."

119. THE WIND-POINTS AND COMPASS-POINTS.—This subject of the bearing of the centre leads me to remark here that the sailor will often find himself puzzled, at first, by the difference between the WIND-CIRCLE-POINTS on our Card, and his Compass-points. I mean by wind-circle-points those points of his Compass *at* which he finds marked on our Horn Card, the wind which is blowing; and these are different in each Hemisphere. Thus the North wind-circle-point is at the Western Compass-point in the Northern Hemisphere, and at the East Compass-point in the Southern Hemisphere. The following is a table of these points arranged with the Compass-points in the

* The cases of these ships will be subsequently given, (pp. 148, 157, 249.)

Part III. § 119.] Table of Cyclone Points.

middle, and the wind-circle-points on both sides, but as the word wind-circle-point is somewhat unmanageable and German-looking, and *wind-point* leads the mind to think as usual of the point *from which* the wind is blowing, I have preferred our word Cyclone, and headed the table on each side with the word CYCLONE POINT, so that the sailor cannot, I hope, misunderstand it.

In the Northern Hemisphere the Cyclone Point of—	Is at the Compass Point at—	In the Southern Hemisphere the Cyclone Point of—
EAST.	NORTH.	WEST.
E. b. S.	N. b. E.	W. b. N.
E. S. E.	N. N. E.	W. N. W.
S. E. b. E.	N. E. b. N.	N. W. b. W.
S. E.	N. E.	N. W.
S. E. b. S.	N. E. b. E.	N. W. b. N.
S. S. E.	E. N. E.	N. N. W.
S. b. E.	E. b. N.	N. b. W.
SOUTH.	EAST.	NORTH.
S. b. W.	E. b. S.	N. b. E.
S. S. W.	E. S. E.	N. N. E.
S. W. b. S.	S. E. b. E.	N. E. b. N.
S. W.	S. E.	N. E.
S. W. b. W.	S. E. b. S.	N. E. b. E.
W. S. W.	S. S. E.	E. N. E.
W. b. S.	S. b. E	E. b. N.
WEST.	SOUTH.	EAST.
W. b. N.	S. b. W.	E. b. S.
W. N. W.	S. S. W.	E. S. E.
N. W. b. W.	S. W. b. S.	S. E. b. E.
N. W.	S. W.	S. E.
N. W. b. N.	S. W. b. W.	S. E. b. S.
N. N. W.	W. S. W.	S. S. E.
N. b. W.	W. b. S.	S. b. E.
NORTH.	WEST.	SOUTH.
N. b. E.	W. b. N.	S. b. W.
N. N. E.	W. N. W.	S. S. W.
N. E. b. N.	N. W. b. W.	S. W. b. S.
N. E.	N. W.	S. W.
N. E. b. E.	N. W. b. N.	S. W. b. W.
E. N. E.	N. N. W.	W. S. W.
E. b. N.	N. b. W.	W. b. S.
EAST.	NORTH.	WEST.

It is almost superfluous to remark here that the Cyclone point is always *eight points* distant from the compass point, but to the right or left of it (looking towards the centre of the circle), according to the hemisphere in which the ship is. Thus the E. N. E. Cyclone point in the Northern hemisphere is at the N. N. W. compass point or eight points to the right hand, and in the Southern Hemisphere it is at the S. S. E. compass point or eight points to the left hand.

120. PROBABILITY OF THE INCURVING OF THE WINDS IN A CYCLONE. I should mention here that, though for convenience' sake I have spoken of the wind as blowing in a circle, and the foregoing tables are calculated on that supposition, yet it is by no means certain that it is a *true* circle, or that even *if* the whole body of the storm be circular, the winds within it blow every where in exactly concentric circles. Mr. REDFIELD on this subject says, in a recent memoir, (American Journal of Science and the Arts, Second Series, No. I. p. 14.)

"When, in 1830, I first attempted to establish by direct evidence the rotative character of gales or tempests, I had only to encounter the then prevailing idea of a general rectilinear movement in these winds. Hence I have deemed it sufficient to describe the rotation in general terms, not doubting that on different sides of a rotatory storm, as in common rains or sluggish storms, might be found any course of wind, from the rotative to the rectilinear, together with varying conditions as regards clouds and rain.

"But I have never been able to conceive, that the wind in violent storms moves only in circles. On the contrary, a vortical movement, approaching to that which may be seen in all lesser vortices, aerial or aqueous, appears to be an essential element of their violent and long continued action, of their increased energy towards the centre or axis, and of the accompanying rain. In conformity with this view the storm figure on my Chart of the storm of 1830, was directed to be engraved in spiral or involute lines, but this point was yielded for the convenience of the engraver.

* * * *

"The common idea of rotation in circles, however, is sufficiently correct for practical purposes and for the construction of diagrams, whether for the use of mariners, or for determining between a rotative and a general rectilinear wind, on one hand, or the lately alleged centripetal winds on the other. The degree of vorticular inclination in violent storms must be subject, locally, to great variations; but it is not probable that on an average of the different sides, it ever comes near to forty-five degrees from the tangent of a circle, and that such average inclination ever exceeds two points of the compass, may well be doubted."

121. And while Mr. REDFIELD was writing and publishing this in America, I was so fortunate as to be *proving* it in India, for one storm at least, that of the *Charles Heddle,* which vessel in a Cyclone off the Mauritius scudded round and round for five successive days! In a Memoir, the Thirteenth of my Series, published in the Journal of the Asiatic Society of Bengal, Vol. XIV. p. 703, in which this

PART III. § 121.]   *Incurving of the Winds in a Cyclone.*   109

storm and the *Charles Heddle's* manœuvres are analysed in close detail,* it is I think distinctly enough shewn that the average *resultant curve*† of the winds for five days was an incurving and flattened spiral, like the figure below, in which (without reference to the storm wave) the ship may in that time have made a curved course from *x* to *y*, and we may imagine that at the centre the incurving of the winds was fully that of the arrows which I have marked there.

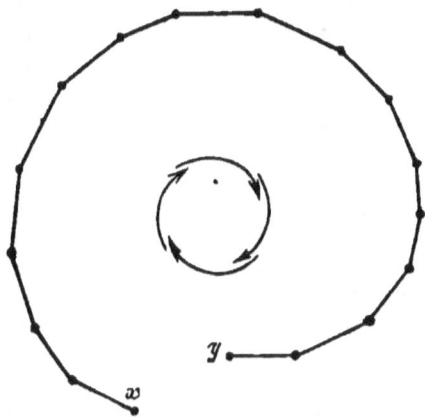

* The *Charles Heddle's* log and its analysis, with the Chart, have been reprinted in the Nautical Magazine for 1846.

† I explain these terms. By the total *resultant* of the winds is meant in meteorology the calculating each separate wind during the number of hours it blows in a given time, its direction being in nautical language ' a course,' and the time or number of hours ' a distance ;' the strength being always supposed the same (or this may also be used), and all these courses and distances, (direction and time,) may make a traverse table, of which as usual one course and distance is the result. Thus, if in 24 hours we have 9 hours of N. E. and 15 of S. W. wind, the resultant is 6 of S. W. wind; or the whole atmosphere of the place may be supposed to have moved for 6 hours to the N. E., if the strength of the two winds was always equal. This is the *resultant* of the wind. If instead of the traverse table we project the directions of the wind for courses, and the hours it blew for distances, we shall have a line of some kind, which in this case is a curve, and this is the *resultant curve* of the wind. Now in the run of a vessel, scudding under bare poles, her run per hour may be supposed to be an indication of the strength of the wind, but then the course and distance shewn by log becomes the resultant, indicating from which quarter also the resultant wind blew.

122. And I have said in reference to this result, that it appears to me not improbable that, while at some point in the whole zone of a Cyclone the wind may be blowing in a true circle, it may on the inner or central side of that part be a series of *in*curving spirals,* (spirals of vorticular inclination, Mr. REDFIELD calls them,) and that on the outer part there may be a centrifugal (*i. e.* a curving *off from* the centre) tendency. I have farther said, and this may not be out of place here—

" If (for the sake of hypothesis only) we admit this incurving of the winds, it follows that there may be also, not a single incurving of the same rate throughout the whole breadth of the storm, but that the incurving may be much more excessive, and amount to two or three points when the centre is nearly approached, and even be so violent at the centre as to prevent ships drifting out of it! just like the vortex of a whirlpool or a tide-eddy; which last we know will often give a boat's crew a heavy pull, or a ship much trouble, before they get out of them. Does it not seem that we have here the explanation of how some ships, as in the case of the *Runnimede* and *Briton*, in my last (XIIth) Memoir, may be blown and drifted round and round, without drifting out of the fatal centre which we should look for them, nautically, to do, and which other ships there is no doubt really do. An excessive incurving of the winds towards the centre, like the wind arrows at the centre of the figure is one, and one very likely mode of accounting for vessels remaining in this hopeless state; and moreover it may assist us in supposing how some dismal losses have occurred, whilst other ships in company have escaped. It adds also a most powerful argument, if any were wanting, for every precaution to avoid the centres—and for every one who can contribute to these researches to do so."

123. I have farther, in relation to this question, expressed a view that it may be possible that when Cyclones are increasing in diameter or dilating, the spiral is a *di*verging one, or that the arrows curve outwardly, and when it contracts (see p. 94 for the contracting and dilating of storms) the spiral is a *con*verging or incurving one. We do not yet *know* this, but it is probable, or possible; and the seaman when using the foregoing table and his Storm Card should bear in mind that the wind-arrows and bearing of the centre are not strictly mathematical lines and curves, but may vary a little; though usually not so much, unless very near the centre, as greatly to affect his estimate and management.

Mr. REDFIELD, we have seen, thinks it within possibility that the

* "*Incurvating*" is, I know, the English-Latin word, but Incurving is a much more manageable one for sailors.

incurving* may amount to two points, or 22° near the centre, and my own views, as expressed above, would require fully this. I have in the Memoir quoted, shewn how this incurving may affect our calculations of the bearing (the *exact* bearing) of the centre, and when we arrive at that part of our lessons which shews how the track of a storm may be pretty nearly calculated by the veering of the winds, we shall again allude to this subject. For all practical purposes however we may say, for example, that if in the Northern hemisphere, a Cyclone has fairly begun at E. N. E., the centre bears *about* S. S. E. from us, and we have then plenty of room and time to make even half a point, one way or the other, in the bearing, of little or no consequence.

124. I have fortunately the log of one ship in which this incurving of the wind, or what appears to me to be equivalent to it, is marked. It is that of the *Thalia*, Captain BIDEN, Track *g* on Chart No. II., which vessel in Lat. 28° to 31° South, Long. 55° E., experienced a furious hurricane-Cyclone from the West to the N. W., having run as long as she could carry sail to the S. W. with a heavy and increasing N. W.† gale. She then laid to at 2 A.M. on the 4th April, when her storm sails, mizen mast, boats, wheel, compasses, &c. were in a short time carried away, rudder-head split, and other severe damage done, so that she was no doubt close to the centre. She now found the wind "*alternating four or five points, viz., from N. W. to W. b. S. and vice versa!*"

The next day, the 5th, the wind is marked as "*alternating* West to N. N. W.;" on the 6th as "*alternating* N. W. to S. W.;" on the 7th at 1 P.M. Nautical time, wind N. W.; 5, W. N. W.; 10, West; 5 A.M. W.S.W., the sky completely overcast, the sea one sheet of foam. On the 8th, the hurricane is said to be abating, the wind being West at noon; at 9 P.M. W. N. W.; and at midnight and following morning veering to S. W., South, and South East, when the gale broke up.

During 4¼ days the ship was drifted to the S. E. b. E. 248 miles, which would give a mean drift of little more than two miles an hour in a direct line, but she probably made twice as much distance per

---

* He uses the words "vorticular inclination" to express this, but I shall prefer incurving as the simplest of the two for the plain seaman.

† Her Barometer had been unfortunately destroyed by an accident.

hour, since, by the alternating of the winds, she must have drifted over a succession of curves.

Since she was hove to, it is evident that she was drifted nearly with the centre, and as the gale from N. W. abated at from West to S.W. ending at South and S. E., that she drifted as it were *through* the body of the Cyclone, while *it* moved slowly away from her, and that she was constantly, so to speak, during the 4th, 5th and 6th, while the wind is said to be alternating—by which I understand veering several points in an hour or two—whirled back into the more central parts of the vortex as it passed on. In the curved wind-arrows at the centre of our figure at p. 109 the arrow at the upper part exactly shews a wind, along the curve of which, if we suppose a ship to be slowly driven, this would by her drift and the motion of the vortex give her winds alternating between West and N. W. or thereabout, for we must recollect that all these data in weather of such severity are necessarily vague and imperfect. The *Thalia's* Log indeed notes as follows:—

" The alternations of the wind (as noticed) were most frequent on the 5th and 6th.* The spray from the waves beating over us as thick as rain, the changes of the wind could only be correctly observed by the effect on the ship's course, or as she lay to by Compass."

We are nevertheless in this, as on so many other points, greatly indebted to Captain BIDEN for these notes, and I doubt not that intelligent seamen will now see the importance of marking correctly the veerings and oscillations of the wind, so as to furnish us with a clue to the right understanding of this deeply interesting and highly important fact in our science—the incurving of the winds towards the centre.

We have other and strong examples of this alternating of the winds close to, or at, the centre of Cyclones, of which I will quote but a few.

In the terrific Cyclone in which the *True Briton* and *Runnimede* were thrown on the Eastern Andamans, (see p. 56) the Log of the former vessel, while drifting with the vortex in a helpless state, says, " wind variable from N. E. to E. S. E.," and again " ship heading from S. E. to North, and wind blowing all round the Compass."

125. In the *Thomas Grenville's* Cyclone, to be subsequently mentioned, it is noted just after the passage of the centre, that the wind was " blowing in awful gusts from East to N. N. E."

---

* She had run close into the centre, before broaching to and losing her mizen-mast.

In the *Northumberland's* log, also to be subsequently mentioned, the wind close to the centre is noted as veering from South to East, the vessel being kept as much as possible before it.

A still more recent instance is the case of the ship *Caledonia*, from China to Bombay, which ship 'chased,' overtook, and ran into a small Cyclone, No. XIV. on Chart No. III. (see Jour. As. Soc. Beng. Vol. XIV.) till she reached the calm centre, and for three hours before she did so, her log states, "Heavy gale, *wind shifting from South to S. E.*" and "increasing gale, *wind continually shifting from South to S. E. and back again ;*" and these are exactly the alternations which a ship running into the vortex of a Cyclone in the Northern hemisphere, on its Eastern side, *should* have from incurving winds.

126. Mr. THOM, p. 103, describing the Mauritius Cyclone of 1840, says,

"As the vortex approached, and the wind began to be marked by lulls at one moment and fierce *spiral gusts** at the next, we observed several of these (Casuarina) trees, as *if suddenly twisted round*, broken off in an instant, and borne away for some distance on the breast of the tempest."

It would seem also that this incurving takes place sometimes at a distance from the centre, for in the Log of the ship *Sophia Fraser*, running to the S. W. b. W. in the verge, and within the body of a Cyclone in 12° to 15° South and from 81° to 78° East (track *pp.* on Chart No. II.) the wind is noted as veering from E. S. E. to N. E. in the squalls and then back to E. S. E. When the ship, unable to run any longer, hove to, the wind appears to have become steadier as the Cyclone left her.

127. Perhaps a further proof, or rather an indication, of this, may be found in the remarkable instances, of which there are many and recent ones, of ships being surrounded or having their decks covered during the passage of the calm centre of Cyclones, in the neighbourhood of the land, with land and aquatic birds, butterflies, horse-flies, &c. Now within a Cyclone these animals must be incapable of doing more than keeping themselves in the air; as a good swimmer in a strong current or eddy can only keep himself above water, but cannot stem or cross it. They must therefore, we may suppose, be gradually carried inwards by the incurving tendency of the winds, and at the

* Italics are mine—H. P.

centre are kept there because they cannot fly out of it, and when it reaches the ship they of course make a last effort to reach her as a resting place. If the wind blew in true circles they would be scattered all over the body of the Cyclone. In Part V. where the dust whirlwinds of India are mentioned, I have detailed one instance in which I myself *saw* the incurving of the wind most distinctly marked.

128. FLATTENING IN OF THE CIRCLE OF A CYCLONE WHEN APPROACHING HIGH LAND. Another consideration must be noted here. When the tracks of Cyclones approach the land at right angles or nearly so, and especially if there be a considerable extent of high land, it seems very probable that the vortex may be *flattened in*, as it were, and the winds thus blow, for the land side of it at least, in an oval.

In the following figure A represents a *Norte* "Norther" of the Gulf of Mexico, which we may suppose generated in the Gulf of Mexico

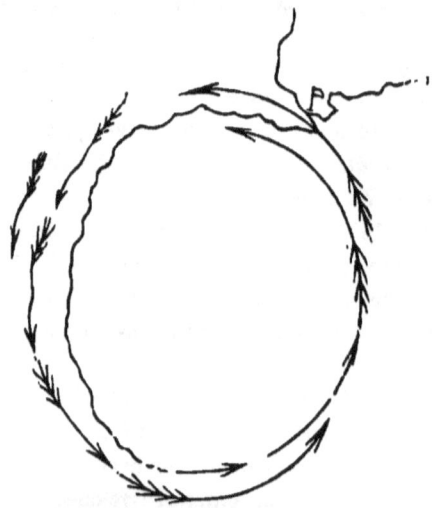

or Caribbean Sea, and travelling up from the E. S. E. or S. E. till it strikes upon the shores of Texas and the Northern part of Mexico, become thereby so flattened in, as to produce a much larger proportion of Northerly winds than would prevail in the same Cyclone in a wide open ocean. The figure B in the next page is an instance of the same kind, where a Cyclone crossing the Bay of Bengal, in so low a Latitude as 6° and 7° North, (and this has occurred, see p. 49,) strikes on the high land of Ceylon. The wavy arrows are drawn in both figures,

PART III. § 129.] *Ascertaining the Tracks of Cyclones.* 115

to show what may be (or *must* be at the surface) the effects of chains of hills or mountains, and if these cross the track of the storm at right angles still greater deviations must arise.

We do not, it should be remembered, *know* that all or any part of this takes place, but it is evidently of high probability that it does, and the seaman therefore should duly allow for it, if near the land, in his various considerations. We shall by and by see how this probability must not be forgotten when putting to sea from a roadstead or harbour.

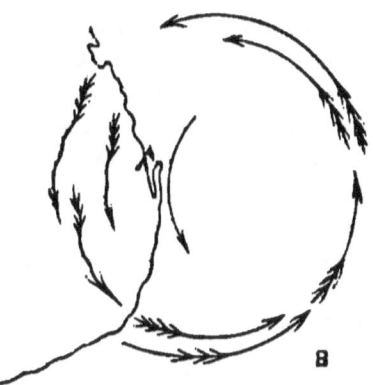

129. ASCERTAINING THE TRACK OF A CYCLONE. The seaman now I presume knows how to estimate the bearing of the centre from him? and I have detailed in Part II. the usual tracks of storms, at p. 8 the law of their revolution, as laid down by Mr. REDFIELD, and now so amply demonstrated for both hemispheres; but as there are still great tracts of the globe, though much frequented by shipping, in which we are ignorant of what may be the tracks of Cyclones, it may be as well to give here the method by which a careful Commander may, approximately at least, estimate what is the track of an approaching Cyclone, or of one towards which he thinks he is running. He will find it useful to do this, not only in parts where the tracks of Cyclones are unknown, but also where they are pretty well ascertained. He should always, in a word, look on the centre of the hurricane as a privateer, or a pirate, or an enemy of superior force, and make his calculations for avoiding its neighbourhood. He must not forget, that if he has *his* course and drift, the storm has also a course of *its* own, and brings

with it currents, see pp. 12 and 18 (often strong ones,) to both of which he must attend. Recollecting this, he may now farther use his storm card, with his log, to judge how the Cyclone itself is travelling, in the following manner.

If it be supposed that the Cyclone is travelling from the S. E. to the N. W., it will be seen that, supposing the ship in any part of the circle, this will give different changes of the wind to what would occur if it were travelling, say from E. N. E. to the W. S. W. An allowance being made for the ship's track between any two changes of wind, he will find that the line between the two points occupied by the centre of the circle, will lie in a direction which is *nearly* that of the track of the storm; and that in all which follows, he will only do what he does when sailing in fine weather along the shores of which he has no exact chart: he takes the bearings of two head-lands, guesses their distance, and from his run lays down a sketch of the trending of the coast.

130. The track is estimated by projection, as in the diagram, p. 118.

Draw a small line through the ship's place at A, in the direction of the wind, which we will call N. E., and another A B, from the same point, perpendicular to it, or S. E., which represents about the bearing of the centre of the Cyclone from the ship, at that time, in the Northern Hemisphere.* We can only *guess* at the distance, which we do by estimating it from the violence of the wind, the rapidity of its changes, and the fall of the barometer.† I should say, that for a strong gale, which would allow a good merchant ship to carry her close-reefed topsails and foresail, we might allow 200 miles. For a hard gale, in which the foresail could scarcely be borne, 150 or 100 miles, and for a very severe gale, a still shorter distance. The veering of the wind, the increase of its force, and the fall of the barometer, are of course more rapid the nearer you are to the centre, and some storms are also more violent and travel faster than others.

* True, and not compass points of the wind, should be used when the variation is considerable, if it be intended to consult the chart, or the track estimated from the compass, wind and course should be corrected for variation.

† In Part V. I have endeavoured to lay down rules for making the *rate of fall* of the Barometer a measure of the distance of the centre; with what success the sailor will judge.

Let us then put down 150 miles on this S. E. line, or from A to B, for our present distance from the centre of the Cyclone, and in six hours afterwards let us suppose the ship to have made 54 miles on a South course, bringing her to C, and that the wind is increasing fast, but is still at N. E., with the barometer falling, and every other appearance of bad weather.

Mark off this run of 54' on a South line, and as the wind is still at N. E., draw a S. E. line as before, which points again towards the centre. We have to consider now that we are probably nearer to it than before, for we know that *it* also, in these six hours, has been travelling to the Westward between 8 and 16 miles at the least, and the barometer has continued to fall, and the squalls are much more violent and frequent. Taking 12 miles an hour, or 72 miles for the six hours in the compasses, we find that this distance from B will strike upon the S. E. line (which is the perpendicular to the wind's course, and now the bearing of the centre) at D, which we may thus take to be the new place of the centre. This, it will be seen, gives the Cyclone a W. N. W. course, which is a likely one, and places the ship now at only 58 miles from its centre. It is clear, therefore, that we are thus running in upon it, and though our distances are mere guesses, they are, for the Bay of Bengal and the China or Caribbean Sea, *very strong probabilities*, because of the continued fall of the barometer, and the great mass of evidence which exists to prove, that at certain seasons almost all their violent storms are Cyclones, and move from the Eastward to the Westward.

I have placed on the Southern half of the Diagram, a case wherein another ship E, in the same tempest, at the same time, but at a distance of 220 miles from the ship at A, may have the wind at N. W. first, and steering N. E. 54 miles, to F, bring the wind to West, because the centre of the Cyclone, travelling as we suppose, bears then about 82 miles North from her. From evidence published in my Fourth Memoir, Jour. As. Soc. Beng. Vol. X., there seems little doubt that the unfortunate ship *Golconda*, which foundered with 300 Madras troops on board, must have run from the south side of a hurricane into its centre, in the tyfoon Cyclones of 22nd to 24th September, 1840, in the China Sea, as I have supposed this vessel to be doing. See p. 97, § 104.

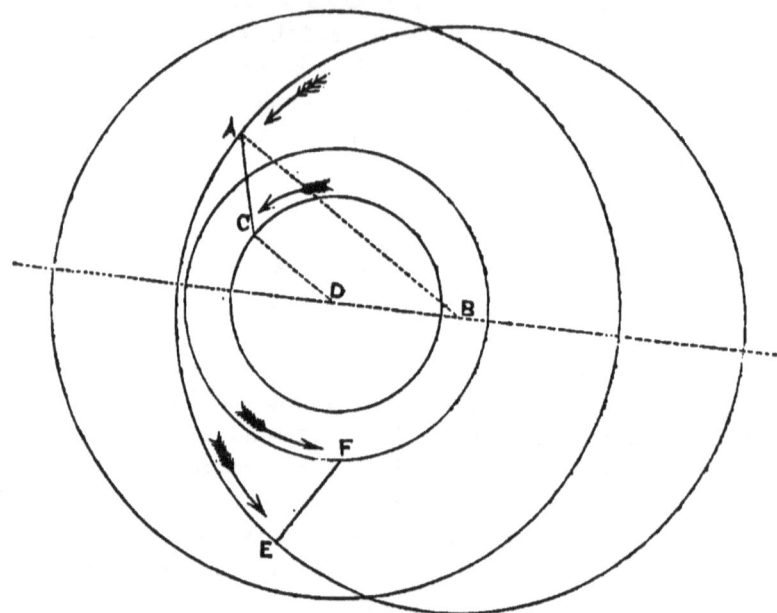

Scale 100' to an inch.

131. Again : we may suppose off the coast of North America, a ship in Lat. 40° North, with a Cyclone evidently setting in at S. E. b. S., and that in a run of 6 hours, at 10 miles per hour, or 60 miles, to the W. N. W., the wind has only veered two points, and become S. E. b. E., but has increased in strength very considerably and rapidly, and with a falling barometer. The projection of this will give us a Cyclone coming up from the S. Westward, as usual thereabouts, and we are thus right in its track. As it may in such a case be impossible or hazardous to scud on account of sea and wind, and the dread of the dangerous shift, we have in such a case to consider *what will be the proper tack to lie on ?* which will be shewn in the next section.

132. In consulting his charts, and judging therefrom, and from his log, and the veering of the wind, what the probable track of a Cyclone may be, the seaman must bear in mind, that, first we have no ground for any certainty that the tracks of one year will be *exactly* those of

PART III. § 133.] *Track ascertained from the shift.* 119

another; and next that while the tracks of some Cyclones on all the charts are laid down, so to say from the logs of a whole fleet of ships; others, such as those near the Channel, some in the Southern Indian Ocean, &c. are necessarily laid down from the logs of one or, at most, two vessels, and this often without either having experienced the shift by the passage of the centre over her, but in all cases from there being sufficient evidence to shew that the gale *was* a Cyclone travelling at *about* a certain distance from the ship on about a certain track. Our evidence for the tracks is, in fact, of all degrees of certainty from careful estimation, based on reasonable proofs, up to what may fairly be called mathematical certainty.

133. The track of a Cyclone may be judged of also from the shift, though some seamen may be embarrassed in considering accounts of storms by others, or from their own logs, to judge what *was* the track of the storm from the shift of the wind at the centre, but nothing is more simple. It may be done by the eye with the Horn Card, and a little attention to the table of Cyclone-points at p. 107; but if exactness is required, it should be done by projection. Thus, I suppose in the Southern hemisphere, that a ship, lying to, has the shift from S. E. to North. See Fig. I.

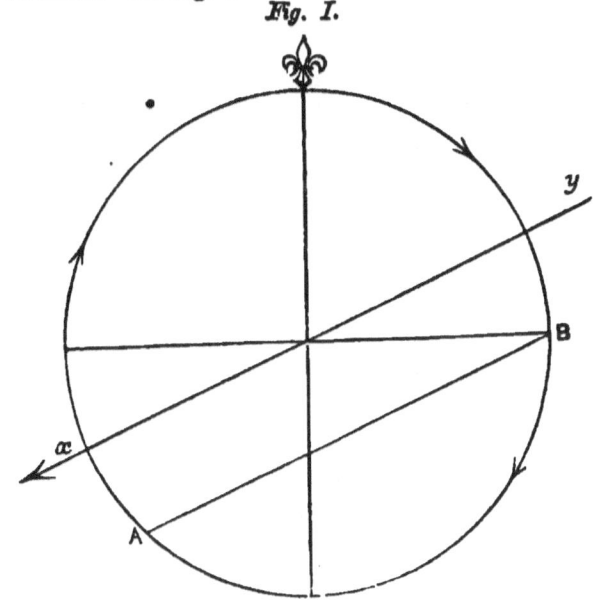

Fig. I.

Strike any circle, cross it with diameters, and mark on it the S. E. Cyclone or *wind-point* (see page 107) of the Storm card for that hemisphere, which will be on the lower part of the left hand side, at A or in the S. W. quadrant of the circle; then mark the North *Cyclone point*, which will be where the diameter to the right touches the circumference at B; join these two, and draw a parallel *x, y*, through the centre of the circle. Measure off the angle B. *y*, or its complement from the fleur de lis to *y*, and it will give the compass track; which in this case is *from* E. N. E. to W. S. W. or from *y* to *x*, because the shift being *from* the S. E. to the North, the Cyclone must have been moving to the left, or from B to A to strike the ship first at A.

Again: suppose in the Northern Atlantic a shift from S. E. b. E. to N. b. W. (See Fig. II.)

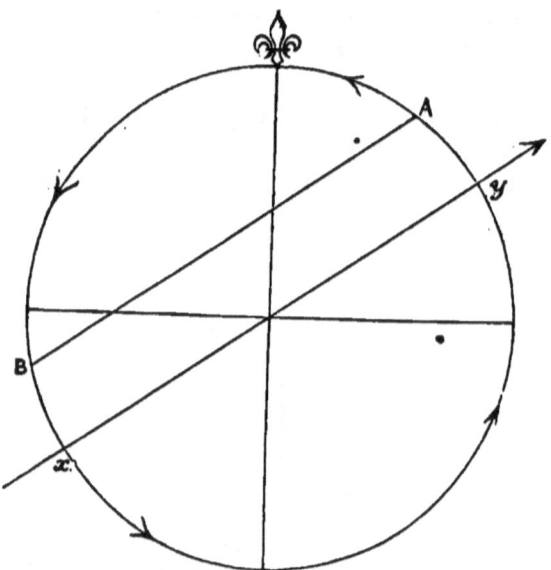

Here, as before, A is the *Cyclone point* of S. E. b. E. (or compass N. E. b. N.,) in the Northern hemisphere, and B of N. b. W. (or compass W. b. S.) Join them, and draw the parallel *x y* through the centre of your circle, and the angle between the *fleur de lis* and *y* is the compass rhumb of the Cyclone's track; and as the wind, this time, was at first blowing from the S. E. b. E. you see the Cyclone must have been *travelling up* from the left hand or from B up to A, or on a track *from* the S. 55° West to the N. 55° E. or from x to y. I need not add that the seaman in these essays must of course take the average direction of his winds when the Cyclone is at full strength, before and after the shift. He will also understand how, without any sudden shift, when the wind has only *veered* rapidly with him, so as for any one point to settle for some hours at another, it was travelling on a given track, and he was to the North or South of the centre, though no distinct calm interval occurred, which is often the case.

134. It may appear at first sight to the old seaman, as well as to the novice and landsman, that we are a little hasty in deducing our opinions of the track from the shift of wind; but I am fortunately enabled to shew that in one instance, certainly, where the shifts have no doubt been most accurately marked, the track is perfectly defined by the various shifts of wind of three ships exposed to the same Cyclone in sight of each other, but in different parts of the same calm centre!

This remarkable and highly interesting fact I derive from a series of logs sent out to me by the Honourable the Court of Directors of the E. I. Company, which contains amongst others, the logs of the Fleet under convoy of H. M. Sloop *Swift*, homeward bound from China, in the Northern Pacific Ocean in July, 1797. In this Fleet, as quoted by Mr. REDFIELD, from Lynn's Star Tables, the H. C. S. *Buccleugh* was in severe distress, and the *Swift* foundered* in the second of the Cyclones which they encountered on the 2nd July (track $c\ c$ on the Chart.) In the third of these, at noon on the 8th of July, the H. C. Ships *Cuffnells, Duke of Buccleugh* and *Taunton Castle* were in company and in sight of each other, in a tyfoon (Cyclone) from the N. Eastward, when a lull took place, and at 2 P. M. a shift of wind, which I give below as in the three logs before me, and I note how such a shift for each single ship would give the track of the Cyclone.

* And we can not only almost mark the exact spot where she did so, but we can say now *to a certainty* by what error (for want of this our new knowledge) these brave hearts were lost to their country! The whole fleet were scudding across the S. Western quadrant of a slow moving Cyclone under their foresails, and the *Swift* was nearest to the centre. The Indiamen mostly broached to and lost topmasts, rudders, lower masts, &c. &c. The *Swift* was never more heard of, and melancholy to relate, she had on board the crew of H. M. S. *Providence*, wrecked on a coral reef some time before, so that altogether she is supposed to have gone down with nearly 400 souls on board!

Ship	Wind at Noon.	Shifted at about 2 P. M. after a lull to W. b. N.	Shewing a track to the
Cuffnells.	N. E.	Veering rapidly to S. S. E.	W. b. N.
Buocleugh.	North	S. W.	W. N. W.
Taunton Castle.	N. N. E.	S. S. W.	W. N. W.

135. If the young mariner desires a lesson for practice, this is a capital one. He should strike a circle to represent the vortex of the Cyclone, say of 15 or 20 miles diameter on any scale. Then marking *Cuffnells* at the N. W. *compass*, which is the N. E. *Cyclone* point on one side; and at the E. N. E. compass, which is the S. S. E. Cyclone point, and so on of the rest, and drawing the parallels as just described, he will have before him at once the proof of how truly the tale of the track would have been told by each separate ship (to be between W. b. N. and W. N. W. accurately) in spite of their different changes, and how nearly we may rely upon a well kept log to indicate it.* The apparent anomaly of the *Cuffnell's* log is rather a confirmation, since it shews, either, that there were, what we have just described, irregular gusts at the centre, or that she had drifted for a time into the Southern part of the circle. He will farther observe in the table of references to our Chart No. IV. that the Fleet was to the Northward of these three ships, and the track (*d d*) is confirmed by the evidence of their logs to have been about North 65° West, on the average of *three* days from the 8th to the 11th or 12th. And I may add, that in more than one instance where we have deduced the tracks from the shift of wind experienced by a single ship, subsequent information has shewn us to be perfectly correct.

136. Colonel REID'S RULE FOR THE PROPER TACK TO LIE TO ON. This is one of those great, but simple, results for which every seaman owes to this distinguished officer a deep debt of gratitude; and every year and every new investigation proves the utility and beauty of the rule. I have already explained that the track or course of a storm is considered like that of a river, and that looking to the mouth of the river, we call the banks of the river right and left banks, according as they then lie; so looking *to* the quarter *towards which* the Cyclone is moving, we call the *right* side of the storm all that half of the sup-

---

* And the logs of the East India Company's ships were in no respect inferior to the best of those in the Royal Navy.

posed circle, or ellipse, on our right hand, and the *left* side all that on our left. I cannot do better here for my reader than to extract nearly what Col. REID has said at p. 530 of the second edition of his work.

'*Rules for laying Ships to in Hurricanes.*—That tack on which a ship should be laid to in a hurricane has hitherto been a problem to be solved; and is one which seamen have long considered important to have explained.

' In these tempests, when a vessel is lying to and the wind veers by the ship's head, she is in danger of getting storm-way, even when no sail is set; for in a hurricane, the wind's force upon the ship's masts and yards alone will produce this effect, should the wind veer ahead, and it is supposed that vessels have often foundered from this cause.

' When the wind veers aft as it is called, or by the stern, this danger is avoided, and a ship then comes up to the wind instead of having to break off from it.

' If great storms obey fixed laws, and the explanation given of them in this work be the true one, then the rule for laying a ship to, follows like the corollary to a problem already solved. In order to define the two sides of a storm, that side will be called the right hand semicircle, which is on the right of a storm's course, as we look in the direction in which it is moving, just as we speak of the right bank of a river. The rule for laying a ship to, will be, *when in the right hand semicircle, to heave to on the starboard tack, and when in the left hand semicircle, on the port tack in both hemispheres.*'

' The first of the two figures inserted here is intended to represent one of the West Indian hurricanes moving from the S. E. b. S. to the N. W. b. N. in the direction of the great arrow drawn across it. The commander of a ship can ascertain what part of a circular storm he is falling into by observing how the wind begins to veer. Thus, in the figure, the ship which falls into the right hand semicircle, would receive the wind at first about E. b. N., but it would soon veer to East as the storm passed onward, and supposing her lying to. The ship which falls into the left hand half of the storm would receive the wind at first at N. E., but with this latter ship instead of veering towards East, it would veer towards North.

' The explanation of the rule will be best made out by attentively inspecting the two figures. In both, *the black ships are on the proper tacks;* the white ships being on the wrong ones.

' The second figure is intended to represent one of those hurricanes in South latitude which pass near Mauritius, proceeding to the South-westward. The whirlwind is supposed to be passing over the vessel in the direction of the spear head. It will be seen that the black ships are always coming up, and the white ships always breaking off; and that they are on opposite tacks, on opposite sides of the circles.

' If hurricanes were to move in the opposite course to that which they have hitherto been found to follow, then would the rule be reversed, for the white ships would come up, and the black ships break off.'*

---

* I have already (p. 115,) explained that the application of the law depends entirely on the knowledge of the track, and this, as here stated, he will constantly see exemplified. H. P.

124   *Rules for laying Ships to in Hurricanes.*   [PART III. § 136.

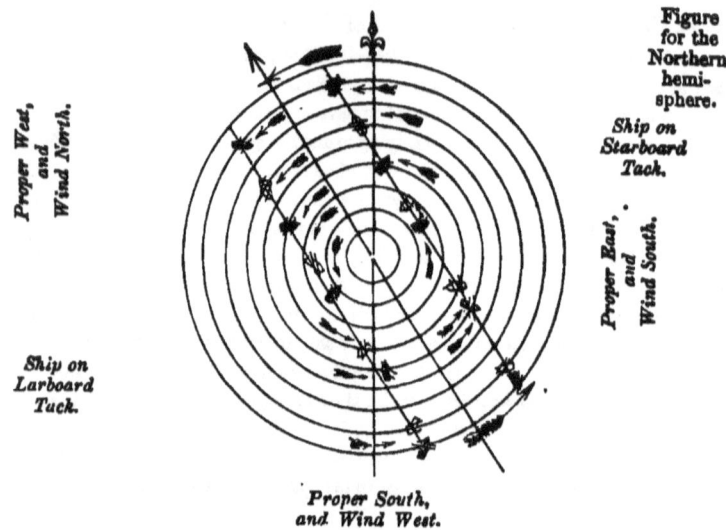

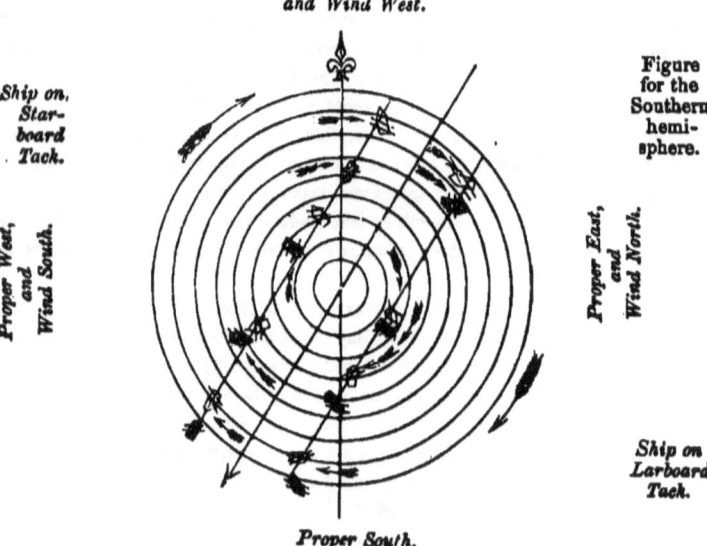

PART III. § 137.] *Rodney's Prizes. Admiral Graves' disaster.* 125

137. The ever memorable loss of the prizes taken by RODNEY, on the 1st of April, 1782, together with an immense number of merchantmen, and nearly all the men-of-war convoying the fleet, should not be passed over here, as affording a truly awful lesson of the importance of our science. From Mr. REDFIELD's "Memoir" in the United States' Naval Magazine, and a Memoir of Admiral GRAVES, before me, which I think has been copied from the U. S. Journal, it appears that H. M. S. *Ramilies, Canada,* and *Centaur,* of 74 guns each, with the *Pallas* Frigate, and the *Ville de Paris* of 110 guns, *Glorieux* and *Hector* of 74, *Ardent* and *Caton,* of 64 guns each, prizes, and a convoy which, even after those for New York had separated, and the *Ardent, Pallas* and *Hector* put back, still amounted to ninety-two or ninety-three sail, were overtaken by a hurricane-Cyclone, on the 16th of September, 1782, which increased rapidly from E. S. E. The fleet fully prepared for bad weather, hove to, but unfortunately on the starboard *(which was the wrong)* tack, for at 2 A.M. on the 17th, when in about Lat. 42$\frac{1}{2}$° North, Long. 48$\frac{1}{2}$° West, the whole fleet were taken aback by a shift of wind, evidently of terrific violence, to N. N. W.*
The *Ramilies,* Admiral GRAVES' flag-ship, lost her main, mizen, and fore-top-masts, was pooped, and apparently in danger of going down stern foremost; and the following day shewed that many of the men-of-war and of the merchantmen also had been as ill treated, for there were "signals of distress everywhere." The Cyclone continued at N. W., and before it left the helpless fleet, the whole of the men-of-war, except the *Canada,* had foundered or were abandoned and destroyed; and so large a proportion of the merchantmen that this is supposed to be the greatest naval disaster we have upon record. Upwards of three thousand seamen alone are computed to have perished by it!

The track of the Cyclone appears from the shift, E. S. E. to N. W.† to have been one travelling to the N. E. b. N. about five degrees to the Eastward of the termination of Track B on Chart I., and though it is true the men-of-war, as well as the prizes, were in a deplorable state, yet no doubt the tremendous shocks and consequences of dismasting, pooping, and the like, added not a little to their danger; for

* Observe that this is nearly our Fig. II. at p. 120.

† The first gust mentioned as from N. N. W. I take to have been an incurving wind at the centre, and that these are the average points.

it is said of the *Centaur*, that the Admiral was thrown out of his cot by the shock! If, with the wind at E. S. E., sail had been carried to the North but some fifty or a hundred miles, the centre would have passed by without reaching them, and heaving to on the *larboard* tack, for they were on the *left* side of the track, would have then given them a common gale, drawing aft till it ended with one at W. N. W. or a fair wind. The young sailor should make another lesson of this.

138. SCUDDING OR HEAVING TO. Every seaman of course begins by considering this question with reference to his sea-room. In what follows, I shall suppose throughout that he *has* sea-room, and that either because he has a fair wind in the Cyclone, or because he wants to profit by it, or wants to get out of its way, or has a leaky or crank ship, making lying to dangerous, he desires to scud, or wishes to know what he had better do. He has also of course by this time a correct notion of how the centre of the Cyclone bears from him, and for many parts of the world what are the usual tracks, or he has calculated pretty nearly what the track of the one he is engaged in is, from his previous run and the veering of the wind, as shewn at p. 117. He has now to consider if he is, with relation to these *two* data, the bearing of the centre of the Cyclone and its track—

A. *Behind it.*
B. *On either side of it, at about right angles to its track.*
C. *Before its track.*

Now (A.) his precautions must be directed if *behind* the Cyclone. *In scudding,**

(1.) Not to run into the heart of it.
(2.) Not to run too far *before* and then gradually across it.
(3.) Not to lose the advantages he may derive from it.

In heaving to,

(1.) To do so on the right tack.
(2.) To be careful when bearing up not to overtake it again, and therefore——
(3.) Not to bear up too soon.

* Sailors will note that I use the word scudding in this paragraph, and indeed elsewhere, rather loosely, and sometimes almost improperly, as nothing but sailing with the wind right aft, or two or three points on the quarter, should in strictness be called scudding, but for shortness sake I have not been particular.

Or (B.) his precaution *if on one side* of a Cyclone at about a right angle to its track must be,

In scudding,

(1.) To scud *with* it, but not to edge too far into it—*i. e.* to keep at a safe distance from the centre.

(2.) Not to over-run it, and be thus led to cross into, and before it, if he has not full sea-room, and time to do so.

In heaving to, to do so on the right tack.

And (C.) his precautions when *before* the track of the storm must be in this, which is the most difficult case, for it is that of a Cyclone "coming down" upon him, (see p. 91.)

In scudding,

(1.) To be careful that he can do so in a direction which will take him out of the way of the centre.

(2.) That he has full *time* (allowing always at least from ten or fifteen to twenty miles an hour for the Cyclone's rate of travelling,) to get out of the way of the centre.

(3.) That if already in part, as it were, out of the way of the focus (*i. e.* when there is every probability that it will pass *near* but not over him) he does not by scudding, put himself more fully in the way of it.

(4.) To see if, by hauling up (should wind and sea, and the condition and qualities of his ship allow of this), and *then* heaving to when the Cyclone by approaching becomes of full strength, he cannot get farther towards its border, and thus more surely at a distance from the dangerous central focus.

In heaving to,

(1.) To heave to on the right tack for the shift and for "coming up" to the wind as it changes, instead of breaking off.

(2.) When the calm centre reaches the ship to keep the vessel's head the right way to avoid being taken aback, if he can do so.

139. The following diagram of a Cyclone in the Northern Hemisphere, travelling to the E. N. E. as in the Northern Atlantic, will shew this more distinctly. In it the ships *A, H, I, M,* on the outer circle at the four cardinal points are pursuing safe tracks *round* the Cyclone; the ship *M* having carefully, but at some risk of course, to cross in front of it.

In the next circle, *O, T, U,* and *V,* finding themselves too far in, are all doing their best to get *out* of the Cyclone circle, as they understand their position, and have plenty of sea room.

But *W, X, Y,* and *Z* are all " carrying on " to get farther *into* it; and while *W, Y,* and *Z* are standing right in to cut off the centre, if they can reach it in time, *X* is chasing it as hard as wind and sea will allow him, and will run on till he broaches to, and is dismasted or upset.

We may call *S* one of that class of the old school who still affirm that " you should always heave to in a hurricane," and we see that he has done so—on the wrong tack, and will just drift into the centre as it passes, and be taken aback by the shift from the North.

140. If the page be held up with the face of the cut to the light, it will then represent on the reverse a Cyclone in the Southern Hemisphere, travelling to the W. N. W.,* the letters distinguishing the ships being chosen so as to appear rightly placed when so looked at also.

141. It will perhaps be useful if we try to give an extract from the imaginary logs of each of these ships, and this will not be so much pure invention as the reader may at first suppose; for I believe that, either from my own materials (published and unpublished) or those of other writers on the science, I could produce instances of every kind of management and *mis*management which I shall here set down.

LOG OF *A (bound to the Southward).* " Finding the wind, sea, and squalls increasing rapidly with a falling Barometer, and all the signs of a Cyclone, of which the centre is to the South of us, bore up to allow it to pass," &c.

LOG OF *I (bound to N. E.).* " We have a Cyclone evidently passing to the North of us. Kept away East to allow it to pass, so as not to approach too near to the centre."

LOG OF *M (bound to the North).* " After all due calculation I am of opinion that we are so far in the direct track of the Cyclone to the W. S. W. of us, that we have yet time safely to cross in front of it. All hands called, head braces stretched along, good hands to the helm, hatches battened down," &c. &c.

* Though this is an unusual track there.

PART III. § 141.]   *Imaginary Logs.*   129

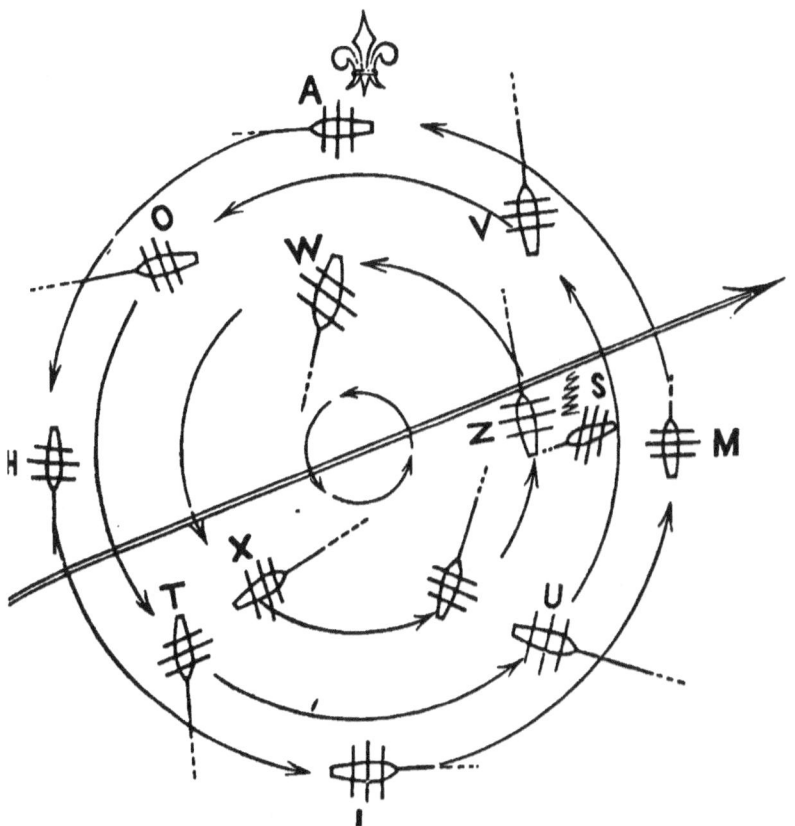

Log of *H* (*bound to the E. S. E.*). "As we are evidently overtaking the Cyclone, bore up South to get well on the S. Western quadrant before hauling up, so as not to plunge farther into it. Tremendous confused sea while crossing; no doubt from the wake of the Cyclone."

Log of *T* (*bound to the E. N. E.*). "Gale and squalls becoming terrific, with a fearful sea and Barometer falling; bore up to run farther out of the Cyclone circle."

Log of *X* (*bound also to the E. N. E.*). "Gale and squalls excessively severe, but the vessel behaving beautifully and steering to a

K

quarter of a point ————. At 1 A.M., in a terrific squall, the wind shifted to S.S.E., which brought us by the lee. Before the helm could be put over and the head yards braced round, we were driving astern at a frightful rate against the former sea; paid off, but in a few minutes had lost main and mizen masts," &c.

Log of *F* (*bound to the N. E.*). "Tremendous gales and sea. At 9 A.M. it fell calm with a sea which was rising in mountain peaks on all sides of us, and running in every direction. Ship perfectly unmanageable. At 11 A.M. a terrific burst of a hurricane came down with a roaring noise like thunder, taking us flat aback. Bowsprit and foremast went in one heavy plunge, and finding the vessel would not pay off, and was pressed down with the lee gunwales far below the water, cut away the main mast," &c. &c.

Log of Z (*bound to the N. N. W.*). "Finding it impossible to run any longer, though dead before the wind, hove to on the starboard tack. At 11 A.M. a sudden shift of wind from S. S. E. to N. N. W. taking us flat aback, lost topmasts, boats, booms, and every thing on deck."

Log of W (*bound to the S. S. W.*). "Running under close-reefed main and fore topsail and main trysail, broached to by a heavy quartering sea; lost topmasts, sprung main and foremasts, 3 feet water in the hold. Head of the rudder split, and two of the rudder pintles gone," &c. &c.

Log of S (*bound to the South Eastward*). "Gale very severe; every sign of an approaching hurricane. Hove to, &c. In a few hours calm with shift throwing the vessel on her beam ends against a heavy sea, and obliging us to cut away mizen mast," &c.

142. The mariner (I mean the careful man who conscientiously endeavours to do his duty to his country or his owners) will now, I hope, easily see how simple it is, first to ascertain his position with respect to the bearing of the centre, and then what line of management it will be best for him to adopt; and he will no more consider the time as lost in these brief calculations than he would that which at the approach of a dark stormy night, near a dangerous passage, he devotes to weighing carefully whether it will be better to stand on, or to heave to, or work to windward, or anchor.

143. I desire to be understood by those who may read and consider all this, and indeed for the whole of this work, that I quite appreciate, and fully allow for, the differences between the smart and well-found man-of-war, with her well-disciplined and numerous crew, and her tried and trained officers, and the ill-found, half-manned merchantman, with her best hands on the sick list; or (I grieve to say) too often her grossly insubordinate, or incapable officers and ship's company: and I am well aware how safely the commander of the first can calculate upon his officers, his men, and his ship, to the hour and the minute, and how little the last can venture upon the chance that

in two or three hours he may have far worse, or better, weather; but writing for all, I explain as clearly as I can the danger and the prudent course, and each must calculate for himself what his ship, his officers, and his crew can bear and do in the hour of need, if it comes.

144. THE USE OF THE TRANSPARENT HORN CARDS. To explain this, I suppose I have to shew them for the first time to a plain Seaman, or to a very young Officer who knows nothing about them, and by following with a little patience the steps laid down in the following lessons, there are none who will not, I trust, be able quickly, not only to understand it themselves, but to explain it to others.

145. Clearly to understand the Horn Cards, the learner should first consider one of them attentively, when he will see that in the middle the eight points of his common compass cards are laid down to guide his eye, but that the *circles* represent the winds as they may be blowing in a Cyclone, *and that these winds are always as tangents to the compass points.* Thus, in the Northern Hemisphere the N. E. wind is marked at the N. West point of the compass, and so on. It is this confounding the wind points (or wind tangents) with the compass points on the storm-card, because they are both on the same card, which sometimes confuses a learner.

The first and most useful lesson the plain seaman can give himself in this new science is, to obtain the habit of considering that all winds *may* be, and many certainly are, *not blowing in straight lines but in curves.* This is not so easy; for we are all habituated from our childhood to consider the wind both as blowing, "where it listeth," and also in a straight line from one point to another,* so that we are a little puzzled, when we must think of the (apparently) straight lines shewn by our dog-vanes and weather cocks, or by the trim of the sails of a whole fleet, or the furious gusts of a squall or hurricane, as only parts of a great curve.

The best way to form a good notion of this, is to take a common map of any kind, on a large scale, say on that projection on which the parallels of latitude are great curves, and put a dot on one of the pa-

---

* Just as we talk of the *level* of the sea, many landsmen using the word as if it applied to the horizontal and not a curved surface of about eight inches to a mile.

PART III. § 145.] *Use of the Horn Cards.* 133

rallels for your ship's place. Now, putting the hands, or two pieces of paper, so as to cover a parallel of latitude all but an inch, or half an inch, on each side of your dot-ship, you will readily see then how the folks on board of her might suppose and affirm *their* wind to be a straight line, just as you do at sea of *your* wind; and yet you can see that it is part of a great curve which *might* be also part of a circle of several hundred miles in diameter. It is indeed supposed by Professor DOVE and Mr. REDFIELD that *all* winds form circuits, which would make them all curve in some part, if not in the whole of their courses.*

The next lesson to give one-self is this. When you have obtained the habit of recollecting that the wind, though apparently blowing in a straight line for you, is, or may be, really blowing in a curve forming part of a circle or ellipse, to consider then *on what part of the circle* or ellipse you are? Say the wind is at North; you may suppose according to the hemisphere you are in, that you are on the East, or the West side of a circle; and so on of every other wind, as in the tables given in pages 105 and 107.

Making marks for ships' positions on a sheet of paper or an old Chart, and putting the Transparent Card over them, so as to see by it *how the centre bears* from the ship, is the next lesson, and will be found highly useful, for it is not every seaman who has the habit of considering his present wind a minute tangent line only, and the centre of a storm lying, according to the hemisphere he is in, at a perpendicular from that tangent. The *habit* of thus considering the wind, is what will give the plain seaman a complete understanding of the theory. He can always refer to the table in p. 105 to see if he is correct, and he will find it to be a sort of "boxing the compass at

---

* If we suppose a Cyclone of 300 miles in diameter, and thus (in round numbers) a little more than 900 miles in circumference, and 32 ships placed round the circumference, each with a wind one point different from its neighbour, they would still be 29 miles apart. Many of the gradual changes of the wind from quarterly to a-beam, close-hauled, and breaking off from the course, are probably owing to these circuital winds as the ship runs on, though in fine weather. *There is no doubt that in tropical countries the wind is often blowing in circular vortices though not rising to the force of a storm.* This remark, which I now print in Italics, closed the note in the first edition; and again, while I was writing this in India, Col. REID was collecting undoubted proofs of it at Bermuda. See Part VI. Miscellanea. (p. 367.)

right angles," which every one has not got. If whenever you look at the dog-vane or a weathercock you make a rule of considering the wind also as a tangent line to a circle, in either hemisphere, you will quickly master this difficulty. "How is the wind, Mr. —— ?" "S. S. W., Sir," (*and if it is part of a Cyclone in the*
Northern } *hemisphere the centre bears* { W.N.W. } *of us.*)
Southern   { E.S.E. }

The next lesson is to move the Storm Cards, each in their own hemisphere—for with them *you have the hurricane in your hand*, recollect—over an Island, or a fixed mark of any sort to represent a ship lying to, on a Chart, and to notice how the effect of a vortex of any size, if so moving, would produce the changes of wind, of which we read; supposing always that we move the Storm Card along the right *track :*[*] As across the Leeward Islands from the S.E.; along the Coast of North America from the S.S.W. or S.W.; in the China Sea from between N.E. and S.E. to the S.W. and N.W. &c.; for, as will be easily seen, *if the track of the Storm varies, the changes of the wind also vary*, and if in either hemisphere a storm came, for instance, from the West instead of the East, the changes would be exactly opposite.[†]

The seaman should next mark off a ship's place, real or supposed, with her run or drift for any given number of hours, and then, moving up the Storm Card near, or across this, observe how the motion of the Storm *and* this run or drift induce different, or more rapid, changes of the wind, or take a ship out of, or run her headlong into a Storm.

---

[*] The seaman and student will find it convenient to have two bullets flattened and covered with silk and a piece of narrow red tape about 18 inches long, stitched to the inner side of each; these may be laid down in any direction on a Chart, and will represent the track of the focus of a storm very distinctly as a sort of "Line of mischief," over which the Card may be conveniently moved at pleasure, and the changes perfectly noted.

[†] He should next mark off *two* ships' places, and move his Storm Card so as to pass the centre between them, noting how the wind will *veer* with each ship, though the whole hurricane is *turning* according to the Law of its hemisphere. If he sets these down on a piece of paper he will see at once how the contradictory accounts of winds veering and "backing round," as it is called, are perfectly true, though till now inexplicable. See Diagram at p. 14.

PART III. § 147.]   *Use of Cards.   Practical Lessons.*   135

These preliminary lessons may appear trivial, but they are not so; and the seaman who will give them a little attention, will find great advantage from so doing. I need, I hope, remind none that when bad weather is coming on they will have other matters to attend to than studying *then* the applications of the Science; and that it is the cool and patient consideration of emergencies in fine weather and at leisure that makes the able and ready seaman when a crisis arrives.

If a small hole is drilled at the centre of the horn card, just large enough to admit the point of a pencil and a mark be made on a sheet of paper to represent an island or a coast town, or a ship hove to, and the wind's place at the outer circle of the card, be put over this, while a pencil point is held in the hole, it will be seen how in moving the card so as to make the successive changes of wind pass over the Island (or *near* it allowing for the drift if a ship is supposed) the pencil point will trace out the track of the Cyclone's centre.

146. I now proceed to apply all this by practical lessons in various parts of the world for the uses of the Storm Cards on Horn, which will be found in the pockets of the covers; one being for the Northern, the other for the Southern hemisphere.

The following rules must be borne in mind:

1.—The card may be supposed to represent any sized storm from 20 to 500 miles in diameter, and as many more circles may be supposed to be added to it as may be necessary to suit the scale of the chart.

2.—The *fleur de lis* must always be kept on the *magnetic* meridian when using it.

3.—It is always to be placed so, that the *wind's place* (or Cyclone Point) is over the *ship's place*.

4.—It is to be moved as required along the known, or estimated, track of the storm.

147. SPECIAL EXAMPLES FOR THE USE OF THE STORM CARDS. FOR THE WEST INDIES.—Mark off the place of a ship in Lat. 16° North, Long. 50° West, and suppose her bound to Trinidad.

The weather and barometer indicate the approach of a hurricane or severe gale, and the wind is at E. b. N.

Now place the Card so that part of the outer circle which lies between East and E. N. E. may lie over the ship's place, and it will be seen that if this is a Cyclone* the centre of it bears S. b. E. or *thereabouts* from you.†

Next you know (see p. 28 and Chart I.) that it is hereabouts that the West Indian, Atlantic and American coast storms seem to take their rise, and that they travel from hence, as far as yet known, on a track to the Northward of West, but rarely so far so as North-west. W. b. N. or W. N. W. then, is probably about its track, and fifteen or eighteen miles its rate of *sailing* in that direction.

I suppose your sea to be not yet *very* heavy, so that with the wind at E. b. N. and every chance of its becoming more Northerly, your ship sound, and crew hardy and smart, with plenty of good helmsmen amongst them, the temptation to run on is very great; but you see also in a moment that, if you do, you will cross before, within, or close behind, or at the very centre of the Cyclone, and that you will thus, as it were, thrust yourself, if not into the jaws of destruction, certainly into a useless peril.‡ Now as no man runs his ship headlong into a waterspout, when backing his main-yard for ten minutes can keep him out of it, your prudent course is to lie to, and allow the Cyclone to travel past you, and the rise of your barometer and the gradual veering of the wind to the Southward of East will shew you that it has done so, and all you have then to be careful of is, when standing on, not to *overtake* the Cyclone, should it by any chance be a slow moving one: see p. 126 for your various considerations and chances.§

* The fall of the Barometer and the appearance of the weather are in all common cases sufficient to distinguish it from the trade rising to the strength of a gale, as trade winds and monsoons sometimes do.

† *Thereabouts*, because there may be some irregularity or incurving of the wind; see p. 108.

‡ You will see also that supposing even that you get safely through or pass the centre by "dashing at the hurricane," all you gain on one side of the storm circle you lose on the other, as the winds become Westerly!

§ It requires, indeed, some little confidence of this kind of knowledge for a Commander to bring himself to what is called "throwing away a fair wind;" and more than one has told me, after being severely damaged, and though they suspected from the Law of Storms that they were running into mischief, yet they could not resolve to lose a chance where others might push on and laugh at them; and many, like myself, may have heard between 30 and 40 years ago, "the Captain's *Barometer-*

148. Suppose your ship in 16° North and 72° West with all appearance of a Cyclone, the wind at W. N. W., and she is bound to Grenada. Mark off her place, and lay the *W. N. W.* of the Horn Card over it; and you see at once that is a storm of which the centre lies about N. N. E. from you, *i. e.* between you and the shores of St. Domingo, which it is probably then ravaging; and you moreover see that, for you, it is a fair wind by which you may safely and surely profit; hauling first a little to the Southward to be sure of a good offing from the more violent and dangerous part, and also to get room to keep away a little if the sea should be heavy when you get on the S. Eastern quadrant of the Cyclone, when the wind, as you see, will haul to the S. W.

149. Suppose you are bound from New Orleans to the Havannah, and that at a day's sail from the first port you begin to have a falling Barometer, a heavy sea, squalls, gloomy weather, &c. and "half a gale" from E. N. E.

With the old knowledge we should all, certainly, "carry on" to this, and keeping the wind about a point before the beam, push on for our port under double reefs, with a close look out on the steerage. But place the Storm Card on the chart, and you will see that if you do so you will be rapidly plunging into the Cyclone.

But you need not therefore heave to, for you will note that the hurricanes hereabouts all travel between the E. N. E. or N. Easterly to North and even N. Westerly. It is clear then that you *can profit by it by sailing round it*, i. e. keeping away West, W. S. W., S. W., S. S. W., and South, as your Barometer guides you, and thus make a curve round the heel of the storm, and which will not be a large one; for as *it* is probably flying ten or twenty miles an hour on *its* track, while you are, in the beginning of the storm, and if the track be one to the Eastward of the meridian, going wholly the other way, the wind

---

gales of wind" grumbled over, when reefing top-sails by its warning; and even his Lunars and Chronometers sneered at! It seems that, with seamen especially, all useful knowledge has to force its way through this dislike of novelties. One would give something to know what the good old first-rate observers with the Astrolabe and Quarter Staff and Davis' Quadrant, said of the "looking glass gimcrack," which we call Hadley's Quadrant?

will veer at first so rapidly, if the track is to the Eastward of the meridian, that you will be surprised to find how soon your Easterly gale, is at North, N. W. and West, just as your Storm Card shews. If however it should be one of the Cyclones which, like *M*, *L* or *E* have tracks to the Westward of the meridian, it is evident that this may be merely scudding *with* the storm, if its track is very Westerly, or crossing the track before the centre, and perhaps much too near for safety. This will be quickly known by the veering or steadiness of the wind (see p. 116), and if it is supposed that the track *is* one of these, the plan is to heave to on the starboard tack, being in the right hand semi-circle.

150. IN THE NORTHERN ATLANTIC. A ship from Boston bound to Bermuda at 200 miles S. E. of Cape Cod, meeting a N. Easterly gale with a falling Barometer and other indications of a hurricane Cyclone will clearly understand also why she has her choice of running into the heart of the Cyclone, at the risk of being dismasted or upset, if she persists on her direct course as long as she can stand on, or of sailing round the Western edge of it, and thus at the expence of a little curve in her track making a fair wind for her whole voyage.

151. Colonel REID has given in his work the following: which I copy entire, as affording the seaman both for that and other quarters of the globe an able instance of how we may in time profit by this enlarged view of the practical utility of the rules of our new science. Every reflecting and experienced seaman must recollect his disappointments (and often his losses, at least of time if not of money) at finding his fair wind gradually become a foul one, and not a few may recollect by how many theories and assumptions they have endeavoured to account for these apparently capricious veerings.

A NOTE.

On the winds as influencing the tracks sailed by Bermuda Vessels; and on the advantage which may be derived from sailing on Curved Courses when meeting with Progressive Revolving Winds.

"In high latitudes the prevailing atmospheric currents, when undisturbed, are westerly, particularly in the winter season. As storms and gales revolve by a fixed law, and we are able by observation to distinguish revolving gales from steady-blowing winds, voyages may be shortened by taking advantage of them.

"The indications of a Progressive Revolving Gale are a descending barometer with a regularly veering wind, or with the wind changing suddenly to the opposite point.

"In the Northern hemisphere Storms revolve from right to left, thus,

"In the Southern hemisphere Storms revolve from left to right, thus,

"The indication of a steady-blowing wind which will not revolve, but blow in a straight-line direction, is a high barometer remaining stationary. When the steady wind blows from either pole, according to the side of the equator, the atmosphere will be both dry and cool. An increase of warmth and atmospheric moisture, are indications of the approach of a Progressive Revolving Wind.

"*Sailing from Bermuda to New York.*—The first half of a revolving gale is a fair wind from Bermuda to New York, because in it the wind blows from the *east*; but the last half is a fair wind from New York to Bermuda. During the winter season, most of the gales which pass along the coast of North America are Revolving Gales. Vessels from Bermuda bound to New York, should put to sea when the *north-west* wind which is the conclusion of a passing gale is becoming moderate, and the barometer is rising to its usual level. The probability is, more particularly in the winter season, that after a short calm, the next succeeding wind will be *easterly*, the first part of a fresh Revolving Wind coming up from the South West quarter.

"A ship at Bermuda bound to New York or the Chesapeake, might sail whilst the wind is still *west*, and blowing hard, provided the barometer indicate, that this West wind is owing to a Revolving Gale which will veer to the *northward*. But as the usual track which gales follow in this hemisphere is Northerly or North-easterly, such a ship should be steered to the Southward. As the wind at *west* veers towards *north-west* and *north*, the vessel would come up, and at last make a course to the westward, ready to take advantage of the *east* wind, at the setting in of the next Revolving Gale.

"*Sailing from New York to Bermuda.*—A vessel at New York and bound to Bermuda, at the time when a Revolving Wind is passing along the North American coast, should not wait in port for the westerly wind, but sail as soon as the first portion of the Gale has passed by, and the N. E. wind is veering towards *north*; provided it should not blow too hard. For the *north* wind will veer to the *westward*, and become every hour fairer for the voyage to Bermuda.

"*Sailing between Halifax and Bermuda.*—A great number of gales pass along the coast of North America, following nearly similar tracks, and in the same winter season make the voyage between Bermuda and Halifax very boisterous. These gales by revolving as extended whirlwinds, give a *northerly* wind along the shore of the American Continent, and a *southerly* wind on the Whirlwind's opposite side far out in the Atlantic. In sailing from Halifax to Bermuda, it is desirable for this reason to keep to the westward, as affording a better chance of having a wind blowing at *north*, instead of one at *south*; as well as because the current of the Gulf Stream sets vessels to the eastward.

"*From Barbadoes to Bermuda.*—When vessels coming from Barbadoes or its neighbouring West India Islands, sail to Bermuda on a direct course, they sometimes fall to the eastward of it, and find it very difficult to make Bermuda when westerly winds prevail. They should therefore take advantage of the trade winds, to make the 68° or 70° of West Longitude, before they leave the 25° of latitude.

"*Sailing from England to Bermuda.*—On a ship leaving England for Bermuda, instead of steering a direct course for the destined port, or following the usual practice of seeking for the trade winds, it may be found a better course, on the setting in of an *easterly* wind to steer west, and if the wind should veer by the *south* towards the *west*, to continue on the Port tack, until, by changing, the ship could lie its course. If the wind should continue to veer to *north*, and as it sometimes does even to the *eastward of north*, a ship upon the starboard tack might be allowed to come up with her head to the westward of her direct course. On both tacks she would have sailed on *curved lines*, the object of which would be, to carry her to the westward against the prevailing wind and currents. There is reason for believing that many of the Revolving Winds of the winter season originate within the tropics; and that ships seeking for the steady trade winds, even further south than the tropic at that period of the year, will frequently be disappointed. How near to the Equator the revolving winds originate, in the winter season, is an important point not yet sufficiently observed. The quickest voyage from England to Bermuda therefore, may perhaps be made, by sailing on a course composed of many curved lines, which cannot be previously laid down, but which must be determined by the Winds met with on the voyage. This principle of taking advantage of the changes of Revolving Winds, by sailing on curved lines, is applicable to high latitudes in both hemispheres, when ships are sailing westerly." W. R.

Government House, Bermuda, 21st March, 1846.

152. While correcting this, I read the account of the *Great Western* Steamer committing the old blunder of *steaming into* a Cyclone in the Atlantic Ocean, in which she was nearly lost. She had it beginning at S. W. and veering gradually to N. N. E. Her drift in distance steamed, in the interval of 36 hours is not given, but it is clear that she was on the South side of a Cyclone travelling about from West to East, taking her to have steamed and drifted about 100 miles to the N. W.

In the same Cyclone a Royal Mail Steamer from the West Indies to Bermuda having met the Cyclone on her track, slacked her speed to allow the centre to pass her, and made a fair wind of its southern quadrants to her port.

153. A vessel from Europe, bound, say to New York, meets with a strong S. S. Easterly gale and falling barometer, about the meridian of Bermuda. Here the seaman will observe, by the tracks and his Storm Card, that he is on the *Eastern side* of the Cyclone, which is travelling *towards* him on an E. N. E. or North Easterly course, and that if he stands on he will inevitably meet it. To run off to the N. W. till he has brought the wind to at least E. N. E. or N. E. and his barometer is rising, would be the means of getting out of the way of the centre; but it is very uncertain what are the sizes of the Cyclones hereabouts, and the distance he might have to run; most seamen will probably therefore prefer heaving to and allowing the centre to pass them, when the wind will become a fair one.* See what has been said before at p. 30 on the tracks and for the sizes of the Atlantic Cyclones during the winter gales, see also Chart No. I.

154. COASTS AND SEAS OF EUROPE. If I venture to give any examples for these parts, is is rather with the hope of exciting the attention of seamen to the subject, than as founded upon any positive knowledge which we as yet have. It is *probable*, as before stated, (p. 34) that the great Atlantic Cyclones sometimes reach the coasts

---

* It has been noted before that these situations, when a ship is directly on the average track of the Cyclone, are those of the greatest difficulty on two accounts. It is difficult to say beforehand, unless great attention has been paid to the run and veering of the wind, on *which* side of the track the vessel is; and the rate of travelling is here a serious question on which everything depends, and it is difficult therefore to judge if by running off for a few hours we shall really be getting further *out* of the way. In the instance given above, the centre bears about W. S. W. and the track may be due E. N. E. *or* N. E. If it be only E. N. E. there might be time to cross in front and round the Northern verge of the storm; but if it be travelling N.E. the wind and sea may become too heavy to allow of standing on, and the run made would then only have brought the vessel directly into the path of the focus; with the encumbrance of her sail if she meets it unexpectedly. Nevertheless there may be many cases, such as urgent service, chasing, and the like, in which the sails, and even the chances of the masts going with them, are not worth considering, and it is therefore proper that the seaman should have all the resources and the risks before him, for it is not impossible that a case might occur in which he might by clever manœuvring inflict a share, at least, of the latter on an enemy. Every experienced officer can tell how often in coast cruising the chase escapes by her superior knowledge of the winds and currents.

of Spain and Portugal, and that their Northern verges extend to the coast of Ireland as shewn on Chart No. I.

We have also authenticated examples of true circular Cyclones passing over Great Britain, and of small, hurricane-like storms, both as to force and veering, prevailing between the Chops of the Channel and Madeira, but of the storms of the inland seas, such as the Mediterranean, Ægean, Baltic, &c. we have no defined accounts, though their suddenness and severity render it most desirable to learn if they be Cyclones or not. I shall thus merely adduce cases in which the seaman may use the Storm Card "to consider with," and then if his experience should justify it, guide himself by the help which they afford.

155. I suppose a vessel bound into the Channel in Lat: 47° North and Long. 20° West, with a smart gale which has veered from East to E. N. E. and a sufficient fall of the barometer to induce the supposition, that this may be a Cyclone travelling towards the Channel or Bay of Biscay.

The Storm Card will shew forthwith that the centre bears S. S. E. of the ship, and though with the wind at E. N. E., if the ship can stand on the port tack, S. E. would offer most advantage, this may be leading her to keep company with the Cyclone and get worse weather; it may be therefore well to consider if standing 50 or 100 miles on the starboard tack, or due North, may not be the safer plan? and the barometer here will soon shew if the distance from the centre has increased.

If heaving to be found necessary from stress of weather, then the card will also shew that being on the port side of the track of the Cyclone, which may have come up from the Bermudas, and in all probability is travelling to the E. N. E. or N. E. the port tack is the one to heave to upon, and that the wind will draw aft and not forward. It will also, by moving it along an average track, warn against the error of running again into the Cyclone when it has passed by; which may be avoided by keeping a little to the Southward, and if possible on the South-western quadrant of the Cyclone, where the wind is from N. W. to West. An outward bound vessel, from Europe to America, would of course manage quite differently, her business being to profit all she can by being on the favourable side of the Cyclone,

PART III. § 157.]   *Examples; Narrow Seas of Europe.*   143

and to avoid (especially if it be a slow moving one) running into the heart of it, where the violence of the wind, though fair, and the heavy sea must expose her to broaching to if she approaches it too closely.

156. If we place a Storm Card between the Coast of Ireland and Cape Finisterre, we shall understand as before said (p. 34), how a Cyclone of 600 or 800 miles in diameter may fill up the whole Bay of Biscay and the entrance of the Channel, and while it is a Westerly storm on the coasts of Spain, be a Southerly one for those of France, and South-Easterly and Easterly for the Channel and coast of Ireland. A ship from Cape Clear bound to Gibraltar might, in some cases, profit by making a somewhat curved track, so as to pass *behind* the centre, and have full room to take advantage of the North Westerly and Westerly gales where the diameter of the Cyclone was not excessive.

157. In the narrow seas of Europe we have usually too little sea-room to do much in the way of profiting by Cyclones if they occur; but the time to heave to, the tack to heave to upon, and often the direction in which to steer, so that, all things permitting, their excessive violence may be best avoided, are matters of great importance to the seaman; for they may at least be to him the difference of arriving at his port with a damaged cargo and a strained and leaky ship, or safe and sound when others have been torn to pieces. Thus in a N. W. storm in the Gulf of Lyons, if we call it a Cyclone, we see that we are on its S. W. quadrant, and that to stand a little to the South would probably carry us from the more violent parts. See tracks *a* and *U* on Chart No. I.*

Again in the sudden and violent storms of the Black Sea, a ship having left the Channel of Constantinople and being bound to Odessa, might if meeting with a violent Cyclone *travelling to the Westward,* advantageously steer, at first, to the Eastward and then haul up gradually to the North, so as to carry a fair wind up to her port, while if she stood at first to the Northward she might get damaged by the strength of the storm, and driven considerably to the Westward. If bound however to Trebizond and the storm was travelling *to the Eastward,* she might also carry a fair wind the whole way by

* See p. 86 for NELSON's management in these gales.

running with the Southern border of it. We do not know that the storms of the Black Sea are Cyclones, nor what are their tracks, but no intelligent seaman can look at a few old logs* of such storms, with the assistance of the Horn Card without detecting by the veering of the wind and his track, if they were so, and how they travelled.

CYCLONES OF THE BLACK SEA. I have mentioned at p. 86 the possibility that the "gales" of the Black Sea may at times be Cyclones, and that they are so I find confirmed by a letter from Captain SILVER of the Brig *Tuscan*, of Portsmouth, who gives me the abridged details of one which seems to have been travelling up from the S. West to the N. East; the wind having veered with him as he ran up from the Bosphorus towards Odessa, whither he was bound, from S. S. W. to W. N. W., North West and North. He has not sent me his detailed log, nor the height of the barometer, but there seems no reason for doubting that this was a true Cyclone.

158. SOUTHERN INDIAN OCEAN.— Off the Cape of Good Hope and to the Eastward of it, as far as Cape Leuwin and Bass's Straits, ships bound to India and Australia are often detained for some days, and even for above a week, by strong North-Easterly, Easterly, and South-Easterly gales, in which little or no progress can be made, and they are for the most part lying to. Now these, as explained at p. 40, are very probably the Southern quadrants of great Cyclones which are passing to the Eastward, North of the vessel, and she may in all likelihood, by standing a degree or two to the South, get out of their influence into the regular Westerly gales prevailing there.

159. For as there is no sort of doubt that the West India Cyclones of say 200 or 300 miles in diameter at most, expand in their progress along the coasts of North America, and become great Cyclones of even 1000 miles or more in diameter as they travel across the Atlantic in high latitudes, we may easily conceive the same thing to take place in both hemispheres, giving rise to Easterly or Westerly gales, according as the ship is on the North or South side of the circle.

Now let us consider the case of a heavy Westerly gale for a ship in Lat. 38° South, and between the Cape and St. Paul's and Amsterdam. If we suppose a polygon of 32 sides for the 32 points of the com-

* Unfortunately none are obtainable in India, where I am writing.

pass, and each side of 96 miles, this will be about equal to a circle or Cyclone of a little more than 1000 miles in diameter.

If our Cyclone is travelling at the rate of 15 miles an hour from West to East, and our ship is on the Northern verge of it, we will take it that she is scudding at the rate of 11 knots due East with a heavy Westerly gale, as it is called.

But as the storm is moving at the rate of four miles an hour faster than the ship, its centre will in the 24 hours have made 96 miles more Easting than the ship, and this in our supposed polygon of winds, and we allow no incurving, will have left the ship at the point where the wind is W. by S., and we may suppose also less strong, for the centre of the Cyclone is more distant than it was from her, so that for the next 24 hours she may scud not more than 10 knots, and thus for this day the Cyclone will gain 120 miles upon her, and thus she will gradually fall out of the storm circle. If the ship's or Cyclone's track be different, and approach to or diverge from each other, the results will of course vary.

As just noted, the heavy Easterly gales also at times experienced in these latitudes, and which so much detain outward bound ships, are probably often referable to the same cause, namely, the ship being overtaken by the Southern half of a great Cyclone, travelling to the Eastward, which may therefore begin at East, and end at S.E.; but if travelling slowly (or if, as seems sometimes to occur, followed by another of the same kind), it may take several days to pass over the ship, as its chord from the N.E. to the S.E. point would be about 680 miles, which even at 16 miles an hour (leaving out all account of the ship's drift) would require nearly two days.

If this view is a correct one, the remedy is to stand to the Southward, to get out of the storm circle into the usual stream of the Westerly gales, and with a homeward bound ship meeting a Westerly gale, which she had good reason to suppose part of a rotatory one, to stand to the Northward. Every seaman will see at once how this applies in the Northern hemisphere, as, for example, in the gales of the Atlantic and in other parts of the world, if Cyclones, and by placing his Storm Card on the Chart, he may often find that he may save himself some days of "thrashing and beating," and his owner some wear and

L

tear, by going a little round about: verifying perhaps the homely old proverb,

> "What can't be done by pushing and striving,
> May often be done by a little contriving."

I should perhaps explain here that I do not by this intend to counsel more than a temporary run to the Southward to disengage the vessel from the influence of the Cyclone. She can afterwards haul up a point or two so as to regain any parallel which her commander may conceive the most advantageous one for running down his Easting. I suggest this course only as preferable by far to wasting time in lying to "till the gale blows over." See at p. 40 the remarks of Captain ERSKINE, R. N. on the passage of H. M. S. *Havannah* under his command.

160. Ships lying at the Bell Buoy at the Mauritius (Port Louis), and having every indication of an approaching hurricane, must consider if it is one like our tracts *a* and *c* Chart No. II., passing directly over the Island, or to the Northward and Southward of it. If passing directly over it, the wind, as will be seen by the Storm Card, will remain steady at S. E. till it shifts to the N. W., probably after a calm interval. If passing to the North Westward of the Island, the winds will veer to the S. E., East, and N. E., and if passing to the Southward and South Eastward of the Island, the winds from South will veer to the S. W., West, and N. W., so as to make the Island a dangerous lee shore in a few hours if a good offing be not run for. It is in this knowledge of what the winds *will be* (certainly) in a few hours, that the seaman's safety, or his avoidance of serious damage, often lies, and it is to the absence of it that mischief is in innumerable cases to be traced.

Mr. THOM, p. 229, quotes the case of a ship which "in the hurricane of 1840 slipped with the gale at S. E., hove to, and drifted from the island, but the next day was carried back by the N. W. wind, and thrown on the reefs at the entrance of Grand River, at the place she had left."*

---

* Mr. THOM quotes also the case of the *Stag*, but, by an oversight, condemns her for running off from the Bell Buoy to the W.S.W. He forgot that, in a heavy gale the sailor often cannot choose, but must *venture* only to run with the wind

161. Suppose yourself in 10° South and 84° East Long. with a strorg gale at W. S. W. and falling barometer; place your Southern Storm Card as before, and you will see, (as you are now in the Southern hemisphere) that the Westerly winds place you on the *Northern* side of the Cyclone and that a W. S. W. wind makes the centre bear about S. S. E. from you.

Now if you are bound to the Northward you may profit by this, for all research as yet shews that the storm tracks are to the W. S. Westward hereabouts,* and thus the Cyclone is leaving you, while you are leaving it.

If bound to the Southward or South-Westward, however, (as to Europe, Australia, or the Mauritius) the case is exactly reversed, and you might, as you perceive, with a very little "carrying on to it," plunge headlong into a Cyclone of unknown fury. Your safe and proper plan is, if so bound, to steer off before the wind, or even N.E. till you raise your barometer a little. The wind you will find draw to West (the centre now bears South of you) and gradually to the Northward of West, when you haul up a point in an hour or more, *as your barometer* and the wind advise you, till hauling South and S.W. you have *sailed round* the storm, and have now literally—for it is almost an Americanism—" got to the other side of it," where it is a fair wind for you.

162. This simple manœuvre of sailing round a storm is especially useful here; for though, of course, a vessel may heave to and allow the Cyclone to leave her, yet the usual tracks are so exactly along or two, or at most three points on the quarter; and then with extreme care for fear of broaching to, which the smartest vessels with the best helmsmen will at times do in spite of every care. Every sailor knows that in a fleet, for instance, all scudding with the same gale, while some ships will steer with one spare hand at the helm to help if wanted, others of the same size, merchantmen or men-of-war, require three or four men at the helm and an officer to watch them; and withal will sometimes tear the wheel from their hands, and perhaps seriously hurt them if the relieving tackles are not carefully attended to.

* In one instance traced by me, (Eleventh Memoir, Jour. As. Soc. Beng. Vol. XIV.) for a short time, and very slowly to the N. W., but standing North or N. N E. would still carry a ship rapidly out of the vortex in such a case, as her barometer would quickly shew. See Track VIII. Chart II.

rather *across* the homeward bound routes, that after losing from six to twenty-four hours in heaving to, a ship would in most instances find, when she bore up, that in a few hours she was again coming down upon the body of the Cyclone and have again to heave to; so that in the end she would lose far more time than by sailing round it as I propose. I am not *supposing* this, but writing from facts. The ship *Orient*, Captain WALES, from Calcutta to the Mauritius, hove to to avoid running into a Cyclone (Track *i i* on the Chart) and bore up to proceed, when she again in a few hours found she was getting into a worse weather, and again hove to; repeating this bearing up and heaving to three times, an excellent instance of careful management.

The transport *Maria Somes*, with troops on board, committed the fatal error of running headlong into another hurricane close to the *Orient's* track, was dismasted and nearly foundering, *and suffocated fourteen persons for want of air during the tempest!* having all her hatches closed.*

163. Suppose yourself bound to India or the Straits of Sunda, say in 20° S. and 75° East, and that the Trade from fine clear weather and a high barometer becomes stormy, the barometer falls and the wind is at East or E. b. S.† This, as you will instantly see by the Storm Card and the Chart, is in all probability a Cyclone crossing ahead of you, and by heaving to for a few hours it will pass you, travelling to the W. S. W. just as a water-spout might do. You will then probably have the wind from the N. E and North, as your Storm Card will shew you, and by standing to the Eastward your barometer will rise, and you will quickly find the Trade again.

164. The most recent instance in this latitude of this kind of careful management which has occurred, I give a little in detail, as being highly instructive. It is that of the ship *Earl of Hardwicke*, Captain WELLER, from England bound to Calcutta, who was good enough to place his well-kept log and a private note-book both at my disposal.

The *Hardwicke* was, on the 26th December, in 28° 42' South, Long. 80° 46' East, with her Barometer at 29.95, and with light, and lat-

* See Sixteenth Memoir, Jour. As. Soc. Beng. Vol. XVII.

† And especially in these cases with a heavy cross sea from the *North* East or more Northerly.

terly strong breezes S. E. and North. Up to this time squally, thick, heavy, wild looking weather, upper clouds coming from N. W., the next stratum N. E. and the lower scud, with the wind, fast from S. E. At midnight from 10 knots ran into a dead calm. Breeze was renewed again, and next day she was in 26° 14′ South, 81° 5′ East, with Barometer at 30.00, and confused sea, heaviest from S. W., wind East to E. S. E. and a strong Trade throughout. On the 28th in 22° 37′ South, 81° 00′ East, and Barometer 29.95; and on the 29th in Lat. 19° South. Long. 81° 00′ East, Barometer, 29.71, strong Trade still, but squally and a confused sea, and barometer falling, made preparations for bad weather: upper clouds from N. E.

On the 30th in 17° 0′ South, 81° 41′ East, Barometer 29.75. To 8 A. M, running 6 and 8 knots to the Northward, but appearances threatening,* hove to, dense lurid atmosphere, very peculiar appearance at sunset the last two evenings. P.M. continued dark appearance to the North Westward, ran twice to the North and found the wind increasing, and drawing to the Eastward with thick weather, but always fine when going South; kept her South till it should clear off a little; a thick lurid appearance over the heavens, the sun only shewing as through a dense veil with heavy leaden-looking clouds to the North and N. W.

Captain WELLER's private note is: At 4 A. M. barometer not falling any more, made more sail to the Northward, weather became more equally with thick weather and heavy rain. At 8 A. M. a heavy squall from the N. E., shortened sail to close-reefed main top-sail, light Easterly air with a heavy arch to the Northward, which kept nearly in the same position till noon, ship drawing to the Southward 3 knots. At 1 P.M. made sail again to the North and East. As we advanced, the weather became thick and squally until 4, when in a smart squall with thick rainy weather, not able to see fifty yards from the ship, wore to the South, and shortened sail to close-reefed fore and main top-sails, the weather clearing a little, but an immense mass of heavy leaden-looking clouds, and over the whole of the heavens a very murky threatening appearance; sun at setting gave the whole a red lurid appearance,

---

* These cloud notes are of high interest; perhaps, as we shall subsequently see, of higher interest than we suppose.

and every thing on board had a red tint. At 8 P. M. a fresh gale S.E. Although the sun and moon were visible during the day, yet they were only seen as through a thick veil. 30th December, Noon Lat. 16° 26′ South, Long. 81° 35′ East, Barometer 29.80. After midnight the stars began to shew, and the thick lurid haze went off, and blue sky was visible at daylight, but still a heavy leaden appearance to the Northward with a heavy confused swell, heaviest from the East.

In this capital instance of good management, there can be no doubt, that Captain WELLER, with the sacrifice of a few hours' run, avoided running into a Cyclone on the South Eastern border of which he was, and of which he saw the body in the bank and arch of clouds to the North and N. W. of him; as distinctly as one might see the smoke of a fleet of enemy's steamers, and stand back not to run in amongst them. The red lurid tint of the atmosphere is a peculiar characteristic of the approaching Cyclone in the Southern hemisphere, as we shall subsequently see.

165. If however in this position (§ 163) you are bound to the Mauritius or the Cape, you will see that you may clearly profit by it, steering only as far to the South as not to let it gradually approach you, and thus become too violent to run in with safety. For this your barometer and keeping the wind to the Northward of East will be sure guides. If the wind is to the *Southward* of East, you are *before* (to the West of) the meridian of the Cyclone, and should haul more to the Southward out of its way.

166. But if in the Latitude and Longitude just stated, 20° South and 75° East, I have the wind South?

You are then right *before* the path of the tempest (Cyclone), and taking into consideration also the state and qualities of your ship, as well as the distance of the centre, as you estimate it from the barometer, you must manage to get out of the way of the terrific centre at all events.

The best plan is to steer off N. W. or N. N. W.* till the wind is at least S. S. W. and the barometer has decidedly risen, when you may

* I give these apparently loose limits, because while a man-of-war or fine-built merchantman, well steered, might run on for a long time N. W. with a gale at South, others could not venture to do so without great risk of broaching to.

haul North and either heave to or run on, as you are bound. Note here, as will be subsequently shewn when we come to speak of the barometer, that your barometer will be proportionately higher *before* the storm than when on either side, or behind it. And also that as the Cyclones here are from 200 to 400 miles in breadth, you must run at least one-third of that distance to be well out of harm's way.

The foregoing examples will also apply to the Cyclones of the Straits of Madagascar and those between Cape St. Mary and the Natal Coast, which are all in the Southern hemisphere, and have all their tracks, *so far as we yet know*,* to the W. S. W. or Southward of West.

But about the Latitudes of Bourbon and Mauritius, there seems no doubt that the tracks of the storms *recurve* to the South, (see Chart and at p. 46 what is said there of the *Serpent's* Cyclone;) and hereabouts the seaman must warily apply all his skill and attention to ascertain the track before he heaves to, and be assured that if he has allowed the storm to pass him he does not run into it again, by crossing the apex of the curve.

167. THE ARABIAN SEA. We are here again in the Northern hemisphere, and must recollect that all the changes are different from those we have considered in the foregoing paragraphs. In the section on the Tracks of Cyclones I have adverted fully (p. 52) to the little that is known of the tracks of the Cyclones in this sea, and the possibility of their curving to the Northward towards the coasts of Persia and Arabia, which should be borne in mind; as also that a few degrees from the Malabar coast, the Cyclones appear at times to be sudden and violent, though of small extent.

Taking a Cyclone to begin at N. W. with a ship bound to Bombay in Lat. 13° North, Long. 67° East, it will be clear that she should not stand on upon her North Easterly course, but run off E. S. E., hauling up East and gradually N. E. as the wind hauls to the S. W. which it will do, and this *may* be the best plan even for a vessel bound to Aden or the Persian Gulf instead of lying to. At all events she should not heave to with the wind at N. W., for should the course of the Cyclone

---

* And this caution especially applies to those to the Southward, where it is quite possible there may be a second recurving track, analogous to that off the Mauritius mentioned in the next paragraph.

be very Westerly and its diameter small, the centre would pass very close to her. To run off to the S. E. or S. W. if the season admits, so as to raise her barometer will be more prudent, and she may regain her former position by running partially round the Cyclone and taking advantage of the South Easterly breezes by which these Cyclones are usually followed.

168. IN THE BAY OF BENGAL. Supposing our ship in Lat. 14° North and Long. 98° East, or not far from the middle of the Bay of Bengal, with a strong squally gale at E. N. E., a falling barometer, and other indications of bad weather. We take the Northern Card, and placing it so that the ship's place may fall, say on the outer circle just between " Wind N. E." and " Wind East," we have it of course at the place of the wind, E. N. E., and we see immediately that the centre of the storm bears S. S. E. from us.

Next, we know that the Cyclones of the Bay of Bengal come generally from about the E. S.E., and travel to the W. N. W., so that we may say with tolerable confidence, " This is a Cyclone coming up from the Andamans and travelling towards Coringa," and if we are bound down the Bay or to the Straits of Malacca, we may add, "and if I run on now, I shall run into the heart of it, for I shall in a few hours cross its track, before, upon, or close behind its centre, according to its rate of travelling." As no careful seaman would of course run his ship headlong into a waterspout, neither would he now hesitate, I suppose, to heave to for a few hours, when he will find that the wind has drawn gradually to East, and then to the Southward of East, *when his barometer will begin to rise*, and he may safely stand on. It is probable that in doing so he will cross a spot where the Cyclone has left its traces in a heavy and confused sea. He will see also by moving his Storm Card to the W. N. W. *how* it is, that (allowing always for his drift or run), the wind will in a Cyclone change *as I have here described it:* and indeed, what I have related above is, with some little difference of position, the case of a troop ship, the *Nusserath Shaw*, in April, 1840, as shewn in my Third Memoir in the Journal of the Asiatic Society of Bengal, Vol. IX, but with this difference, that her Captain, being in all probability desirous of profiting by the " fine fair wind," unfortunately ran down 100 miles towards the track of the Cyclone,

PART III. § 170.]   *Lessons; Bay of Bengal.*   153

which travelled up about 180 miles in the same time; and as the ship crossed it close to the centre she was dismasted, narrowly escaped foundering, and was obliged to put back at a ruinous expense to repair the damages! Several other vessels suffered from the same error, and two at least, the brigs *Freak* and *Vectis*, had to put back to Calcutta for repairs, after being near foundering.

169. Let us suppose now, with the card in the same place, that our ship is in the same Longitude as before, 89°, but in Latitude 11° or 11° 30′ North, with the same signs of bad weather, falling barometer, &c. and the wind at W. N. W. Placing the card as before *with the wind's place upon the ship's place*, we see that now, if bound South or to the Straits of Malacca, we may safely and surely profit by the fair gale; for the farther we get to the South or S. E., the more distant we are from the dangerous centre. On the contrary, if bound to the Northward, we should be standing on into danger, if we ventured, especially with a falling barometer, to "carry on," since we should then clearly be nearing the centre, which now bears N. N. E. from us.

170. Assuming a ship bound to Calcutta in the month of October, in Lat. 18° N. Long. 88°, with a strong gale at N. b. E., and a falling barometer. The Storm Card will shew that this is probably a Cyclone coming up from the E. S. E. or S. E., and to lie to, would be to lie in its track as if waiting for it, and that in a position in which if disabled there is a dangerous lee shore. A short run to the S.W. will raise the barometer sufficiently to assure the seaman that he is out of the way of the violent parts of it, when as the wind hauls to the N. W., West, and S. W., he can soon sweep round its Southern and Eastern border, and regain his lost distance without straining a rope-yarn. It will no doubt be some time before this plan, which will probably be called "turning tail to one's port for a gale of wind," will be adopted; and many will prefer taking their chance, subjecting their ship to the wear and tear of three years in twenty-four hours, (to say nothing of loss and risks which underwriters pay); and those who do so, will of course not think it worth while to consider of the right tack to heave to upon; *but this will be done once only.* Those who have passed through a "real hurricane-Cyclone" in the Bay of Bengal or West Indies, or a "genuine tyfoon" in the China Seas, will not mind sacri-

ficing a little time to getting out of their way; and still less when they consider that, as before said, all that is gained in the run into, is lost in the drift *out of*, the fatal circle!

171. Supposing the ship in 12° North and 90° East; but the wind from the S. W. or S. E. ? This may safely be profited by, for we have as yet no example here of a Cyclone moving to the Eastward of the meridian; and the barometer and the Storm Card will fully advise if it should be one of those (apparently rare ones) which move up on a N. N. W. course towards the Sand-Heads, so as to prevent the seaman from crossing into it, or approaching the Sand-Heads too nearly in bad weather where no Pilot is procurable.

172. In my Eighteenth Memoir (Jour. As. Soc. Bengal, Vol. XVIII. p. 912,) I have given detailed rules with the reasons on which they are grounded for the management of a ship at (within), or near the Sand-Heads on the approach of a Cyclone, of which the summary is as follows.

"To sum up these rules then. For the ships to the Northward (*i. e.* on the right hand side of the track) of the Cyclone when the gale, with a falling Barometer, is at S. E. to E. S. E., they may, if they have sea-room, stand to the N. E. to allow it to pass them comfortably, heaving to of course on the starboard tack when far enough out of its way. Between E. S. E. and East they may, if the Barometer be not too low, and they have sea-room to give the shores of Orissa a wide berth, with a stout ship and good helmsmen, venture to cross in front of the Cyclone, or if this be not advisable, heave to on the starboard tack, when they have made an offing if in an outward-bound ship, or before they run too far in, if inward bound.

"With the wind between East and N. E. to North. I have already shewn, that crossing may almost always be safely adopted for the ships to the right (Northward) of the Cyclone path, and for those to the left (Southward) of it that they must not run into it, fancying it a fine fair wind, or too far up if in its rear, as that is wholly useless, and may bring them into soundings with a heavy Southerly gale; and this much sooner than they expect, if the storm-wave and current are strong and if the track is one near the meridian.* All will I hope recollect that

* In the *Loudon's* Cyclone of October, 1832, noticed by Colonel Reid, the ship *Albion*, commanded by Captain N. McLeod, one of the oldest commanders to Calcutta,

PART III. § 173.]   *Rules at the Sand-Heads.*   155

they have first to consider what their ships and the crew can do, next what their best course is, and lastly in what position they may be if dismasting, or even loss of topmasts should occur; and this may happen from broaching to, or from sheer hard blowing, in the best ship. With an on-shore gale, the resource of anchoring in the open ocean, which all the Sand-Heads anchoring, and most of that of the coast of Orissa, is, becomes one to which no good seaman would desire to be reduced if he can avoid it.

"Finally, I need not remark that in all this, both as to the expected Cyclone and what is to be done to avoid it, much must depend on the judgment guided by careful observation. Thus for instance a sea from the Eastward crossing a Southerly one with a bank of clouds to the E.S.E. is a strong sign of a Cyclone, though it may be blowing fresh from the South at the time ; and, again, if the bank is heavy to the S.S.E., that it may come up from that quarter. In a word, the vigilant seaman will watch every thing and despise no indication, and the dull and the careless will see nothing and be always too late."

About Ceylon and the Southern parts of the Bay, the same rules hold good, as also for the Cyclones which occur in Madras roads, Coringa, and other anchorages on the coast. To the management in these cases I shall advert in the Section upon that advisable in Roadsteads and Harbours.

173. IN THE CHINA SEA. Suppose yourself in about Lat. 20° North, Long. 115° East, with the wind varying between West and N. W., a falling barometer and other signs of bad weather, being in the tyfoon months withal, and bound to Canton.

Place the Storm Card on the chart, and it will shew you, that in all human probability, and with reference to what has been said at p. 57 of the tracks in that sea, this is a tyfoon-Cyclone coming up from E S. Eastward or down from the North Eastward, and that if you stand on upon your direct course, though you may indeed make 50 or 60 miles of latitude, you will be in the very track and heart of it;

ran up with a terrific Southerly Hurricane, and making every allowance hove to as he thought 60 or 70 miles to the S. E. of the Pilot Station, but at the same moment found himself in 13 fathoms water only ! Fortunately they were enabled to beat off, but another hour's run would have been destruction to them.

but that by steering away to the South Eastward, and hauling up gradually as the wind and your barometer guide, you will have allowed the storm to pass you, and have only made a little curve, as in the case of a head-wind. The case which I have supposed here, *really happened* between the 27th and 28th September, 1809, when a fleet of four of the East India Company's China ships, by standing on 64 miles to the Northward, ran headlong into a tremendous tyfoon, in which the *True Briton*, a new ship, foundered, and the other three were not far from it; See Sixth Memoir, "Storms of the China Sea," Jour. As. Soc. Beng. Vol. XI. and track V upon Chart No. IV; the Fleet being at noon of the 27th in Lat. 19° 49' North; Long. 114° 43' East.

174. Again, in the same Lat. and Long., with a strong Easterly gale, falling barometer, and all other indications of a tyfoon, and in the month of July, and your ship bound *down* the China sea or to Manilla. The probabilities (or may we say at once certainties?) here, are as before, and you may commit the same error by standing on; the centre of the Cyclone being South of you; but it is travelling more or less to the Westward or North Westward. If wind and sea yet allow, you may stand off to the N. N. E., when as your barometer rises you will be able to lay more Easterly, as the wind will haul to the S. E. and South, and with a very little sacrifice of distance you will allow it to pass you, with but little more than the violence of a common gale for a few hours. If you stand on, the chances of your meeting the centre, and even of being involved in it *when disabled* are most imminent; and this last is perhaps the *most* dangerous situation in which a ship with sea-room can be placed.*

In 16° North, 113½° East, with the wind at W. b. N. and all the usual indications of a tyfoon (Cyclone) and bound to China.

* The centre of a hurricane or tyfoon (for the tyfoon is yet more furious than the hurricane, using the words now to designate force only), is always an awful situation for the best found, manned, and perfectly prepared ship, even were she the smartest and stoutest frigate in the navy. Now if we supposed a disabled, and consequently unmanageable ship in this position, with a mast or two hanging by the lee chainplates, and thumping at her bottom or rudder, and this in the midst of "the pyramidal sea" which can only be likened to a boiling cauldron, flying mountains high in all directions, the most reckless sailor, or landsman, may suppose it worth while to avoid the chances of such a purgatory.

PART III. § 176.]   *Examples; Golconda and Pluto.*   157

If you stand on you get into the heart of mischief, but by running away to the E. N. E. and gradually North, which is about your course, you will avoid it, and perhaps really lose no time at all in the end.

175. One case, which affords three lessons at once, has occurred in this sea, and should not be omitted. It was the subject of a special Memoir* by me, in consequence of a reference from the Marine authorities at Calcutta, of which the object was to ascertain as nearly as possible on what day the transport ship *Golconda* was lost. The results of the inquiry shewed that of three ships which must have been close together in a double tyfoon in September, 1839, Tracks XXVI. and XXVII. on our chart—one, the *Thetis* of London, perfectly aware of her position, and her Captain well acquainted with the Law of Storms, hove to at the right time and place, and sustained no damage. The second, the *Thetis* of Calcutta, ran on, evidently in ignorance, till she could run no longer, lost her mainmast which crushed her pumps in its fall,† had three feet water in her hold, and narrowly escaped foundering; and the third, the *Golconda*, which with 300 Madras troops, their officers, followers, and her crew, had nearly 400 souls on board, ran up, and was no doubt caught at the meeting of the two Cyclones and foundered! see p. 97 § 104.

176. The most recent instance of serious mischief, detriment to the public service, and useless risk of valuable lives and property by disregard of the rules of our science, occurred in this sea June 28th and 29th, 1846; in which month the Honourable Company's War Steamer *Pluto*, bound from Hong Kong to join the force off Borneo, having the wind from the Eastward in the S. W. monsoon and in the tyfoon months, with a falling barometer and every other indication of a tyfoon-Cyclone, steamed directly on her course, making one of South, 32° West, while the tyfoon was travelling up to the North, 30° West. She of course met the calm centre and shift, with the pyramidal sea, and indescribable fury of the winds found there. She lost her funnel,

* IVth Memoir, Jour. As. Soc. Beng. Vol. X.
† Builders, ship-owners and underwriters should note this accident, and place the pumps so as not to subject them to this serious risk, for it must have occurred before, and doubtless have occasioned the foundering of ships.

rudder, &c. &c. and was almost foundering when the Cyclone left her. She then put back, and being of course nearly unmanageable drifted on the rocks at Hong Kong, but was saved by the boats of the *Vesta* frigate fortunately lying there.*

177. IN THE NORTHERN AND SOUTHERN PACIFIC OCEAN. We have usually here abundant sea-room, but our difficulty will lie, till more knowledge is collected, in judging of the tracks of the Cyclones. The careful seaman will not fail to apply here with due care the method I have given (at p. 115), for finding what the approximate track is, and guided by that, and by analogies drawn from what is known of other parts of the world, will rarely, I should suppose, be unable to avoid the worst parts of the Cyclones he may meet with. He can confer no greater obligation on nautical science than by carefully registering his own and collecting the experience of others, to enable us to lay down rules for him here with as much certainty as in other parts. He will find in Chart IV. a few tracks in the Northern Pacific; and at p. 121, a striking instance is given of doing exactly what should *not* be done, and the mischief arising out of the error. For the Southern Pacific, and for the present, I can only recommend to him the careful study of what is said of the Southern Indian Ocean, both as to tracks and stationary storms, and of the little information collected, p. 73 to 81, as well as close attention to his barometer, the winds, and his run by log at the approach of bad weather, so as to enable him to calculate roughly, (for this is all that is usually required) the probable track of the threatening Cyclone. It is *probable* that the tracks may approach to those of the Indian Ocean, but the groups of Islands, particularly when high, may occasion great variations.

178. I have not in these examples for practice, referred at all to the incurving of the winds, because it will perhaps rarely be required that the seaman should consider it, since when he gets to where it is of considerable amount he is in all probability too far involved in the Cyclone to make it a consideration influencing his management; and

* See Chart No. IV. Track *j*. I drew up a detailed investigation of this error, which the Government of India did me the honour to lithograph, with the chart accompanying it for the instructing of its Steam service. See also Naut. Mag. for January 1847, p. 12, for a capital lesson in this sea.

if he has, either while lying to or scudding, the winds "alternating," as described at p. 111, he will take the mean as the average direction of his wind, as in fact he often does in fine weather, when the wind is a little variable. In his examination of old logs, however, and especially those of Fleets or of numbers of detached ships, he will find this element of much importance in settling correctly the place of the centre.

179. RECAPITULATION OF VARIOUS CASES OF ERROR AND OF GOOD MANAGEMENT. I have thought it useful, as I proceeded, to adduce striking examples of good or erroneous management, and consequent advantage, loss, or mischief in the preceding sections; so that the precept and the warning against neglecting its practice might go together, and be thus perhaps more forcibly impressed on the mind than if separated by a different arrangement. I merely recapitulate here, then, what has been instanced as to these separate heads:—

HEAVING TO ON THE WRONG TACK. Admiral Graves' Fleet with
the *Ville de Paris* and *Rodney's* prizes, . . Page 125
RUNNING INTO CYCLONES. H.M.S. *Swift*, with E. I. Company's Fleet in the Northern Pacific, . . . 121
The *Great Western Steamer*, Northern Atlantic, . . 140
Maria Somes, in Southern Indian Ocean, . . . 148
French Frigate *La Belle Poule* and Corvette *Le Berceau*,
Southern Indian Ocean, ... 46
Nusserath Shaw, and other ships, Bay of Bengal, . . 152
E. I. Company's Fleet, China Sea, 1809, . . . 156
Transport *Golconda*, China Sea, 1839, . . . 157
H. C. Steamer *Pluto*, China Sea, 1846, . . . 157
PROFITING BY A CYCLONE. Royal Mail Steam Packet in the
Northern Atlantic, ... 140
Ship *Lady Clifford*, Coromandel Coast, . . . 166

180. SHIPS PREVENTED BY THEIR SITUATION IN THE STORM CIRCLE FROM RUNNING INTO THE CENTRE. I mean by this, ships the Commanders of which really would, if they could, have driven their ships headlong into the midst of a furious Cyclone; in which dismasting would have been perhaps the least they could have escaped with! They have not done, and do not do this *knowingly*, but hundreds, there is no doubt, are yet doing and will yet do it for some time.

In fact in every case in which the Mariner meets with that quadrant of the Cyclone which gives him a foul wind, he is, I must say, sadly tempted to wish the wind was "but a few points the other way," yet these few points, obliging him to lie to, are in fact his safeguard from the mischief he would plunge into if he could: An example will make this quite clear.

On our Chart No. III. will be seen, marked XII. the Cyclone of the *Briton* and *Runnimede*, which was indubitably destruction to those two fine ships; and by placing the Storm Card over the track, it will at once be seen that a ship about the Island of Narcondam must have had the wind Easterly, and one a little to the East of it, the wind well to the Southward of East and South Easterly. Now at the very time the *Runnimede* and *Briton* were driving helpless and dismasted, round and round the centre and finally on shore, the ship *Prince Albert, bound to the Southward,* was lying to off Narcondam, with a gale at S. E. and was thus prevented from running farther down towards the vortex.

181. PROOFS OF THE ACCURACY OF THE RULES AND DEDUCTIONS AFTER PUBLICATION. The seaman who has not studied our new science, will scarcely believe that by means of it, when clearly understood, nothing is more easy than to predict beforehand or to declare after the event, what winds and weather ships under given circumstances have had or may expect: and yet not only may this be done almost with certainty for single ships and Cyclones in given positions, but, which is perhaps still less credible at first sight, we have found that long after the track of a Cyclone has been traced, and the chart lithographed and published, other ship's logs of the same storm have been obtained, and when their position has been carefully ascertained and laid down, it is found that they had the wind nearly or exactly as the Storm Chart shewed, and this even where the Storm Circle was laid down by the logs of two ships only or even of one!

Now the strongest possible proof of the truth of any natural law is, that it will shew, and predict as it were beforehand, what *has* happened, or *would* happen under certain circumstances; and in these cases the prediction is made, and even printed and published, that a Cyclone of a certain size *must* have passed on a certain track, at a given rate on a given day and hour; and *therefore* as it did so, all ships in certain positions *must* have had winds of certain violence, and blowing from such and such points and no other; and so it proves to be. I quote one or two of these cases which have occurred, to shew how satisfactorily the seaman may rely upon the results obtained.

182. In my Third Memoir* I had lithographed my chart, and, principally from the log of a single vessel, the *Nusserath Shaw*, (which, however, had the centre passing close to her) placed the track of two days as there laid down. I then obtained the log of another ship, the *Marion*, which also lost her mizen-mast in the same hurricane (Cyclone), and in placing her on the Chart it was found, not only that the track of the hurricane had been correctly marked, but moreover that its rate of travelling, which had been assigned by two ships' logs to be on an average for the 24 hours $14\frac{1}{4}$ miles an hour, was for six or more hours, with the *Marion*, $16\frac{1}{2}$ miles per hour, so that its true rate was probably about $15\frac{1}{2}$.

Mr. THOM also, p. 342, refers to a case, that of the Ship *Patriot*, in which, after his chart for the Mauritius hurricane of 1840 was constructed and engraved, her log was obtained and found exactly to coincide with what had been laid down.

In a Cyclone of October, 1848, in the Bay of Bengal, I published in the *Calcutta Englishman* of the 16th October, a notice that a Cyclone of considerable intensity had been crossing the Bay during the preceding week, 11th to 14th October, on a track about from Cape Negrais to Point Palmiras. This was deduced merely from the appearances of the weather, the winds and the Barometer; for at no time was it blowing more than a moderate gale in squalls and puffs at Calcutta. It proved afterwards that this was nearly the track of the Cyclone. This is an instance of how easy it is for any seaman with moderate care of his observations of the weather to be fully warned, long before even the outer verge of the Cyclone has reached him, or he has run down upon it.

183. Numerous other instances have occurred to myself and to Mr. REDFIELD, Col. REID, Mr. THOM, and M. BOUSQUET, in which

* Jour. As. Soc. Beng. Vol. IX.

we should have been able to say to the Captain of a ship relating his wind and weather at a given time—

Stop : Did you then run on, or heave to?

Oh! I carried on to it of course.

Very good : then you soon found the winds veering to——and had a larger share of the hurricane than you looked for?

Yes, I lost my mizen-mast, top-masts, boats and sails, and had my decks swept, and four feet water in the hold, *but how the d——l do you know that ?**

Many more cases may be cited in all these sections (179 to 182), but for brevity's sake I have noted but a few.

184. IN ROADSTEADS, HARBOURS, OPEN COASTS, ANCHORAGES, AND RIVERS. Tremendous as the Hurricane and Tyfoon are at sea, the mariner is well aware that there is one situation in which they are invested with tenfold danger; and this is when riding in open anchorages, or where caught by them on lee shores. To get an offing is then his only chance ; and yet how often do we hear of a dozen of ships putting to sea, of which three or four are never heard of, and three or four return dismasted, or otherwise so seriously damaged as to be condemned. No doubt this often rises from ships being ill prepared, too light, or badly manned; but there is no doubt also, as we shall soon see, that it often arises *from steering the wrong course!* The mariner looks only to gain an offing as fast as he can, and is not dreaming that on the course he runs, he is steering perhaps in a direct line to meet the furious focus of the Cyclone, and the shift of wind it brings with it ; and this too perhaps in a light ship† and with top-gallant masts on end and top-hamper in the way, not having time to

---

* I have really had more than one such conversation, under circumstances which (and this created the astonishment) almost excluded the possibility of my knowing by any chance what had happened to the ship; as speaking of an old gale of 10 or 15 years' date, and the wonder was always that we could predict how the wind veered.

† It is not long ago since a very fine teak-built East Indiaman was utterly dismasted, nearly lost, and so strained that she sold for a fifth of her insured value, or thereabouts, in consequence of putting to sea from Madras Roads in an approaching Cyclone with top-gallant masts on end—and this with ample time and warning from the Fort to get them down! Had she known which way to steer, she might, even in this state, have got off with the loss of a top-mast or two!

PART III. § 186.]   *Examples ; Jamaica.*   163

get them down, till it was no longer safe to send men aloft. Capt. LANGFORD in the paper quoted (p. 2) speaks fully of this, and advises the practice of running to sea in the West India hurricanes; and seamen well know in how many parts of the world it is necessary to ride in certain months, with the expectation of being obliged to slip and run to sea at very short notice. The question I now propose to examine is, what ought to be the course steered, *for the greatest distance from the land may not always be the best offing.*

185. Let us, as an instance, suppose a ship lying at, or becalmed close in with, any part of the South Coast of Jamaica, between Portland Point and South Point Negril, with every indication of a hurricane, and the wind at N. E. blowing already strong. From the average of the tracks (p. 28) he will see that the hurricane is probably coming up from the E. b. S. or E. S. E., and his Storm Card (which I repeat is the hurricane *ready-made* for him to manœuvre with)* will shew him that though his present situation may be now a tolerably safe anchorage, or that it is a good breeze for running to sea with, yet remaining is out of the question, for the Cyclone will shift to the S.W. or dead on shore, and if he runs to sea with his North Easterly wind without due caution, he may find the Cyclone veer rapidly to the North and N. W., when if an accident should occur, (even should he avoid the centre) he may be in a disabled state with the Cays of the Pedro Bank under his lee. By running out at first well to the Westward, and then to the Southward before he heaves to, he may cross in front of the storm without damage, if its rate and time allow him ; or at all events, if the track of the Cyclone lies well to the West, and he can keep on the North or right hand side of it, obtain a clear offing West of Point Negril as the wind draws to the East, S. E., and South, where it will leave him.

186. Captain LANGFORD in his paper (Phil. Trans. for Oct. 1698,) speaks of " the benefit and experience he has had in foretelling hurricanes" as follows:

"That whereas heretofore they were so dreaded that all ships were afraid to go to sea, and did rather choose to stay in the roads at anchor than to run the hazard of the

* And he will find it as advantageous as the gambler does the looking-glass when he can see his adversary's cards in it.

M 2

merciless sea, though never ship escaped at anchor, but was cast a shore oftentimes twenty to thirty yards from the wash of the shore and the vessels set whole.——"

He then gives directions for preparing a ship for sea, and continues:—

"And they may, having their ships in readiness, stay in the road till the storm begins, which is always first at North, so to the N. W. till *he* comes round to the S. E. and then *his* fury is over. So with the North wind they may come away to the South, to get themselves sea-room for drift, if the S. W. wind, where *he* blows very fiercely; by these means I have by God's blessing preserved myself in two hurricanes at sea, and in three on shore, and have had great advantages by it."

He relates also an instance of preserving his own ship from damage, and at Nevis, in 1667, of a fleet under Sir John Berry in the *Coronation*, putting to sea by his advice before a great hurricane, and coming back again safe in four or five days, "to the great admiration of the French, then our enemies." And he relates that a Capucin Friar in his sermon to the French, said,—

"You may now see your wickedness in praying for a hurricane to destroy the "*English* fleet, when you see they are all come back safe, and we have not a house "left to serve God in, nor for our own convenience, nor forts, nor ammunition left "to defend ourselves against these preserved enemies."

It is curious to find 150 years ago, so striking an illustration of the advantage of our science on so considerable a scale; and I trust we shall again find, if the misfortune of war is forced upon us, that now, as then, the science and skill will be on the British side. Let every seaman then who can contribute a line to it remember, that his notice *may* be important enough to influence the fate of a ship or a fleet!

187. Let us take the case of Madras roads, a Coast which lies North and South. Here the Cyclones, which are at times of such fury, that no vessel can lie with the remotest hope of riding them out, and the surf breaks in nine fathoms water,* which is in from four or four and half miles from the shore, come in from East to S. E.; that is, their track is *from* that quarter, or perhaps from E. S. E. on an average; see Chart III. The Cyclones usually begin from N. E. or N. N. E. but at times from North.

* In the hurricane of 1809, when the *Dover* frigate was lost, it did this; and the wreck of a vessel which had been blown up in 8 fathoms water, twenty years before, was hove on shore! The expedition to Java, of probably 70 or 80 Transports and Men-of-War, had sailed but a few days before, but steering to the Southward fortunately escaped.

PART III. § 187.] *Examples; Madras Roads.* 165

The Storm Chart will shew that without allowance for any "flattening in," as explained at p. 114, these winds indicate Cyclones coming up from the S. E. to East. We take the average as before and say E. S. E., and we will also allow the Cyclones to have an average diameter of 200 miles; that is, that when we slip, its centre is 200 miles E. S. E. from the anchorage. We have the wind N. N. E.

Now to run out E. S. E. and as long as we can carry on, with the wind about abeam or a little abaft it, has been the usual rule, but the Storm Chart will forthwith shew that this is running out to meet the focus. If S. E. be adopted, it will be seen by a little calculation, that in about 16 or 18 miles, or in the first two hours run, the wind will have hauled to the N. b. E. or more Northerly, and that in an hour or two more it will be to the Westward of North, though probably the barometer may still fall, when the course can be made more Southerly to raise the barometer by increasing the distance from the centre; and then Easterly to "sail round the heel of the Cyclone," so as almost to be in with the anchorage at the first fine weather. All this of course if the wind be not too violent, the sea too heavy, or the ship crank, or steering too wildly or dangerously. At all events if heaving to *must* be adopted (on the port tack of course as being well in the left hand side of the circle), every seaman can see that he has thus avoided two dangers, the being disabled by the fury and shifts of the centre, and the being drifted into the Northern or right hand semicircle, where the winds and current of the storm are all more or less from the Eastward, and therefore *towards* the shore, while on the other half they are as much *off* it: so that even should a misfortune have befallen him, he has a clear drift to sea while rigging his jury spars to get back again.

By the following extract from an official report to the Master Attendant I am enabled to corroborate exactly the view I have there taken of the proper course to be steered when running to sea. Captain A. STUART, commanding the ship *Baboo*, says—

> "With regard to casual storms or other incidents, I can say but little; with exception of the gale of the 21st May at Madras, when in company, and after the *General Kydd*, I put to sea. We had two very severe nights of it, but fortunately I got well to the Southward of Madras, and returned to my anchorage on the 24th.

"During the gale the wind with us was most severe from North and N. W.; at the latter point it blew with fearful violence, then moderated and veered round to S. W. in the afternoon of the 22nd ; with which I made sail, being then distant from Madras 118 miles, bearing N. N. W., and arrived without any damage."

The *General Kydd* is the ship alluded to in the note, at p. 162, and she ran out so as directly to meet with the centre. The *Baboo* "fortunately got well to the Southward of Madras," *i.e.* steered far enough to the South to avoid the centre, though in passing the S.W. quadrant of it, she had, as we see, the gale excessively violent, being then nearest to, and close upon the centre. Some of the other vessels which slipped from the roads also escaped the centre, but the unfortunate barque *Braemar* being, like the *General Kydd*, too far to the North, was laid on her beam ends, and only righted by cutting away her masts, the wreck of which carrying away her rudder. she drifted in an unmanageable state on shore, and was wrecked near False Point Divy.

188. Captain MILLER, of the ship *Lady Clifford*, has given us a capital instance of this management. In the Bay of Bengal Cyclone marked VIII. on our Chart, he was lying at Nagore, an open roadstead, and as the centre passed close to Pondicherry he had the winds Westerly. Being bound to Madras, and well acquainted with the law of Storms, he weighed on the 24th October, with a gale at W. N. W. and stood out to the N. E., ran round the heel of the storm, and on the 26th anchored in Madras Roads, having carried the wind, veering in 16 hours from W. N. W. to S. E.

189. IN RIVERS. Ships in large rivers may be able to derive much benefit from weather shores in Cyclones, and it may often become with steamers, boats, and small craft, and at times with ships in the smaller rivers, of great importance to know that the shift or veerings of wind will take place in such or such a direction rather than in another; for by this knowledge they can often run at the commencement of a Cyclone for an anchorage, at which they can lie sheltered through the whole of it ; or they can, if anchored on a weather shore, profit by the lull to get over to the other, before the shift of wind comes on, which would convert their former shelter into a dangerous lee shore.

190. Let us, as an example, take the Hooghly or Canton river,

both much frequented and running about North and South towards their mouths, and suppose ourselves in them exposed to a Cyclone crossing them from East to West, in a launch with treasure on board, or in a steamer.

It is clear that for the first half of the tempest, if the wind is to the Eastward of North, the centre will pass to the South of the boat, and as the wind will be N. E. and East, and South Easterly, the Eastern shore or any bank sheltering from these quarters is safe; but that if obliged to anchor to the South of a bank or island, when the centre is passing exactly over the boat's position, the latter part of the Cyclone may drive her on shore, or sink her at her anchors. The Card also shews how, if the wind is to the Westward of North, it will veer to the West and S. W., and thus perhaps render an apparently safe berth really dangerous, if due precaution be not taken.*

To the last edition of Sir William REID's work, The Law of Storms, is added an appendix with figures, shewing that ships mooring in roads or harbours where Cyclones prevail, should so lay their anchors as to ride with an open hawse as the wind veers, and this indeed is usually done by the pilots from long practical experience, as in the Downs, and other roads and harbours at home.

In those parts of the world, however, where no good pilots, if any, are to be had, and if mooring be desirable there, it will be highly useful to consider, as Sir William REID advises, how the anchors may be laid so as to have the best bower if possible to ride by on the first part of the gale, and both anchors with an open hawse on its veerings and close. But this again must depend on the usual tracks of the Cyclone, if known in these quarters, and it is very difficult to lay down any positive rules, for we have again the tides often interfering, and always to be considered in fine weather. Few ships would moor when bad weather was decidedly coming on, even if they could do so, unless in very confined anchorages, as they would probably prefer riding with a long scope, with a second anchor dropped under foot to veer upon if the first should not hold.

191. In a word, the Law of Storms will here, as in so many other cases, *fore-warn* the mariner of what is to come; "and fore-warned

---

* See also p. 172 for the effects of storm-waves in rivers.

is fore-armed." The tales which we have all read of the wind's "*unfortunately shifting to the opposite quarter*," when "the boat (or ship) was driven on shore and all hands perished," will be changed into "During the lull, (or when the wind had veered to ——) the boat (or ship) in due anticipation of the latter part of the Cyclone, changed her berth to the —— shore, where she safely rode out the remainder of it without damage." And let me add, that every officer and commander of whom this shall be said or written, will be held in that degree of professional esteem which is the rightful meed of such careful and scientific management of the lives and property entrusted to his charge, as contrasted with the fatalism, the fool-hardiness, or the helplessness of ignorance.

PART IV.

1. OTHER PHÆNOMENA CONNECTED WITH CYCLONES; STORM-WAVES AND STORM-CURRENTS.—2. INSTANCES OF THESE AND OF THEIR PROBABLE EFFECTS.—3. INUNDATIONS FROM THE STORM-WAVE.— 4. PYRAMIDAL AND CROSS SEAS EXPERIENCED IN CYCLONES, AND SWELL FELT AT GREAT DISTANCES FROM THEM.—5. NOISE OF CYCLONES.—6. PASSAGE OF THE CENTRE OF CYCLONES.—7. SIZES OF THE CENTRAL SPACE OR LULL.—8. ELECTRICITY, MAGNETISM, ELECTRO-MAGNETISM, AND EARTHQUAKES.—9. ARCHED SQUALLS, TORNADOS, ETC.

192. STORM-WAVE AND STORM-CURRENT. These two phænomena, of which the very names will be probably new to most of my readers, certainly exist, and often to an extent little dreamed of. They have also, I have no doubt, frequently occasioned losses which were quite unaccountable at the time, and thus, as being in effect a sort of *unknown, capricious, and sudden currents arising in the midst of tempests*, demand the most serious attention of the careful mariner: I begin with the Storm-wave.

If any of my readers have ever observed the passage of a whirlwind or water-spout across a harbour, a river or a lake, they may have noticed that it raises the water above the level as it travels along, and that light floating bodies, such as pieces of wood, or even sometimes the wrecks of boats upset in its passage, are often brought from the middle of the river or from the other bank to the shore where it lands. This is in miniature a Storm *Wave*.

193. Colonel REID (p. 523, 2nd edit.) says—

"If a revolving power, like a whirlwind, were the only one exerted, it might be expected that the level of the water would be diminished at the centre of the vortex, though heaped up towards the verge of the storm. But it may be possible that a wave of a round or oval form, moving onward like a tidal-wave, but at the rate of the storm's progress, may accompany the storm in its course, and that its height may depend on the degree of atmospheric pressure, modified by the revolving power of the wind. The impulse in the direction of the storm's course being given, and maintained for a few hundred miles, currents very similar to the ordinary currents of the tidal-wave might be created, so that if the effect produced by such a wave is added to the spring tides, it may assist in causing these inundations in flat lands which often occur in violent storms. It will, therefore, be very desirable to note the height to which the tides rise on the leeward side of islands, particularly those lying at a distance from and influenced by continents."

194. Every seaman has also seen that under a water-spout the sea is boiling and foaming, and rising up, and travelling along with the spout, in a space of perhaps some two hundred yards or more in diameter,* and he can suppose that, if a boat could live within it, it might be carried along as far as the water-spout travelled, which would then be also a miniature representation of what we suppose to take place in the case of the Storm-wave; which, *at present*, for we want information greatly on this as on many other points, we suppose to be somewhat as follows:

As there is no doubt that the atmospheric pressure† when diminished in any one particular point of the globe occasions a temporary

---

* The black column of a small water-spout where narrowest, will often subtend a horizontal angle of a degree at least, one would suppose; though I do not know that it has ever been measured. An angle of 1°, at a mile distant, requires the object to be about 31 yards in diameter.

† Most seamen are acquainted with the meaning of this term; such as are not, will find it briefly explained in the following section, where the Barometer is treated of.

rise of the waters under this point, which are pressed up by all the rest, and that at the centre of Cyclones a diminution of pressure which varies from an inch, to two and two and a half inches of mercury occurs, so there is no doubt that at this peculiar part of the centre the water is raised a little more than a foot for every inch of fall of the mercury, or about two feet in ordinary Cyclones, and proportionably over the whole area of them. The incurving of the wind, already explained (p. 108), must also have a tendency to keep up and support a mass of water at, and towards this part, and we assume that this mass of water, or at all events a body floating upon it, is borne by the conjoined action of the wind *and* this wave, onwards in the direction of the path of the storm. This is what we denominate the STORM-WAVE, and what I have quoted and said is the THEORY or theoretical description of it. We shall come forthwith to the proofs I have been able to collect of the fact of this phænomenon so occurring, and with more of these we shall obtain in the end a LAW for its action in all cases, and our estimates of its effects will then be more certain.

195. Sir HENRY DE LA BECHE, in his "Geological Report on Cornwall, Devon and West Somerset," p. 11, has the following remarks, which I will not attempt to abridge, confirming so exactly as they do what has been just said. I have however marked a few passages in italics.

"Mr. WALKER* has observed, with respect to the influence of the pressure of the atmosphere upon the tidal waters on the shores of Cornwall and Devon, that a fall of one inch of the mercury in the barometer corresponds to a rise of sixteen inches in the level of the sea, more than would otherwise happen at the same time, under the other general conditions; a rise in the barometer of one inch marking a corresponding fall in the sea-level of sixteen inches. This he has found to be the usual rate of such alterations in level; but very sudden changes in the pressure of the atmosphere, are accompanied by elevations and depressions equal to twenty inches of sea-water for one inch of mercury in the barometer. Regarding the whole pressure of the atmosphere over the globe as a constant quantity, all local changes in its weight merely transfer a part of the whole pressure from one place to another; and hence he concludes that the subjacent water only flows into or is displaced from those areas where, for the time, the atmospheric pressure is either less or greater than its mean state, in accordance with the laws which would govern the indication of

* Assistant Master Attendant H. M. Dock Yard, Devonport.

two fluids situated in the manner of the atmosphere and sea.* We might account for the difference observed by Mr. WALKER in the amount of depression or elevation of sea-level produced by sudden changes in atmospheric pressure, by considering that a sudden impulse given to the particles of water, either by suddenly increased or diminished weight in the atmosphere would cause a perpendicular rise or fall *in the manner of a wave*, beyond the height or depth strictly due to the mere change of weight itself.†

" As regards the influence of the winds on the mean level upon the south coast of Cornwall and Devon, Mr. WALKER observes that East and West winds scarcely affect it, but that Southerly winds raise the sea above it from one to ten inches, and off shore winds depress the water beneath it as much according to their force. On the morning of the 29th November, 1836, when the velocity of the wind was estimated at about one hundred feet per second, the sea at Plymouth was raised three feet six and a half inches above the mean level, the greatest height above the equilibrium level he has seen. The hurricane began at S. W., and the barometer was very low; therefore this great increase in height is due both to the wind and diminished atmospheric pressure. A gale of wind from the Southward, a low barometer, and a high spring-tide concurring, cause damage and inundations on the southern coast of Cornwall and Devon. From the form of the Bristol Channel, and the absence of a free passage for the waters, such as exists at the Straits of Dover in the English Channel, Westerly winds force up and sustain a great body of water, thereby raising the sea above the mean level several feet. It appears from an account of the great storm of the 26th November, 1703, that the tide flowed over the top of Chepstow bridge, inundating all the low lands on both sides of the Severn, washing away farmyards, drowning cattle, &c. and it is worthy

* Walker, MSS.

† A circumstance connected with this subject, of considerable practical value, has been noticed by Mr. WALKER during his long-continued observations. He has found that changes in the height of the water's surface, resulting from changes in the pressure of the atmosphere, are often noticed on a good tide-gauge *before* the barometer gives notice of any change. Perhaps something may be due in these cases observed by Mr. WALKER to the friction of the mercury in the barometer-tube, as it is well known that in taking careful barometrical observations it is necessary to tap the instrument frequently and carefully, to obtain the measure of the true weight of the atmosphere at a given time and place. The practical value of the observation is, however, not the less, be the cause of the phænomenon what it may; for if tide-gauges at important dockyards shew that a sudden change of sea-level has taken place, indicative of suddenly decreased atmospheric weight, before the barometer has given notice of the same change, all that time which elapses between the notices given by the tide-gauge and the barometer is so much gained; and those engaged with shipping know the value of even a few minutes before the burst of an approaching hurricane.

of remark, that the barometer is recorded to have then fallen lower than had ever been previously noticed.*

"It will be obvious that while in a hurricane, such as that of November, 1836, noticed by Mr. WALKER, the level of the sea was raised on the South coast of Cornwall and Devon, *it was also depressed on the North coast of those counties;* so that the difference in the sea-level on the two coasts thus caused, would be the sum of elevation and depression produced on each coast respectively."

196. In the Nautical Magazine for August, 1848, Mr. K. B. Martin, Harbour-master at Ramsgate, gives a very carefully detailed account of a sudden rise and fall of the tidal column at that port in August, 1846, which took place three times in unequal undulations during a severe thunder-storm, and just at the moment of a severe discharge of what is termed the return stroke of the electric matter, or its flashing *from* the earth to the clouds. He describes the flash as "crimsoned! spiral! and intense," and as ascending "from the earth to the heavy cloud which hung over us." Upon carrying the tidal diagram from the tide-gauge to London, it was found at the Admiralty that the phænomenon of the positive state of the electricity of the earth's surface, had been noticed there on the same day during the storm. In June, 1848, also during a thunder-storm, the sea, as shewn by the tidal gauge, rose to a height of four feet in ten minutes with undulations as before, but to a greater extent. The return flash was not observed in this instance.

In reference to what will be subsequently said of the possibly electric nature of Cyclones, these observations are of much interest.

197. In a brief note forwarded to me by Captain HUTCHINSON, of the Barque *Mandane* of Liverpool, from an old Mediterranean Commander, it is stated that the rise of the water above its usual level is the first and surest indication of a gale in most of the parts of that sea, especially upon the Italian coasts.

Where, as would often appear to be the case with tropical Cyclones, the track follows the course of a river, or crosses a country intersected by rivers, great and sudden rises of the waters, analogous to the Bores of the Hooghly or the Pororoca of the Amazon, appear to take place, which tear boats and small vessels from their anchors and shore-fasts, and capsize them in a moment. A friend, writing of the effects of

* Walker, MSS.

these rises, says that the force of an ordinary gale or tyfoon is not sufficient to account for them, and he instances the October Cyclone of 1832, in which he was proceeding down the river, but fortunately escaping found that whole fleets of boats had been sunk at the different halting-places, and that the mischief was universally said to be done in a single moment *by a wave supposed to be occasioned by the instantaneous change of wind from N. East to nearly the opposite point.*

In the Nautical Magazine for September, 1854, is a letter from Mr. COODE, in which, from observations made by a sensitive and self-registering tide-guage at the Portland Breakwater, that gentleman has satisfactorily ascertained, and this in the entire absence of any swell, that in every electric storm (thunder-storm) for the last four years there has been a very perceptible amount of variation in the level of the sea. A diagram accompanied the paper, but this has not been given, the result shewn by it only being stated in a note, which is "the depression and elevation of the surface of the tidal wave from its regular curve to the extent of three inches each way." This leaves much to be desired, but, however, the fact appears indubitable.

198. The STORM-CURRENTS we assume to be a succession of currents analogous to those which we see round the sides of a whirlpool, and to be produced by the forces of the various winds blowing round in the area of the Cyclone.

It will be seen then, that the Storm-wave and Storm-current give rise in every storm to two sets of forces (currents) independent of that of the wind, acting upon a ship; the one carrying her bodily onward on the track of the storm, and the other drifting her round the circumference of that part of the Storm circle in which she may be.

On that side of the Storm circle where the winds agree, or nearly so, with the track of the storm, both may be acting together; on the opposite side they will counteract each other, and on other parts of the circumference they will act more or less across, or diagonally: thus, taking, as the simplest case, and one nearly that of Madras Roads, a storm travelling from East to West, and striking upon a Coast running North and South, its centre passing over Pondicherry, we should have then, for all ships in the offing, one current, 'the

Storm Wave,' carrying them directly on shore, with greater or less velocity, as they were nearer or further from the centre; and other currents, 'the Storm Currents,' varying in their direction according to the situation of each ship in the Storm Circle, but always agreeing pretty nearly with the direction of the wind.

The current of the Storm *wave* then is setting due West: but that of the Storm *current* West on the North side of the Storm circle, and due *East* at its South side; South at its Western edge, and North at its Eastern side, and so on in all the intermediate directions; and a ship putting to sea from Madras Roads in our supposed case, will be carried right towards the shore by the *Storm-wave*, and to the South-Westward also by the *Storm-current;* but if putting to sea from any place to the Southward of Pondicherry, that is on the Southern side of the Storm-circle, she would be carried one way by the Storm-wave, and the opposite one, or partly so, say to the S. E., East, or N. E., by the Storm-current; so that as to mere Westing, the effect of the one would probably neutralize that of the other. The case of ships on that half of the storm where both forces are against him, should however be borne seriously in mind by the seaman at all times. I have no doubt that it has very often occasioned the loss of vessels, and I proceed to give some instances shewing certainly the existence of this wave and of these currents in various seas, and of their probable effects.

199. INSTANCES OF THE STORM-WAVE AND STORM-CURRENTS. Many seamen, and those old and experienced ones too, will say that they have gone through plenty of gales and hurricanes without meeting with *much* or any current, and will be therefore a little doubtful of the importance which I attach to these; or may even suppose my views exaggerated. It may be worth while therefore to consider first how it is that we have so few accounts of them. I take the reasons of this to be:

(*a*) That *at times* there *is* very little Storm-wave or Storm-current, while, as will be presently shewn, they are at times of extraordinary strength. We do not yet know *why* this is so, but there is no doubt of the fact.

(*b*) That in most merchantmen, and I fear, formerly, even in some

Men-of-War, the log is very badly kept, or in the first not even kept at all in severe Cyclones. When, after a Cyclone, it is found that the ship is very far out of her reckoning, the observations are taken for a new departure, and the unexpected difference of position set down as excessive drift, (the rate of drift never having perhaps been ascertained by the log throughout the gale) or heave of the sea, or as a *current*: in the usual acceptance of this word to signify *any* difference between the log and observation, which the seaman cannot readily account for.*

(c) That sometimes, as before stated, the ship's position in the Storm-circle is such, that the Storm-wave and Storm-current neutralize each other, and then, though both have been strong, the ship has felt no effect from them; and if another vessel differently situated *has* been carried a long way by them, and the two accounts are compared, the one is thought not to have kept so good a reckoning as the other, or the matter is passed over like all other " very unaccountable things" (and truly till now they were such) by the careless sailor.

(d) That in some situations, as in the Gulf Stream, they occur in the localities of known currents, and if they act in the same direction or "thereabouts" are thought to be merely an increased action from the same cause, and are thus again passed over or set down to it.

(e) Both the Storm-wave and Storm-current appear, like the winds, to be of great force at, or near, the centres of Cyclones, but to lose this rapidly as the distance increases from the centre; so that while one ship is carried a wonderful distance by the Storm-wave and Storm-current *and* her nearness to the centre—all three acting the same way—another at a greater distance, and on a different side of the circle, experiences very little of this fatal current. Hence from all these apparent contradictions, the instances related, though perfectly authentic, have come to be held as utterly exaggerated or inexplicable. Like the histories of stones falling from the heavens, they have been

---

* I am not here writing sarcastically, but seriously. How few ships are there which, even now, duly ascertain and take any account of the local Deviations of their compasses, which may amount with some vessels and with various cargoes or armaments, and on certain rhumbs, to as much as one or two points! The effect of all this is set down as *current*.

thought mere travellers' tales, because they could not be readily accounted for.

(*f*) In some cases, and no doubt with some ships more than with others, when vessels are fairly so pressed down by the force of the wind as to be lying with their lee gunwales in the water, and the "lea sea" washing half way up to the weather bulwarks, there is no doubt that they drift much faster to leeward than they suppose; and rarely indeed *can* the log be hove at such times when the seaman is but waiting perhaps for a heavier gust or a deeper lurch to cut away his masts. But in such situations, a stout ship, with a good crew will lie for many hours, particularly if relieved of her topmasts by cutting or carrying away, and it is at such times that the drift by log should be accurately known to ascertain what is really due to the Storm-wave. In the log of the French ship *Le Grand Dusquesne* (16th Memoir, Jour. As. Soc. Bengal, Vol. XVII.) I ascertained that while the leeway and drift were carefully noted, and (the wind being from N. N. E. to N. N. W. throughout the Cyclone) they would have carried the vessel to the South of her reckoning, the difference in Southing was only 12' while that of the Westing, which was due no doubt to the Storm-wave, was 213 miles, which were made in three days at right angles to the drift.

(*g*) And finally, until public attention is properly directed to this and indeed all other parts of our subject, and public means adopted to insure the collection and digestion of the vast masses of knowledge which, in the shape of detached facts, are now, as heretofore, annually lost to science and humanity, there must always be hundreds, if not thousands of notices in existence, but beyond the reach of the labourers in the cause.

The foregoing will account, I hope, to the oldest seaman for the want, hitherto, of definite information, and the frequent absence of any notice of such effects as those we have assumed to arise, frequently if not always, from the action of Cyclones.

200. IN THE WEST INDIES Colonel REID (p. 523, 2nd Ed.) says in reference to the Gulf Stream:

"After the storm of September, 1839, Mr. HURST of the brigantine *Queen Victoria*, (whose place is marked in the chart), found the current of the Gulf-Stream neutralized; and the same commander, on another occasion, found the current running to the Westward, a fact corroborated by other printed statements at the time.

"The storm of 1839, when crossing the Gulf-Stream, was probably five hundred miles in diameter, and a diminished pressure, amounting to a fifteenth part of the atmosphere, at the centre of a moving circle of this extent, seems adequate either to arrest or to accelerate existing currents, or create new ones."

201. Mr. REDFIELD, in his last Memoir, American Journal of Science and Arts, 2nd series, vol. II., gives, p. 169, the abridged log of the barque *Zaida* off the West and N. W. part of the Island of Cuba, in the great Cuba Cyclone of October, 1844; and at the conclusion says:

"Captain CHAPMAN states that during the gale the Gulf Stream current off Cuba *had become changed in its course*, drifting the *Zaida* rapidly to the Westward, so that on the morning of the 7th he found himself off Cape Catoche in Long. 86° 40'. On the 11th October at 3 P. M. he picked up the two survivors of the *Saratoga* off Cape Florida, in Lat. 25° 40'. This position of the piece of wreck appears to shew also the extraordinary check of the surface current of the Florida stream, for these men were drifted off the bank of Bahama as early as the 7th."

202. In a paper published by Dr. PEYSSONNEL (Phil. Trans. for 1756), on the currents of the Antilles, he speaks of what he calls Counter-tides, which he expressly says were very violent at Martinique and the neighbouring islands when they had storms or hurricanes; and again that when St. Eustace (St. Eustatius) was ruined by a dreadful hurricane "coming in a contrary course," on the 1st November, we had here (at Guadaloupe) the most violent counter-tides. He gives also some details, but not sufficiently precise for our purpose, and some of them apparently erroneous or exaggerated, but there can be no doubt that he refers in most cases to true storm-waves and currents, and I trust that attention will be given to this curious but highly interesting part of the phænomena of storms in those quarters.

In a well arranged Manual of Physical Geography, published in Paris in 1839 by M. J. J. N. THIOT, it is said (p. 112), though the authority is not given, that,—

"It is only known that when hurricanes rage at Guadaloupe, the *ras de marée** appears all at once near the coasts of Martinique lying to the S. East (S. S. E.) of the former island, and that reciprocally the gales (*coups de vent*, gales or hurricanes), which are felt at Martinique, give rise to the *ras de marée* at Guadaloupe."

* See p. 192 for the explanation of this term.

The distance between the centres of these two islands is 92 miles, and Dominica lies between them.

203. In Mr. REDFIELD's paper on the Mexican "Norther" and Bermuda Gale of Oct. 1842, (Amer. Jour. Science of 1846, p. 157,) it is mentioned, in Report No. 22, from Cedar Keys, Coast of Florida, Lat. 29° 9′ Long. 82° 56′ West, which was a little to the Northward of the centre of the Cyclone's path, that "the water is stated to have risen 20 feet above low-water mark, and was within 6 feet of covering the island." About 5 or 6 feet are to be allowed for the tide here, so that it was evidently a very remarkable rise. Mr. REDFIELD attributes it to the effects of the Southerly wind on the right side of the Cyclone. The barometer is not given.

204. While the first edition of this work was passing through the press, I received a newspaper notice of the Cyclone and inundation at Grand Cayman in October 1846. No positive heights, from which we could deduce the total rise of the sea are given, but we cannot suppose it, by what is said, to have been less than 15 or perhaps 20 feet, and it appears also to have been a true Storm-wave, which "having risen above the iron-bound coast about midway the island (Savannah) flowed through 'Newlands' into the sea on the other side, forming a kind of rapid, entirely impassable in the usual course."

This is exactly what should occur in a small low island lying in the track of any part of a Storm-wave.

In a Chronological Account of West Indian hurricanes (Naut. Magazine for 1848) is one of a severe Cyclone at Belize, which was travelling from the S. E. b. E. to the N. W. b. W. The account, so far as relating to the present subject and compared with the Annual Register of 1788, is as follows:

"1787, Sept. 23rd, at Belize, Honduras, between the hours of 4 and 5 A. M. a gale came on from the N. N. W. veering to W. N. W. About 10h. it shifted to the S. W. and blew with increased violence. At the same time the sea rose and prevented the running off of the land flood. The low lands were consequently overflowed: not a house, hut, or habitation of any kind on either side was left standing on Belize! One hundred persons perished; dead animals and logs were floating about every where, and eleven square-rigged vessels besides small ones were totally lost."

In the same paper, p. 458, it is stated that in the Antigua Hurricane of 1772 the sea at Santa Cruz rose *seventy feet* above its usual level and carried everything before it.

205. IN THE SOUTHERN INDIAN OCEAN. The brig *Charles Heddle*, to which I have before alluded (pp. 47 and 108) between the 22nd and 27th of February, scudded *round and round* on the track of a Cyclone for five successive days! When her log was carefully worked up, it was found that by it alone her course and distance was *North* 42° East, 111 miles, while by good latitudes and chronometer sights, before and after the Cyclone, she had really made good in these few days a course and distance by observation of *South* 55° West, 366 miles; giving her, thus, a current (the Storm-wave) of South 52° West, 476 miles, or in round numbers (which would require 480 miles) four miles per hour for the 120 hours, or five days. The average distance from the centre at which she scudded round it, is about 42½ miles; the greatest being 62 miles and the least 25 miles. There is no doubt of the carefulness and authenticity of this vessel's log, which was procured for me by the Master Attendant at the Mauritius, Capt. ROYER, nor of her position being nearly correct, for she had just left the Mauritius and returned to it after the Cyclone: it is thus clearly an unquestionable instance of the Storm-wave.

206. Another had just occurred in the case of the ship *Tudor*, Capt. M. J. LAY, which vessel, bound to Calcutta, in Nov. 1846, encountered a severe Cyclone in Lat. 13° South, Long. 83° East, and hove to, to allow the centre to pass to the Northward of her, for 16 hours. Captain LAY had excellent sights before and after the Cyclone, and estimates his drift at 56 miles during the time he was hove to; yet he found that he had been set 133 miles to the West of his dead reckoning, and his estimate would still leave 77 miles of drift in the sixteen hours to be accounted for, and this we can only do by supposing both Storm-wave and Storm-current acting together (as the ship was very near the centre, having her barometer down to 29.20), at the rate of nearly 5 miles per hour.

207. ON THE COASTS OF AUSTRALIA. In the account of the Port Essington Cyclone, p. 63, already alluded to, it is stated that the sea rose ten feet and a half by measurement above the usual high-water mark. This Cyclone was one of great intensity, though of small

extent, and its centre did not pass, it appears, exactly over the settlement, but it is quite probable that the rise was a true Storm-wave.

208. There is also a very clear and remarkable account of the Storm-Currents in the Nautical Magazine for 1841, p. 725, on the Western Coast of Australia, by Commander WICKHAM, of H. M. S. *Beagle*, who says:

"There are other never-failing indications of a Northerly wind, (with which the N. W. storms begin on the coast,) such as the change of the current, which owing to the prevailing Southerly winds usually set to the Northward, but run strong to the Southward during Northerly winds, *frequently* preceding them, and giving more timely notice than the barometer."

Again, after describing a N. W. gale, and these seem to be truly Cyclones, for "they commence at N. N. E., and invariably veer to the Westward, and between W. N. W. and W. S. W. blow the hardest," he says, p. 725, N. M.

"The change of current did not precede the wind, but changed with it. When the gale was strong from N. W. and W. N. W. the current ran a knot an hour to the S. E., and when the wind changed to S. W. it ran with the same velocity to the N. E."

He also describes a Storm-wave, or rise of the water, p. 722.

"A rising of the water is likewise a certain prognostic of a Northerly wind, and has been invariably noticed at Swan River to precede all gales from that quarter; this of course can only be observed while at anchor on the coast."

It is evident that the "rise of water" is the effect of the distant Storm-wave, and that it at once explains the current. As observations are collected, we shall doubtless find on this, as on more peopled coasts, the points where the centres of the Cyclones come to land, to be those where the Storm-waves and currents are most strongly felt. The sets of Current along the Coromandel Coast are very well known, and at the approach of the Cyclones are often very strong.

209. IN THE BAY OF BENGAL, I have been enabled to ascertain, *certainly*, by means of circular letters and queries addressed to the commanders of the Pilot and Light vessels stationed at the Sand-Heads, that at the approach of, and during gales from the Eastward, or Cyclones, a heavy set of from three to five knots an hour from the Eastward is experienced. If the Storm Card be laid down at the

head of the Bay of Bengal, and moved across it from East to West, or up from the S. W. to the N. E. which are the usual tracks, it will be easily seen how the weaker of these currents are Storm-currents, and the stronger ones probably Storm-*waves*, as occurring when the centres pass near to, or over the Sand-Heads, or farther up, which they not unfrequently do. Nothing, indeed, is so common, as for ships which have put to sea from the Floating Light to gain an offing, to find themselves most unexpectedly, and after large allowance for their drift, in shoal water on the edge of the reef off Point Palmiras; and Arab and other foreign vessels, as well as not unfrequently English ones, unacquainted with this peculiar current in bad weather, are often wrecked thereabouts. In a number of Logs and official reports from the Pilot and Light vessels at the Sand-Heads, collected for the investigation of the Cyclone of October, 1848, in the Bay of Bengal, I find this current mentioned in several logs; the rates assigned to it being from 3 to 4 knots per hour, and the average of the whole 2½ knots. One vessel indeed anchored in 35 fs. water with her kedge, but found that she could not hold on, and shortly after that the current was so strong that the vessel would not steer with a 5 knot breeze.

In the *London's* Cyclone of October, 1832, as quoted at p. 154, the ship *Albion*, Captain McLeod, ran up from 15° North Lat. in about Long. 89½° East on the Eastern verge of this Cyclone, track *i* on Chart III., with a terrific hurricane about South. When they thought they had run as far as they prudently might, they prepared to heave to, but suddenly saw discoloured water ahead, and before sail could be made, were in 8½ fs. on the tails of the Sunderbund reef, but fortunately beat off. Captain McLeod says:—

" I remarked in the foregoing our consternation at seeing discoloured water ahead, and well it might be so, for it is almost incredible that we over-ran the Log 70 miles or upwards in 30 hours, after allowing at the rate of 10 miles an hour under bare poles; (in a merchantman), besides we made a very liberal allowance, in working up the dead reckoning as a precaution to keep ahead of the Log : Could this be the Storm-wave you speak of ?"

I have no doubt it *was* the combined effect of the storm-wave and storm-current.

210. IN THE CHINA SEA. We have several notices here of un-

doubted authenticity, being logs and reports from the Captains of the East India Company's China ships, who were all skilful navigators, and men far above any suspicion of exaggerated or careless reports to the Court of Directors, as they were liable to be very strictly called upon and heavily fined in cases of delay, or damage, or misrepresentation, which last would always be known by the evidence of their numerous officers. Hence these documents are of as great value as the logs of men-of-war, and as the instances are very extraordinary, it is important that the reader should know that they are quite trustworthy before he reasons on them.

The following are instances, in this sea, taken mostly from my Sixth Memoir.

211. HORSBURGH, who had great personal experience, and who made also extensive inquiries amongst those who were well acquainted with the China sea, says, p. 289, speaking of the tyfoons between the Grand Ladrone and Hainan, as coming down from the E. N. E. (such as our tracks I. IV. V. and XI. on Chart IV.), upon ships bound to China,* after describing the changes and shifts of wind, from N. W. and North to N. E. and East, S. E. and South, adds in conclusion: "The current at such times runs strong to the Westward;" and again, "When a gale happens to blow out of the Gulf of Tonking from North Westward and Westward, the current at the same time sets generally to the S. W. or Southward, in the vicinity of the Paracels, or where these gales are experienced." With the Storm Card, and with attention to what has been said in the preceding paragraphs, it will be seen at once why this difference is made. In the first instance, a ship crossing the South Western quadrants of a tyfoon Cyclone—if, as we have described, she does so rapidly with a fair wind and then heaves to at, or close to the centre, finds the Storm-wave and Storm-currents both setting her the same way for the greater part of the Cyclone, while in the latter case as she remains entirely in the Southern quadrants of it, the storm-*wave* is carrying her to the S. W., and the storm *current* to the S. E., so that between the two and her drift, she may find the *current* Southerly and South Westerly, and sometimes no doubt South Easterly.

* They never pursue this route if bound from China homeward.

212. In July, 1780, the H. C. ship *London*, Captain WEBB, was driven at four in the afternoon from her anchors, in Macao Roads,* and was obliged to put to sea with an E. N. E. gale. She stood out with all the sail she could carry, and at six had the Grand Ladrone bearing N. E. At eight she was in 20 fathoms water, and the wind "flew round" to S. E., and at midnight to South. At day-light (or say about 7 A.M. in such weather), they had driven close in with the land in 12 fathoms water, and were saved by their sheet anchor.

"The drift we made was amazing. I imagined it at first only about fifty miles, but to my astonishment, when the gale was over, I found myself as low down as Hainan within the Easternmost† island."

We may suppose Captain WEBB to have made the largest allowance he could for his drift before so reporting, yet this ship must have been drifted, *by the Storm wave alone*, for the wind was at S. E. and South, which would have drifted her to the N. W. and North, upwards of 125 miles in thirteen hours, (for the Taya Islands bear about W. S. W. 175 miles from her position at 6 P.M., and Captain WEBB allows 50 miles, or nearly four miles per hour for the drift.) This gives ten miles an hour for the storm wave! and in a W. S. W. direction, or directly with the track of the Cyclone. Making any allowance we please for her having been perhaps unmanageable, and at times before the wind, though her masts were all standing, and she was well prepared for the Cyclone, having her topgallant-masts on deck, &c.; still this is almost incredible, were it not borne out by other strong instances.

213. In September, 1809, the H. C. ship *Scaleby Castle* was drifted in a tyfoon 111 miles to the Westward of her supposed position (no doubt after every consideration and allowance made), in 48 hours or 2.3 miles per hour. The ship was not in the centre of her Cyclone though not far from it; her barometer falling to 28.30. She was on the Northern side of the Storm-circle, and thus her excessive drift was probably a Storm *current* only.

214. The H. C. S. *Castle Huntley*, 25th to 27th September, 1826, was drifted in a Cyclone which was travelling from the South 77° East to the North 77° West, and this while she was at perhaps 70 miles

* Jour. Asiatic Society, Vol. XI. and Annual Register for 1781.
† Printed *Westernmost* in original, but evidently an error

from its centre, at least 220 miles* in 46 hours, or nearly 5 miles an hour. If we allow the ship's drift to be three miles, there still remain two per hour for the Storm-current.

The rates at which the Storm-wave travels, as shewn in the foregoing paragraphs, may be perhaps thought excessive, but since the former editions of this work have been published, I have found that we have in the British Channel itself a very remarkable confirmation of the incredible velocity with which it can sometimes travel, and carry a vessel along with it. In the third edition of Luke Howard's CLIMATE OF LONDON, published in 1833, p. 254 to 258, quoting from a pamphlet which describes the Great Storm of 1703,† it is stated that—

"A small vessel laden with tin being left in the small port of Helford, near Falmouth, with only a man and two boys on board, drove from her four anchors at midnight, and going to sea made such speed before the wind almost without a sail, that at eight in the morning by the presence of mind of one of the boys she was put into a narrow creek in the Isle of Wight, and the crew and cargo saved."

The direct distance between Helford and Chale Bay in the Isle of Wight is about 160 miles, to which if we add 24 for the curved course and bad steerage, &c., this will be 184 miles in eight hours, or 23 miles an hour, from which if we take 6 miles per hour for the little coaster's run, and 5 miles per hour for a flood tide, there still remains at the lowest a storm-wave of 12 miles an hour to account for.

We have thus, for this sea, clear and undoubted proofs of Storm-waves and Storm-currents of great intensity and considerable dura-

---

* Mr. WISE, an officer of the ship, estimates it at "upwards of 300 miles under bare poles."

† The greatest English Storm of which we have any record. An annual Sermon is preached in London on the 25th November in the Baptist Chapel, Wild Street, in commemoration of it. During the reign of QUEEN ANNE this day was observed as a solemn fast, and a private individual named ROBERT TAYLOR has left a sum of money in the funds of which the interest is appropriated to this commemoration. Fifteen sail of the line, with Admiral BOWATER and all his crew, with several hundred merchantmen were lost. London appeared like a city which had sustained a protracted siege, whole streets being destroyed, and several thousand individuals buried beneath the ruins. 8000 seamen perished, and upwards of 120 persons on land, and it was computed that 25,000 fine timber trees were uprooted in the South of England and Wales. See Annual Register for Nov. 1843, p. 168, and Luke Howard as quoted.

tion, and when we recollect that the China tyfoons often bring with them awful inundations on the coasts,* we can have no doubt that the phenomenon exists much in the form in which I have described it. The close relation of the inundations to the Storm-wave, we shall presently shew. A careful consideration of our Chart No. IV. will shew how a tyfoon-Cyclone coming in, perhaps at least from the Marianas, Bonin Islands, or Loo Choos, if not from farther Eastward, may sweep the Southern coasts of China with a huge wave, of power and extent sufficient to produce these extraordinary effects.

215. I have not adverted here to the Northerly, North Westerly, and North Easterly currents, described by Major RENNELL, KELLY, and others, and commented upon by PURDY in his able Memoir on the Chart of the Atlantic, as common and undoubtedly existing at the entrance of the British Channel, for the simple reason that I consider the whole of the data on which they are calculated as uncertain and probably erroneous. Currents there are no doubt, and strong and most dangerous ones too, even in comparatively fine weather, but as before observed (p. 175, Note), in how many merchantmen in the present day, is the DEVIATION of the compasses duly ascertained, studied, applied to correct the courses, and watched to see if any alteration occurs in it? Until we know what was the Deviation on board of ships said to experience currents in fine weather almost all the "*log book currents*" must be distrusted. Where ships in which the Deviation is known and allowed for, are hove to in a gale, and a due allowance is made for their drift, or in the rare cases of a perfect calm where the current can be tried, we may reasonably allow that the current has existed; but when the course on a considerable distance is really uncertain to 5, 10, or even 20 degrees, we should waste time, had we even the details before us, to endeavour to analyse these data for our purpose, which is to demonstrate if we can, or at all events to put the mariner on his guard against, the *temporary* currents produced by storms.†

* To the Westward of Canton, where there are no large rivers of which the streams, being dammed up by the wind and swollen by the rains, might produce them.

† After the loss of the *Great Liverpool* Steamer, belonging to the Peninsular and Oriental Steam Navigation Company on Cape Finisterre, in 1846, a Committee appointed to investigate the circumstance, reported it as owing to an "unusual in-

216. The Western entrance to Bass' Straits is also a locality in which what we have said of the Coast of China, (all things being re-draught," and the P. and O. Company issued an order to their commanders that, at night and in thick weather they were to give Cape Finisterre (and strange to say Ushant was not alluded to!) a berth of twenty miles. In an article published in the Calcutta '*Englishman*,' I commented upon this order as follows, after explaining what the Storm-wave and Storm-currents were.

" The rule given by the Directors of the P. and O. Company is, that at night and in thick weather, twenty miles berth is to be allowed to Cape Finisterre.

" Now, putting aside temptation to 'turn the corner sharp,' and supposing fully twenty miles to be always allowed, we shall shew that instances may arise in which this is only half enough !

" The great storms of the Atlantic are, there is no doubt, very often, if not always, revolving storms of from 500 to 1000 miles in diameter, and travelling from the track of the Gulf-stream past the Bermudas and Azores to the coasts of Spain and Portugal, the Bay of Biscay, the Channel, or the Northern Ocean. They sometimes preserve their full size, and then *appear* to be gales blowing in nearly one direction only—that is veering but very slowly—and sometimes contract in size, when their duration is shorter.

" When the first case happens, there is probably little or no Storm-wave, or currents on the exterior borders of the storm ; but as it approaches or passes over a spot these may be felt moderately and even strongly. In the case of the *Great Liverpool*, if we take it that the Westerly to S. S. Westerly gales were the Eastern and Southern quadrants of an Atlantic revolving gale, and the insets the storm currents, her case is clear enough : but we are told that, in future, a berth of twenty miles is to be given to Cape Finisterre. In ordinary cases, of a moderate current, this may be enough; but taking the case where the centre of a storm passes near, and to the Northward of that point, the Storm-wave and Storm-currents together may induce a set of at least three miles an hour. We will put it at two only, and allow the steamer to have run up from a sight of the Burlings. It is clear that to put her twenty miles wrong she needs only ten hours run in this Westerly set, the ten hours of an ordinary night in those latitudes, and (here lies the mischief) the commanders may think that they fully take precaution enough if they give the Cape the 20 miles required. If the currents should unfortunately run stronger, or the thick weather lengthen out the nights, or render bearings or distances uncertain or impossible, the chances of another misfortune are much augmented.

" Our view, then, is, that the rule laid down by the P. and O. Company should rather have stood thus :

" And the Board further direct that the ships of the Company, on a voyage be-
" tween England and Gibraltar, both outwards and homewards, and especially with
" reference to the extraordinary inset noted above, and the possibility that in cases
" where the focus of a revolving storm from the Atlantic may be passing near Cape
" Finisterre, this dangerous inset may be much augmented, strictly enjoin that, when

PART IV. § 216.] *Bass' Straits; Channel; Ushant.* 187

versed here as being in the Southern hemisphere, and in a high Southern latitude) might theoretically be expected to hold good; and

"passing this point and the latitude of Ushant, if Southerly, S. Westerly, or
"Westerly gales prevail, or have recently prevailed, and especially if the barometer
"should have fallen, that a wide berth of from thirty to forty miles or more may
"be given in the night and in thick weather."

"We submit with all deference that this rule comes much nearer to the present state of nautical science than the one given. We are aware that experienced seamen usually say, and we believe the 'Directories,' and 'Pilots' also, that when a North Easterly storm is blowing out of the Chops of the Channel and Bay of Biscay, there is a strong current along the Coast of Portugal, and round Cape Finisterre. If there is also a Westerly gale, this would be exactly a Storm-current, such as the authors we have quoted infer to exist, and Mr. REDFIELD, indeed, inclines to think with Professor DOVE, of Berlin, that *all* winds blow in circuits; and certainly there is much to confirm these views. At all events, nothing should be risked for want of a trifling precaution, which, for two or three steamers in the course of the year may cost an hour or two more of steaming, by making rather a wider circuit past these dangerous points."

This letter was duly acknowledged by the Directors, and its suggestions I have no doubt acted upon. That they were correct is shewn by the Log of the P. and O. Company's steamer *Iberia*, given in Mr. MARTIN's work, p. clx., of Appendix to Part I. This vessel experienced, a little to the Northward of Cape Finisterre, an inset of 31 miles in the 24 hours, and Mr. MARTIN rightly remarks, that had she left Southampton a little sooner this current might have put her on shore, like the *Great Liverpool*. The same occurred to the P. and O. Company's steamer *Ganges*, (MARTIN, p. 143,) between Ushant and Cape Finisterre.

I at the same time drew up and addressed to the Lords Commissioners of the Admiralty, a paper of "*Instructions to officers of H. M. Navy, regarding the observations desired on the Storm-wave and Storm-current when in or near soundings, and from the Lat. of Gibraltar to the British Channel*," respectfully suggesting that it might be published, as a means of obtaining some useful data to enable us to ascertain all the conditions of this danger, and offering to analyse the observations. Their Lordships, I am glad to say, have done me the honour to recommend its being printed in the Nautical Almanac, and I trust that all British seamen, at least, will see the great importance of the questions, to them of life or death at times, which may depend on the investigation; and aid me by all the information old or new which they can collect. The Peninsular and Oriental Company, in acknowledging my communication, acquiesce on the part of their committee and for themselves, (and they had the best nautical advice in London to consult) in the utility of the suggestion. The sailors must do the rest for themselves by furnishing material.

Since the first edition of this work I have met with an account of the loss of H.M.S. *Repulse*, on the 9th March, 1800, on a sunken rock, supposed to be the Mace, about

if the reader will first refer to what has been said of the tracks of storms hereabouts, (p. 64) and then placing his Horn-Card on a Chart move it along from the West or W. S. W., to the Eastward, he will see at once that there are good grounds for supposing at least, that with the heavy Westerly or North Westerly gales forming the North and N. W. quadrants of a Cyclone, which has travelled in from the Southern Indian Ocean, strong Storm-currents or a strong Storm-wave, or both acting the same way, might occur. And so it notoriously proves, for the repeated and dreadful wrecks of Emigrant ships off the N. W. point of Van Diemen's Land, with the "narrow escapes" of many, which have just seen their danger in time, are indubitable proofs of it. The following, which is abridged from a newspaper notice, is a direct instance, and it appears to be written by one who understood something of nautical matters :—

"As to the Western entrance into Bass's Straits, the writer of this has reason to be glad that Providence preserved him and the ship in which he sailed from England from the fate of the *Cataraqui*;* as she was in very nearly the same latitude and longitude, and during a dark stormy night, in soundings which, compared with Captain STOKES' Chart, shew her to have passed close to the Harbinger and Navarin Rocks, on which vessels have been previously wrecked.

"That there is probably a set of some kind off the coast, Westward of Cape Otway, by which ships are carried to the Southward, is very probable. Captain Stokes does not, however, allude to it. Something of the kind probably occurred to the *Cataraqui*, for Captain STOKES ascertained, by correspondence with her owners, that her master had his chart on board. When it is considered that the coast slopes down rapidly from the Great Australian Bight to King's Island and the Western side of Tasmania, it is not surprising if there should be under the action of Westerly winds, frequently

---

25 leagues S. East of Ushant; in which it is stated that on Saturday the 9th, it was blowing a very heavy gale, and the ship having shaped a course to carry her far to the Westward of Ushant, owing to the thickness of the weather (not having had an observation for some days) and the different set of the tides, had got near *three degrees!* to the east of her reckoning, to which the loss is attributed. She was at the time she struck, going nearly right before the wind, which was thus we may suppose to the S.W. or Southward, and thus in the front of a Cyclone, if it was one; with the Storm-wave carrying her as I have described. The great rises of the water in the Ports of Brest, and thence to the Southward in bad weather, with certain winds, are exactly analogous to what, we have just seen, occurs elsewhere.

* Wrecked at the entrance of Bass' Straits with 400 Emigrants on board.

blowing a heavy gale, a deflection of the current, setting a ship in dark stormy weather out of the proper course ; and it is a matter of surprise rather than otherwise that but five wrecks (however fatal they may have been) should have taken place off King's Island in twelve years. It is to be hoped that for the future, with lights on Cape Otway and the Kent Group, the navigation of Bass's Straits will be as safe as the British Channel, where, it will be recollected, wrecks occur off St. Catherine's, in the Isle of Wight, from a set off from the Channel Islands, as in the case of the *Clarendon*, West Indiaman, which was cast ashore on a rocky ledge, despite all the lights from the Lizard to the Needles, during a S. W. October gale, which the writer recollects, having been exposed to it, and having lost relatives in the wreck of the ship."—*Sydney Morning Herald, Dec.* 10.

217. The seaman also, adverting to what has been before said that winds may be blowing in true circuits (Cyclone) without rising to the force of a gale or hurricane, should bear in mind that in narrow seas especially, and where perhaps other causes may be acting, *and acting the same way*, so that two moderate causes may produce a strong effect, the Storm-wave and Storm-current may also be acting then to a certain extent, and carrying him far out of his reckoning while he is deprived of observations in dark gloomy weather.

218. The *Nortes* of the Gulf of Mexico, may often be of this description, for storms there which are moderate as compared with hurricane forces, may yet be true Cyclones, and produce *Nortes*, (Northerly gales) in various parts according to their tracks. One coming into the Gulf from the S. W. or South, or from the Pacific Ocean, across Central America, might give a *Norte* on its Western edge, and one coming in like the tracks V. XII. XV. &c. on Chart I. would also give a *Norte* on its Western verge. In the two cases ships would be carried quite different ways by the Storm-wave if it existed, for this would be regulated by the track of the storm. [See also p. 366.]

219. The dreadful loss of the Royal Steam Mail Company's Packet ship *Tweed* on the Alacranes, in 1847, appears to me not improbably a case of the Storm-wave in a moderate gale, as I have just described it.

The following account is given in the *Steam Navigation Gazette*, and *Nautical Standard*, as to wind and weather previous to her striking from the statements of the first and second officers.

"It appears from the statement I have received, (at a brief interview, immediately on their arrival), with Mr. ELLISON and Mr. ONSLOW, the first and second officers,) that the unfortunate *Tweed* left Havannah on the evening of the 9th of February, for Vera Cruz, with a fresh Southerly breeze, that lasted about ten hours. It soon after became thick, and at midnight on the 10th, the wind suddenly shifted to the North, and in a short time it blew a stiff 'Norther,' weather still thick, sun obscured, and no observation. At 3-30 A. M. on the 12th, when the Captain judged the ship was 30 miles to the South of the Alacranes, the look-out on the forecastle cried out 'Breakers a-head.' Captain PARSONS, who was on deck, hearing them at the same moment, immediately ordered the engines to be stopped and reversed, and the helm to be put hard a starboard, which order was instantly obeyed, but the fore-sail, fore-trysail, and fore-topsail being set, and a heavy sea running, she forged a-head and struck."

From the Havannah to the Alacranes, the true course and distance is South 84° West 440 miles, though it becomes somewhat curved while running along the coast of Cuba, but this we will leave out. The vessel seems to have had ten hours run with a Southerly breeze, (under the lee of the coast of Cuba,) and then thick weather till midnight on the 10th, when the shift of wind took place; this gives her, at let us say nine knots an hour from 6 P.M. on the 9th, 30 hours run or 270 miles, which would bring her about opposite to the middle of the channel between Yucatan and Cuba.

Now placing a Storm Card (even upon our small chart) so as to give the *Tweed* a Northerly wind (*Norte*), and supposing the storm to have slowly moved up on a course about parallel to our track XV., we can see how the *Tweed* standing across, but nearly along with it, might keep the Northerly winds till 3-30 A. M. of the 12th, or 15¼ hours longer, and yet be set up 30 miles to the Northward of her supposed position. The error in position might also, it is true, arise from a small unknown deviation of her compasses, but the account of the weather is distinctly that of a moderate Cyclone, and in narrow seas, as before said, the seaman cannot be too careful, or too often recollect that two or more petty causes *acting the same way* may produce a serious amount of error.

Colonel REID's Chart (at p. 399 of his works, 2nd Ed.) of the storm which dispersed the Spanish Fleet under Admiral SOLANO in 1780, is exactly one of a cyclone passing on the track which I describe.

220. INUNDATIONS FROM THE STORM-WAVE. I have shewn the seaman, and I trust satisfactorily, that there exist in many parts of the world, both Storm-waves and Storm-currents, which for him, and

while at sea, are to be regarded and calculated upon as currents; and no doubt we shall find, as our knowledge extends, that they exist in other parts, and moreover obtain more accurate knowledge of them everywhere. I now proceed to shew him that there really exists an actual *wave* or elevation of the sea, as announced theoretically at p. 12, and described at p. 174 and following ones, which rolls in upon the land like a huge *wall of water!* as such a supposed wave should do, causing of course dreadful inundations. He may at first suppose that he has little to do with the inundations on land, if he can keep his ship in safety at sea; but lest this objection should occur to any, I answer first, that it is of much importance to *prove* that there is a real lifted wave; and next that this very wave, while it is a source of great danger—for ships have been swept out of docks and rivers by it—is on the other hand an element of safety, for it has sometimes carried vessels over reefs and banks, so as to land them high and dry, and preserve at least the lives and cargo. Of this the case of the *Briton* and *Runnimede* already alluded to pp. 56, 112, is a striking example, both ships being carried at night, and while in the centre of the vortex of a furious Cyclone, over reefs which are dry at low water, and thrown up amongst the mangroves! Captain HALL, of the *Briton*, estimated that the whole rise of the water must have been 30 feet, judging from marks on shore. If from this we take 10 feet for the tide, we have still 20 feet for the storm-wave!

221. Col. REID in reference to this says, p. 513, 2nd Edition:—

"An anchorage, which would be of sufficient depth in ordinary gales, might prove too shallow during a hurricane, in consequence of the depth of the trough of the sea, from the unusual undulations created by such storms. Instances have been here given, when the effect of hurricanes, blowing into a bay, has been to heap up the water within it for a time; so that vessels which have dragged their anchors during such a crisis, have been carried into places whence they could not float after the storm had passed over.

"The opposite consequence may also occur; such as happened to the *Lark*, surveying schooner, when at anchor off the West coast of Andros Islands, in the Bahamas. 'Owing to the receding water, that vessel struck heavily from 6 to 8 P. M., on the 6th of September, 1838; but floated again on her being raised by the S. E. gale, whilst the wrecking schooner, *Favourite*, 68 miles North of the *Lark*, was left completely dry.'"*

* Lieut. Smith's Report, *Nautical Magazine* for January, 1839, p. 30.

But this sort of inundating wave, as will be seen, must not be confounded with that gradual rise of the sea or rivers which is familiar to our mind as the "*heaping up*" of the water by the effect of wind, and which always occurs gradually. The wave, of which we shall give instances, is one resembling that which so often accompanies or follows earthquakes. In some cases, indeed, the Cyclone, earthquake, and wave of inundation have apparently all occurred together!

222. IN THE WEST INDIES. In the great hurricane of 1772, which devastated St. Christophers, St. Croix, St. Eustatia, Antigua, Dominica, &c. it is stated (Annual Register for 1772) that at Santa-Cruz the sea "*swelled up*" 70 feet* above its usual level, roaring so that it was heard a hundred miles off. It overtook and swept off above 250 persons who ran up to the mountains to save themselves.

223. At Savanna La Mar, which was totally destroyed by the first of the two great October hurricanes in 1780, we find that—

"The gale began on 3rd October, from the S. E. at 1 P. M., abating about eight; the sea during this last period exhibiting a most awful scene. The waves *swelled* to an amazing height, rushed with an impetuosity not to be described on land, and in a few minutes determined the fate of all the houses in the bay. About ten, the waters began to abate, and at that time a smart shock of an earthquake was felt. Three ships were carried so far into the morass, that they could never be got off."†

Colonel REID says, referring to other notices of this storm—

"There seems no reason to doubt, from what we now know of the effect caused by hurricanes, that Savanna La Mar was overwhelmed by the accumulated water of the sea raised solely by the power of the wind."

224. In the second great October hurricane, 1780, which began on the 10th of October of that year, and committed dreadful ravages at Martinique, we find, from a French official report, given p. 344, by Colonel REID, and of which I abridge and translate what is essential to our present subject, that (apparently) about the time that the centre was passing close to the town of St. Pierre and Fort Royal,

"An awful '*raz de marée*'‡ completed the misfortunes under which we were suffer-

---

* So in the Annual Register!

† Abridged from the 2nd Edition of Col. REID's Work, p. 297. The account therein is copied from the *Annual Register*.

‡ Raz (or Ras) de Marée. This word is used by the French to express every sort of sudden swell, or waves, or rollers from the sea, whether occasioned by tides or cur-

ing. It destroyed, in an instant, upwards of 150 houses on the sea shore, of which thirty or forty were newly built; those behind them had mostly their fronts driven in, and the goods contained in them almost totally lost. It was with much difficulty that the inhabitants escaped with their lives."

225. At Barbadoes "the sea rose so high as to destroy the forts," and at several other of the Islands, the same kind of devastation is alluded to. It will be noted that the Cyclone came up from the S.E., and that Bridgetown and the ports of Martinique alluded to in the foregoing report are on the S. W. and West, or leeward sides of their respective Islands, so that the sudden rise of the sea has much more the character of a true elevation of the waters, by the effect of the centre, than of a mere wave or number of waves rolled in upon the land by the horizontal force of the wind; as we see in high spring tides with strong South Westerly and Southerly gales on low parts of the coast of the South and West of England for instance.

226. In the Bermuda storm of 1839, referred to at p. 30, and which was no doubt a true Cyclone, Colonel REID, p. 449, 2nd Edition, says,

"By examination of the South coast of the islands, the sea was found to have risen fully eleven feet higher than the usual tides. It carried boats into fields thirteen feet

---

rents; except the *bores* of rivers, which they distinguish by the word *maorée* or *masseret*. The definitions of "*Raz* or *Ras de marée*" (whence, no doubt, our word "*Race*") are; "*Ras des Courans* (current races) masses of water which at sea in narrow passages, as amongst islands, shoals, rocks or coasts, advance with rapidity. Finally, *Ras de marée* (tide-race) a name given to those swells which, without appearing to be driven by any wind are formed suddenly, increase in a moment, and mark some strange agitation in the waters of the ocean:" so far ROMME in his *Dictionnaire de la Marine*. At Bourbon and the Mauritius, the heavy rollers which, as in many other places as at Madras and St. Helena, rise suddenly and break with tremendous force, are called by the name of *ras de marée*, though the rise and fall of the *tide* there is very small. Mr. THOM, p. 340, in the notes to a French table of Barometric Observations, made at Bourbon during the Mauritius hurricane of 1840 has given in it the remark "*Ras de marée très fort* from the 7th to the 11th, so that no communication could be held with the shipping." The Cyclone did not extend to Bourbon. A tremendous *ras de marée* is also experienced in the harbour of Port Louis, when hurricane-Cyclones pass to the Westward of the Island. In January, 1844, five vessels at once were thrown on the rocks near the Bell Buoy, where they had anchored, or were close in, endeavouring to get into the harbour.

above the usual high-water mark, and removed several rocks, containing by measurement twenty cubic feet, some of them bearing evidence of having been broken off from the beds on which they rested by the surge. On the North and leeward side of the island, and within the chamber of the dock-yard, the water was observed to rise two feet and half higher than the ordinary tides.

"As the weather became fine at Bermuda, and the hurricane proceeded on its course, the Northern reefs of the islands, in their turn, presented a line of white surge from the swell, rolled back by the gale. Vessels as they arrived from the East or the West, reported that they met the wind in conformity with what appears to be a law of nature, in those tempests. Thus the *Jane* coming from Baltimore and the Westward, had the wind Northerly; whilst the schooner *Governor Reid*, from England and the Eastward, had the wind Southerly."

It will be noted that here the great rise was on the windward side of the island, and but a small one to leeward; but we may well suppose that over a large extent of reefs and islands like the Bermudas, lying directly in the track of a Cyclone, the wave might divide and roll off to the East and West, leaving a comparative hollow or curve on the North side of the islands; an effect which we frequently observe on a small scale, amongst reefs and rocky shores, where the waves rolling in are divided by a ledge or single rock; a boat riding to the shoreward of which, will *seem* to sink in her smooth water, lower than the surrounding waves on both sides, though it is really these which are raised above her, and they afterwards unite in the common elevation of the swell a little within shore of her;* the spot at which she is riding being perhaps really elevated, though but a little, above its usual level.

227. Mr. REDFIELD in his first Memoir, published in 1830, describing a Cyclone, of which the centre passed close to New York, says that—

"The gale was from N. E. to East, and then changed to West. More damage was sustained in two hours, than was ever before witnessed in the city, the wind

* I write this with the vivid recollection of the very appearance, having been, now about thirty years ago, severely beaten and bruised in swimming off, through the surf on a coral reef, to help to save a boat under my charge, the two men in which could not, without a third hand, weigh the grapnell, and pull her out in time to round a point of the reef, to get into a safe berth. It was only after repeated trials, and taking advantage of the smooth hollows, that I could reach her.

increasing during the afternoon, and at sunset was a hurricane. *At the time of low water, the wharfs were overflowed, the water having risen thirteen feet in one hour."*

228. In the East Indies, Coringa, on the Coromandel Coast, is frequently subject to inundations from gales blowing from seaward, and the consequent rise of the waters of the river Godavery; but these are gradual, though rapid, rises, and when occasioned by the river, the waters are of course fresh, and when by the sea, salt. I should mention also, that the tracks of the storms in the middle of the Bay of Bengal appear to direct themselves very frequently towards this station and that of Cuttack, (see Chart No. III.,) probably from their being situated at the mouths of the valleys of two considerable rivers, while the rest of the Coast is, as it were, walled with high land at a short distance in the interior.

In December, 1789, Coringa was utterly destroyed by a succession of three great waves, which rolled in upon it, apparently during a hurricane-Cyclone. M. DE LA PLACE, of the French Frigate *La Favorite*, who collected his account on the spot, about 1840, says, Vol. I. p. 285 of his voyage,

"Coringa was destroyed in a single day. A frightful phænomenon reduced it to its present state. In the month of December, 1789, at the moment when a high tide was at its highest point, and that the N. W. wind blowing with fury, accumulated the waters at the head of the bay, the unfortunate inhabitants of Coringa saw with terror three monstrous waves coming in from the sea, and following each other at a short distance. The first, sweeping everything on its passage, brought several feet of water into the town. The second augmented these ravages by inundating all the low country, and the third overwhelmed everything."

The town and twenty thousand of its inhabitants disappeared; vessels at anchor at the mouth of the river were carried into the plains surrounding Yanaon, which suffered considerably. The sea in retiring left heaps of sand and mud, which rendered all search for the property or bodies impossible, and shut up the mouth of the river for large ships. The only trace of the ancient town which now remains, is the house of the Master Attendant and the dockyards surrounding it.

229. So far M. DE LA PLACE, whom I have exactly translated, but he makes the year 1789, and the Cyclone to have occurred in *December*, but in the Annual Register for 1788, is a letter (p. 238)

from Mr. PARSONS to A. DALRYMPLE, Esq. describing an inundation from the Storm-wave of a Cyclone which occurred at Ingeram, five miles South of Coringa in the month of *May*, 1787, from which it would seem that there were two of these dreadful visitations. Mr. PARSONS, whose account I abridge, says—

"From the 17th of May, it blew hard from the N.E. but nothing was apprehended on the 19th, at night it increased to a hard gale, and on the 20th in the morning it was a perfect hurricane, untiling houses, and beating in doors and windows, blowing down walls, &c. A little before eleven it came with violence from the sea, and I saw a multitude of the inhabitants crowding towards my house, crying out that the sea was coming upon us. I cast my eyes in that direction, and saw it approaching with great rapidity, *bearing much the same appearance as the bar (bore) in Bengal River*. I took refuge in the Old Factory, which is built on a high spot and well elevated: so that we were not driven to the Terrace. I think the sea must have risen fifteen feet above its natural level, the wind favoured the subsidence of the water (at 1½ P. M.) by coming to the South where it blew the hardest. At five I got to another house during a lull. It blew very hard the greatest part of the night, and at midnight veered to the Westward, and was so cold I thought we should have perished as we reclined in our chairs. The gale broke up towards the morning. The natives have a tradition that about a century ago the sea ran as high as the tallest Palmira-trees (45 to 50 feet at least). Everything was destroyed with us, but at Coringa and nearer the sea, not more than twenty inhabitants out of the 4000 were saved. At first with them the sea rose gradually with the tide, when it increased they mounted on the roofs of their houses till the sea impelled by a strong Easterly wind rushed in upon them most furiously, when all the houses at the same awful moment gave way. This was seen from the terrace of Mr. Corsar's house, over which the sea sometimes broke, and where the wrecks of vessels were seen drifting past. At Jaggernautporam about a thousand lives were lost, and the inundation extended as far North as Apparah (15 miles N.N.E. from Ingeram on the Coast), but not many lives were lost there. It penetrated about twenty miles inland. It is computed that 20,000 souls and 500,000 cattle perished. The writer further remarks that it was considered very remarkable that the vast tract of low ground, from Gotondy to Bundaramalanka, on the South side of the Godavery, which is often overflowed by the spring tides, suffered very little. He considers that point Godavery and the small low islands near it broke the force of the sea."

Taking the average of the winds to have been from N. E. to South, this was a Cyclone travelling from the E.S.E. to the W.N.W. There was no calm centre, and no sudden shift is spoken of, so that we may suppose the centre to have passed a little to the South of Ingeram,

PART IV. § 231.]   *Inundation of Cuttack and Balasore.*   197

but for our present purpose, this detailed account fully demonstrates the existence of the Storm-wave and the terrific inundations it gives rise to.

230. In 1839, Coringa was visited by another of these inundations, which, by the Collector's Report, (published in my second Memoir, Jour. As. Soc. Beng. Vol. IX. p. 401,) much resembled the one above described, to which indeed the old inhabitants seem to have compared it. There are no accounts of distinct *waves* in this last storm, but the rise of the waters of the sea is described by one writer as *rushing* in with such violence, that "the only houses remaining at Coringa, are a particular large house and three or four other brick built houses. More than 20,000 persons are said to have perished. Vessels were drifted from the docks and rivers, and a large sloop (of 50 to 150 tons burden) carried four miles inland."

A friend says, "I visited Coringa about a month after the Cyclone of 1830, and could relate facts which would be thought travellers' tales. The number of vessels of from 100 to 200 tons that were high and dry miles inland, some bottom up, gave the country the appearance of having been visited by a party of gigantic demons, who had been throwing the huge hulls at one another."

231. Cuttack and Balasore have also been subjected to these inundations, but one of the most frightful of them in recent times occurred in May, 1823, along the Northern shore of Cuttack from Point Palmiras to Kedgeree.

A newspaper correspondent of the *Bengal Hurkaru*, writing from the spot, thinks the high wind not a sufficient cause, but that there might also have been an earthquake,* but as I have traced the Cyclone in from sea at least for two days, (track *g* Chart III.) I see no reason to doubt its being a Storm-wave. It is said in these notes that—

"All the Natives agree in asserting, that the sudden flood consisted of three great waves only, and that these occurred in the short period of one minute, the last wave rising about 9 feet above the highest embankment,† and committing, of course, the most awful mischief in its progress. Upwards of 600 native villages are said to have been destroyed."

* None of the accounts allude even remotely to any shock, and it would have been probably felt at Calcutta where there was no storm.

† The embankments are probably not less than from 10 to 15 feet high.

In October, 1831, on about the same parts of the coast, another storm inundation and wave occurred. In an official report, it is stated, that the sea came in "in a tremendous wave."

232. In 1822, an awful inundation took place at the stations of Burisal and Backergunj, at the mouth of the Megna; but here though 50,000 human beings are said to have perished, the rise of the waters was gradual, and the strength of the inundation is described as coming in with the spring tides. This storm too, which I have accurately traced, as well as the former one, moved very slowly, about 53 miles in the 24 hours; which may account for its not bringing so heavy sea with it. See p. 91 where more details are given. We have also on the Coasts of China, and even from the islands of Hainan and Formosa, frequent accounts of these tremendous inundations, accompanying their tyfoons; and as there are no large rivers West of the Canton River, there seems no reason to doubt that there are frequently Storm-waves, especially from the Tyfoon-Cyclones which travel up from the S. E., as will be seen on our Chart IV.

233. I have also obtained at 60 years of interval, exactly a corroboration, if any were wanting, of this phænomenon given by an eye-witness, a sailor, and one placed in the most favourable situation for observing it. During the remarkable Cyclone of the Bay of Bengal, in October, 1848, (Eighteenth Memoir, Journal Asiat. Soc. Beng. Vol. XVIII.) which *descended* in the middle of the Bay and travelled up to False Point over which the centre passed, Mr. BARCKLEY, the Superintendent of the False Point Light House, after describing the passage of the calm centre and the renewal of the Cyclone from the S. S. E. says—

"The rise of the tide which was about 9 ft. more than usual (entire rise 17 ft.) came in with a rush like the bore.* I saw it come in a heavy foaming surge of a wave, like the surf outside of Plowden's Island; I heard it coming in and went up to see what it was, and from the gallery of the Light House saw it distinctly. I at first thought it was the island being washed away. This bore came in about two o'clock in the morning, when the hurricane had reached its full

---

* The very comparison used by Mr. PARSONS above. But it was high water at 9½ P. M. at False Point, so that this occurred at three quarters ebb.

height. To the North and East the tide rose 19 or 20 feet in all, or about 11 feet above its usual height. Great numbers of lives and much property were destroyed by it."

234. In the Malabar Coast Cyclone of April, 1847, in which the H. C. Steam Frigate *Cleopatra* foundered, great numbers of the inhabitants of Kalpeni and Underoot, two of the Laccadive islands which are low coral banks, lying upon the track of the Cyclone, were swept away by the storm-wave, while other islands at greater distances from the centre suffered but little from this cause. All the islands were visited by a vessel belonging to the British Government which was sent to afford them assistance after the Cyclone, and this statement is taken from her official report.

235. Enough then will have been said to shew the mariner, I think, that the wave really exists in many cases, and is therefore a danger to be borne in mind and provided against. I have not alluded to the well known instances of our sudden rises of tide, and even true inundations (as at St. Petersburgh)* in Europe, because these are probably more frequently the joint effects of tides and the driving up of waters by the wind, than the kind of *disk* or umbel, or pot-lid-shaped rising of the water of the ocean, which we assume to be the true Storm-wave. It is in fact at times difficult so to separate the various causes, which *may* frequently all act together and all the same way, as to say where the one ends and the other begins. If we could meet with an instance in which the wave rolled in during the passage of the

* The St. Petersburgh inundation of November, 1824, appears certainly to have been the effect of a storm, but whether of a true storm-wave at, or near, the centre of a Cyclone seems doubtful. At least I have not access to any documents which would shew it to be clearly such, though its progression is clearly shewn from the S. W. to the N. E., being a violent storm on the Coast of England and Holland on the 15th from the N. W. and W. crossing the North Sea, Norway and Sweden, where whole forests are said to have been prostrated by it, like the Tornados in the West Indies, see § 439, and both in Sweden and Norway *sudden* rises of waters are noticed. At St. Petersburgh, however, on the 19th in the morning, the waters of the Gulf of Finland drove up those of the Neva so rapidly that in *less than five minutes* the whole city was inundated, and the country for miles around, causing a frightful destruction of life and property. The rise of the water at Cronstadt is stated to have been 14 feet. Twelve sail of the line and four frigates are said to have been driven on shore.

calm centre of a Cyclone, that would be so far a proof that we might exclude from that instance all the other causes, as tides, winds, &c. and affirm of it then that it was purely the Storm-wave, or the effect of an earthquake; but as yet, and from the frightful confusion which these calamities must create, we have no accurate accounts of this kind nearer than that of False Point just quoted.

236. PYRAMIDAL AND CROSS SEAS. If the seaman will take both Storm Cards, and reversing one of them so as to have both sets of the wind-arrows pointing the same way, place the one upon the other upon a sheet of paper, he will then see that *a perfectly stationary* Cyclone would be raising a sea, which he might represent by so many rays towards the centre, and which therefore has wider intervals (*i.e.* is a *longer* sea) at the outer part of the vortex than towards the centre, where each wave must interfere with the other.

If he will now suppose his hurricane moving on upon any track he pleases, and slide his upper Card gradually forward in that direction, he will see that it is only the front part of every hurricane which gets "fresh sea" to act upon,* and that every successive wave of the following part, passes over and crosses in a thousand more ways even than those shewn by the circles, the sea formerly left by the front portions of the storm.

This is bad enough already: but if he will suppose moreover that there may be a gradual incurving of the wind arrows, to at least as much as two points at the centre, and then the wind blowing with such fury that nothing can resist its force, he will begin to suspect what has been written and said of the pyramidal sea of a China tyfoon, or Mauritius hurricane, and sometimes of the West Indian once may be no exaggeration,† and the more if he adds to all this the certainty that, at and near the centre, the sea may be raised in a disk or umbel of two feet above the general level by a fall of the barometer, shewing the atmospheric pressure to be diminished so as to compel that

---

* We may suppose too, that, as often occurs, it comes upon a track of sea where only calms and light winds with "long heavy squalls" have prevailed for some days.

† If he will look at the rose engine-turning so common on the back of watch cases, he will understand how every little boss there might represent a sea worked up by the infinite crossings of the forces acting on the waters.

rise* by the pressure of the surrounding ocean. He may farther also consider it as a well established fact that the Cyclone drives on, to a considerable distance before it, a heavy swell peculiarly its own, and which is felt at a great distance before the winds reach the point. See the next section on the swell felt at a distance.

When the sailor has duly considered these facts and allowed them to be possible, he may then, though before incredulous, believe some of the following extracts, and think it worth while to attend to our rules for avoiding such predicaments.

237. Lieut. ARCHER, in his capital account of the loss of the *Phœnix* frigate, re-published by Colonel REID, p. 308, says:

"Who can attempt to describe the appearance of things upon deck! If I was to write for ever I could not give you an idea of it—a total darkness all above, the sea on fire, running as it were in Alps, or Peaks of Teneriffe (mountains are too common an idea); the wind roaring louder than thunder (absolutely no flight of imagination); the whole made more terrible, if possible, by a very uncommon kind of blue lightning."

238. HORSBURGH says of the Tyfoon-Cyclones in the China Seas, that they are—

"Frequently blowing with inconceivable fury, and raising the sea in turbulent pyramids which impinge violently against each other."

239. Mr. THOM, p. 15, speaking from the log of the *Robin Gray*, in the Rodrigues' hurricane, says :—

"It is the sea, however, which is most to be dreaded in rotating gales. It is described as having been a tremendous, cross, confused, outrageous sea, raised in pyramidal heaps by the wind from every point of the compass, and has been compared to the surf breaking on a reef of rocks. In fact it was 'such a sea as gave a ship no chance.' Near the centre of the hurricane a ship is always unmanageable, even if she has not lost masts or rudder ; the lulls and terrific gusts, which follow one another in quick succession, are alone sufficient for this, but when we take into consideration the fierce conflict of raging waters, it is only wonderful how a vessel can live through such an encounter."

240. Captain RUNDLE of the ship *Futtle Rozack*, whose able and scientific log† I printed at length in my Eleventh Memoir, (Jour. As. Soc. Beng. Vol. XIV.) has given in it some remarks on the sea, in a Cyclone and at the approach of one, which are highly illustrative of

* See p. 169 and Part V. at the Section on the barometer.

† A pattern for every sailor who loves his profession, and has the interests of his country, his owners, or his own at heart. With a few hundred such observers afloat, and access to their observations, we should soon be able to track and define the laws for storms all over the world.

what this sea is. The Cyclone of which the Memoir treats, was one in the Southern Indian Ocean, and is the storm track to which I have alluded, p. 48. Its track is that marked VIII. on our Chart, and it is also alluded to at p. 92.

At the approach of the Cyclone Captain RUNDLE says—

"I find the barometer considerably fallen, with an exceedingly long swell from the Southward, and at 7 a high N. N. W. sea meeting the Southerly swell created an exceedingly turbulent sea. In the squalls the sea has a strange appearance, the two seas dashing their crests against each other shoot up to a surprising height, and being caught by the West wind, it is driven in dense foam as high as our tops. The whole horizon has the appearance of ponderous breakers."

When involved in the Cyclone he says—

"P. M., wind N. E., tremendous squalls blowing with inconceivable fury. The sea rising in huge pyramids, yet having no velocity, but rising and falling like a boiling cauldron. I have never seen the like before. I was in the height of the terrible hurricane of September, 1834, in the West Indies, I have been in a Tyfoon in the China Sea, in gales off Cape Horn, the Cape of Good Hope and New Holland, but never saw such a confused and strange sea. I have seen much higher seas, and I am sure wind heavier, but then the sea was regular and the wind steadier.

"Noon, blowing with inconceivable fury at times, with the sea I think more agitated and confused than ever; rising up in monstrous heaps, and falling down again without running in any direction. Noon, laid to again."

Captain B. SPROULE, of the ship *Magellan*, thus describes it from actual observation in the China Sea, in August, 1850:

"I never saw anything to equal the sea. You could not say it was running from any point, but meeting from all quarters—impinging one against the other, and flying into the atmosphere in pyramids of foam, falling again on the spot where they rose."

241. So violent and remarkable are these seas, that repeated instances have occurred in which ships, running or bearing up, so as to cross the track of a Cyclone soon after the centre has passed, have found the sea still so violent in comparison with the force of the wind, and withal so irregular, that they have been in great danger of losing their masts. H. M. S. *Serpent*, with 2½ millions of dollars on board, the last instalment of the China Treaty money, hove to, as related in p. 46, to avoid a small but severe Cyclone in 27° South, 64° East, and when she bore up again, crossed the track of the Cyclone. Commander NEVILL says, "When the barometer rose to 29.80, I wore and bore up, coming into a dreadfully confused sea,

where the centre of this turning gale must have passed. For nearly 24 hours I thought I should have rolled my masts away; but by getting up runners and tackles for the lower masts, and having preventer shrouds for the topmasts previously fitted, I did not lose a spar." A like instance also occurred in the *Golconda's* Cyclone in the China Sea (see pp. 97, 157) to the *Thetis* of London.

Many more such extracts might be given, for they occur very frequently; but these will no doubt suffice to our purpose, which is to shew that the best built, and best found, and best managed ship can rarely pass through the centre of a Cyclone without damage, and probably of a serious kind, if it be only that of the straining and tearing of years in a few hours.

242. SWELL FELT AT A DISTANCE FROM STORMS.—I have adverted in a preceding section of this part to the swell felt on shores and in harbours—such as rollers, surfs, *ras-de-marée*, &c. at the approach of Cyclones; and these are evidence enough that the swell often precedes the gale for a great distance, sometimes, indeed, for as much as 24 hours.

In truth, sailors want no examples of this; but it may be as well to advert here to the fact, that in Cyclones of considerable extent and violence, the swell is often felt as a *double* sea, namely, one preceding the track, mixed up with one such as the wind at the Cyclone point of an ordinary storm would give. Thus, if we suppose that side of a Cyclone to be advancing towards us, which has the wind at South, as in the Northern Atlantic, the wind alone, if a straight-lined wind (see p. 10), would give an East and West *line* of sea, or what we call a Southerly sea; but as the storm is advancing to the East, *it* is also perhaps driving a sea before it in North and South lines, or a sea from the Westward. At a great distance, the two forces will give probably " a confused sea from the South Westward," but nearer the two seas may often be clearly distinguished, and are capital indications.* We might indeed distinguish these two seas by separate

---

* See SECT. 240, Captain RUNDLE's clear descriptions of these crossing seas. Such phænomena must frequently occur; and those who will read the quotation from Lord BACON which I have placed on the title page, will understand why it is now, at least, desirable that they should be always noted.

names, such as the "*Wave of Progression*" for that sea which is driven before the body of the Cyclone, and the "*Cyclonal wave*" for the sea occasioned by the wind on different parts of the storm circle. The position and run of the ship, with the average tracks, and the laws of rotation which have been previously explained, and his barometer, will shew the seaman in most cases when the cross sea is that of an approaching or passing tempest, and when it is that left by one already past, but which he is overtaking. The two seas may often be best distinguished from the masthead, under the varying lights and shadows of a cloudy day.

243. Captain BLAY of the brig *Standard*, an extract from whose log in the Bermuda Cyclone of 1839 (track A on our Chart No. I.) is given, says in his remarks, p. 451 of Colonel REID'S work:

"I remark that I have experienced several hurricanes at sea, and have invariably found that by observing strict attention to the set of the swell previous to the commencement, and even after, a tolerably correct idea may be formed of the direction the wind is likely to take.

"I particularly noticed this in the last two which I experienced; and on the 2nd September, 1838, in a hurricane that commenced at E. N. E., although the sea when I first hove to set from that quarter, I found it afterwards altered its direction, and came from S. E., for some time before the wind shifted to that point.

"I felt so confident from that circumstance that I should have the hardest of the storm from that quarter, that I continued to lay to on the starboard tack, well knowing that when the wind shifted I should head the sea much better, and consequently the vessel would lie safer.

"In the storm of 12th September last, although the sky looked much more dismal in the S. E. than any other direction, the swell gave no indication of the wind coming from that quarter, as it set constantly from the Northward."

Colonel REID in his new work gives abundant instances to shew that the swell is often felt at 10° or 15° of distance, or even farther, and he thinks that the heaviest swell must always be in the line of the track. We must, however, in considering this question, take the prevailing winds, trades, and monsoons into account, and carefully weigh all the circumstances, before we finally judge of a swell or cross swell.

There is another peculiarity which most seamen when it is mentioned will readily call to mind, and of which I have many examples noted in logs which have been sent to me. It is, to describe it generally, that when, in the log book phrase, the ship has "a cross confused sea making her very laboursome and uneasy," from two or three different points, there is also sometimes seen at intervals a heavy, breaking, roller-like surge of a sea amongst the others, much larger and higher than the rest, and altogether of a different character— seeming at times as if some submarine earthquake or other commotion had given rise to it, or as if an iceberg had capsized not far from the ship, and sent its powerful wave across all the others. The suddenness with which these waves appear, and the peculiar heavy manner in which they strike the bow or quarter or beam of a ship in dull weather and light winds, will also be seen to be well worthy of note. In one instance, related to me by a capital observer, the whole of the passengers of a large passenger-ship from India were sitting at table after dinner, in fine weather, with all the ports open, when the first of a series of these Cyclone-rollers, as they may be termed, reached the vessel so as to strike her on the beam and quarter, and it filled half the ship's cabins below, washing even into the cuddy ports of a ship of 1200 tons! Upon examining the Barometer it was found to have fallen considerably since noon, and in the course of the afternoon and night there were clear indications of a passing Cyclone not very far from them. This is another indication of that BAROMETER OF SIGNS to which I am so desirous that sailors should direct their attention. There is no doubt that these Cyclone-rollers are the *resultant** waves of the whole of the complex forces acting on the surface of the ocean in a distant Cyclone, and that they may afford useful, and above all, timely warning of its existence or approach.

244. THE NOISE OF CYCLONES.—There appears no reason to doubt, and I have myself experienced it in one case at sea and in another on shore, that both at the commencement and at the passage of the centres of violent Cyclones, peculiar noises are heard. At the commencement, the wind sometimes rises and falls with a moaning noise, like that heard in old houses in Europe on winter nights, and this in situations both near and far from the land, and independent of the noise made by the wind in the rigging. Captain RUNDLE,†

---

* See page 109 for the explanation of this word.
† See VIII. Memoir, Journal Asiatic Society, Beng. Vol. XII.

who had not seen my Horn Book of Storms for the Indian and China seas, in which I first described it, at the time he wrote, exactly describes it in the following, almost incidental, remark in his notes: "I do not like this gloomy weather, with wind lulling and then coming on again *with a moaning noise;* there either has been or will be bad weather." In the summary appended to the Memoir in which his log is printed, I have, in addition to the description of the noise given above, said that in England,—

"We attribute it to the noise of the wind in the chimneys, or amongst the trees, or on board a ship to the rigging ; yet here there can be no doubt of its being as distinctly heard at sea as the 'roaring and screaming' of the wind in a tyfoon or hurricane certainly is. My present theory to account for it is this. I suppose the storm to be really formed, and to be 'roaring and screaming' at, say 200 miles distance, and that the noise, if not conveyed directly by the wind, may be so reflectively from the clouds, as in the case of thunder-claps. A noise is known on some parts of the Coasts of England by the name of 'the calling of the sea' as occurring in fine weather, and announcing a storm, and also in mountainous countries. All these may be connected, and seamen may render great service to science and to themselves by noting these curious phenomena."

Captain Ross, in his Notes from the Cocos Islands, says:—

"Being a native of a mountainous country on a coast exposed to the full sweep of the North Atlantic gales, these noises have been familiar to me from childhood. Whenever the weather is disposed to be what we call 'angry,' peculiar sounds are elicited by the wind passing among precipitous cliffs, hills, &c., though the wind may not be nearly so strong at that time as at others when no such sounds are produced. I have consequently been in the practice of listening carefully to the tones of the wind amongst the rigging and spars and over the bulwarks, and very rarely indeed have been mistaken in my anticipations of the coming weather."

The instance in which it occurred with me on shore, very remarkably, was in the Calcutta Cyclone of June, 1842, which forms the subject of my Seventh Memoir.*

On the morning of the day on which the centre passed Calcutta, I have noted:

At daylight, "wind rising and falling."

Between daylight and 10 A.M. "Wind rising and falling very remarkably, at varying intervals of 15, 17, and 5 minutes, with the peculiar moaning noise which accompanies high and variable winds."

245. I have also met with logs in which this noise has been noticed.

* Journal Asiatic Society, Beng. Vol. XI.

Colonel REID mentions (p. 36), that amongst the signs noticed at Barbadoes by Mr. GITTENS in 1831, who appears to have been a careful observer, and well acquainted with the signs of an approaching Cyclone, was, "2ndly. The distant roar of the elements, as of the winds rushing through a hollow vault."

In the log of the ship *Ida*, given by Colonel REID, it is remarked at the onset of a hurricane (Cyclone) as follows:

"Fresh gales and squally weather; at 4, handed the fore-top-sail and fore-sail; *at intervals the wind came in gusts then suddenly dying away, and continued so for four hours.*"

No doubt the moaning noise might have been distinguished here by an observant person, if his attention had been directed to the subject.

This moaning noise has also been noticed in a Cyclone between the Cape and Australia, Lat. 41° S., Long. 34° E., by Capt. LEIGHTON, of the barque *Secret*, whose chief officer was also struck by it. It is also noticed in the log of the Dutch ship *Loopuyt*, Capt. VAN WYCK, in an approaching Cyclone in the Northern Pacific, in which it is said that the wind was "increasing gradually, and producing now and then a plaintive, and the next moment a thundering noise." It has also in one instance, in the ship *John Ritson*, Captain RITSON, in the Southern Indian Ocean, been noticed at the close of a Cyclone, when the weather is described as "decidedly improving, sky breaking out clear, but wind still moaning."

246. PASSAGE OF THE CENTRE. The noise which is heard just at the passage of the centre of some of the most violent Cyclones is still more remarkable. All accounts, and we have many of them, agree in describing it as resembling the deafening roar of the most terrific thunder, though no thunder or lightning can be distinguished at the time, and it is a part of the storm itself, for it is at first not heard in the calm interval, but in general gradually increases as the shift comes up to the becalmed vessel. The following extracts from various sources describe it.

247. In the log of the *Exmouth's* Cyclone. (Track *s* on our Chart No. II.) Mr. THOM gives (p. 95) the following passage, after describing a calm which lasted from 11-30 A.M. to 12-30 P.M., "with a

most awful silence," during which the quicksilver disappeared in the tube of the barometer, says,

"12 P.M. 20. The sun made its appearance for a few minutes, and then disappeared, *followed by an awfully hollow and distant rumbling noise*. In a few minutes we received a most terrific gust from the S. S. E., laying the ship completely on her beam-ends, &c. &c."

Captain BIDEN, in his remarks on the log of the H. C. S. *Princess Charlotte of Wales*, says, when close to, or say *at* the centre of her Cyclone (track *e* on Chart No. II.):

"At noon the barometer had fallen suddenly from 29.25 to 29.09. Stationed every one for pumping and baling, prepared axes and secured guns, ports, &c., in the best possible manner. The gusts from noon till 7 P.M. *were like to successive and violent discharges of artillery or the roaring of wild beasts.*"

248. In Mr. REDFIELD's Memoir on the Cuba hurricane of 1844 (Amer. Jour. of Science, p. 359) is the following passage, perfectly confirming the foregoing:

"Captain CATTERMOLE (barque *Charleston*) states, that in the night of the 6th, his barometer had fallen to 28.10 (an incorrect instrument no doubt), *attended by a continued roar in the air:* soon after the hurricane struck the ship with tremendous force from a point East of South, afterwards veering gradually to S. W., West and W. N. W."

249. I have also alluded to the *screaming* of the wind in a Tyfoon. Colonel REID gives (p. 92) a narrative by Mr. MACQUEEN, master of the ship *Rawlins*, in which he says, "The wind representing numberless voices, elevated to the highest tone of screaming."

Captain SMITH, of the ship *Futtle Oheb*, states that he has on two occasions distinctly heard this screaming of the wind. On the first in the Southern Indian Ocean, near the Cocos Islands, the schooner *James* was twice nearly foundering in heavy gusts during a gale, which was preceded by a noise like thunder, and on the second occasion it was "mingled with a sound resembling the scream of a steamer's whistle." He again recognized this noise in a heavy squall on the West Coast of Sumatra, when his Malay crew said it was a sure sign of mischief, being the "Devil's Voice."

250. The violent flaws and gusts of winds at the centre, noted in many logs, are also worthy of close attention. Mr. REDFIELD, speaking of these, says in his Second Memoir, p. 7,

"It is also possible that the vortex or rotative axis of a violent gale or hurricane oscillates in its course with considerable rapidity, in a moving circuit of moderate extent, near the centre of the hurricane, and such an eccentric movement of the vortex may, for ought we know, be essential to the continued activity or force of the hurricane. Such a movement will fully account for the violent flaws or gusts of wind and the intervening lulls or remissions, which are so often experienced towards the heart of a storm or hurricane when in open sea; but of its existence we have no positive evidence."

In my Twenty-fourth Memoir (Jour. A. S. Bengal, Vol. XXIV. pp. 425-28) will be found also some very interesting notes from correspondence with F. CRANK, Esq., Bengal Salt Agency, relative to the Cyclone of May, 1852, of which the centre passed over his station of Bagundee, 39 miles E. b. N. of Calcutta, destroying his brick-built house, and devastating everything in its progress to the Northward.

Mr. REDFIELD is inclined to suppose, that independent of the two principal motions of a Cyclone, its whirling round and moving forward, there may be a third at the centre, which he calls the Axial Oscillation—or, in other words, a pendulum-like movement of the central part of a storm; such, that it does not move forward on a straight line at any time, but that while the body of the storm is whirling round and moving forward, the centre also either moves from side to side, so that it goes forward on a waving line, or it has a small whirling motion of its own, which if traced would form a series of corkscrew spirals, at a greater or less distance from each other; or, in language which sailors will understand, that the central part is constantly "taking a small round turn in itself" as it moves along. Mr. REDFIELD has, he thinks, shewn some of these deviations of the course of the axis (or lull at the centre), without being able to determine them precisely, and I incline to agree with him as to the probability of their existence. This is another of the many points on which we require observations, and especially observations at sea, for to those on shore I attach very little value for such matters. (See pp. 25, 26.)

In my Nineteenth Memoir (Jour. A. S. Beng. Vol. XIX. p. 349) I have shewn, I think satisfactorily, that H. M. Brig *Jumna*, under the command of Lieut. RODNEY, from Bombay to England, on the 23rd April, at 3 A.M. in Lat 9° 50′ S., Long. 85° 35′ E., was between or near the tracks of two larger Cyclones, travelling down to the

S.S.E., while she herself was on the verge of and within the influence of a much smaller one, which *seems* to have been oscillating between the two first-mentioned ones, causing it to describe a singularly waving line of track, and giving the *Jumna*, which was unfortunately running down on its Western edge with a tremendous gale, some singular veerings of the wind. When the three Cyclones met, the fury of the wind became irresistible, and H.M. Ship was upset, but fortunately righted, by cutting away her mainmast. It is remarkable and most instructive, that the *Jumna's* barometer, which was at 3 A.M. at 29.57, only fell at the time of her upsetting to 29.16! The inference from this is that it was *driven up* between the two or three Cyclones, for of their terrific violence there is no question.

251. Mr. REDFIELD, I may mention, supposes, as at least one probable cause of these oscillations of the axis, the different and varying pressures of the atmosphere within and without the Storm circle; but until we have ships sent out to experiment on Storms,* it is probable that many of these questions will remain doubtful. They are fortunately, however, not questions which affect the safety of the seaman or alter his management, which is always directed to keep clear of the centres altogether; but they may throw some light on the steps by which a Cyclone is developed, and its manner of continuing to support itself as it were from its own fury for a time, and perhaps upon the causes of the phænomenon. In all these lights, then, the sailor who desires to aid us should consider that neither he nor the writers on the science can yet pronounce what is, and what is not, important, and he should continue to register carefully *all* that he can collect.

252. There is an appearance of the clouds sometimes seen in hurricanes at the passage of the centre, which I trust in future will be noted and registered when it occurs. It is the following, described

---

* In some readers this will excite a smile. My opinion is, that if we remain at peace it will be done within the next five years. The subject is not certainly of less importance than the surveys of dangerous channels or coasts, or magnetism, or hydrographical positions, or a N.W. passage; and for the honour of our flag we will not suppose that a great naval nation like England will allow any other to precede her in systematic researches on this great national and scientific question.

in my own register of the Calcutta Cyclone of June, 1842, at the time the centre passed over that city. Track VII. on Chart No. III.

"At 3 P.M., 3rd June, calm, scud from East, but very slow and indistinct; a light air from East with drizzling rain.

"At this time I drove out on the Esplanade. The appearance of the sky was very remarkable. In the zenith the haze was so thick that the direction of the scud could not be determined, but to the East and N. E. it was slowly moving as before to the West and S. W., while in the South, from thick heavy masses of clouds, the scud was rising and flying to the North and N. E."

In 1845, on the 7th and 8th March, a severe Cyclone was experienced at the Mauritius, and in the *Cerneen* of the 11th March, some barometrical observations and other remarks were published in that paper, which concluded with the following:

"Finally, the whirlwind was distinctly shewn when, by the progress of the meteor, we were involved in the actual centre of the movement of this system of circular currents: in fact, we then observed all around us winds in all directions, without feeling the effect of any. The clouds at sea were moving to the North and those of the mountains to the South, *while at the Zenith every thing was perfectly stationary*. I should observe that the directions of the wind have not been exactly taken."

These are both instances in which the opposite movements have been distinctly seen, and they are always of much importance, as shewing how clearly the circular motion has been proved, if indeed that can be any longer questioned by the followers of Mr. ESPY and Professor HARE in America; for I do not recollect that any European authority in science has doubted of its existence.

253. THE SIZES OF THE CENTRAL CALM SPACE of Cyclones, or, more properly, taking them as circles, their diameters appear to vary very much. The sailor, however, must be careful not to confound this, the actual measurement of the lull, with the time it may perchance take in passing over his ship; for it is clear that three Cyclones, having respectively central spaces of 5, 10, and 15 miles, would each take half an hour to pass over an anchorage, if their rates of motion were 10, 20, and 30 miles per hour; and at sea it is difficult, on account of the drift or run, and the very few notes usually taken in times of such anxiety and danger from the rolling and shift of wind, to calculate with any precision what the diameter of the lull is. Of the few shore observations we possess, the following are notes.

Mr. THOM, calculating from the time the calm began at one point and reached another at the Mauritius (p. 171), estimates it to have been about 21 miles for a storm moving at the rate of 5 miles per hour.

My own estimate for the Calcutta Cyclone of 1842 is, that in a vortex moving at the rate of 5.3 miles per hour, the breadth of the lull at the centre was 11 miles.

In my Fifteenth Memoir (Jour. As. Soc. Beng. Vol. XVII. p. 27) investigating the Cyclone, on the Malabar Coast, of April, 1847, in which the H. C. War Steamer *Cleopatra* foundered, I have from pretty good data estimated the diameter of the calm centre at $18\frac{1}{2}$ miles, when that of the more violent part of the Cyclone was from 150 to 180 miles, and its rate of travelling about 9.2 miles per hour.

254. It is possible, and in this view, as will be presently seen, Mr. THOM partly coincides with me, that at least at the commencement of Cyclones, if not after their full formation, the calm space or lull at the centre is so wide, in proportion to the whole extent of the storm, that fully one-third of its breadth is occupied by the calm! I think, upon good grounds, that I have traced something of this kind in one very clear instance, of which the details, which I extract and abridge from my Fourteenth Memoir,* are briefly as follows:

The ship *Caledonia,* Captain BURN, of 1000 tons, from China to Bombay, overtook the Cyclone marked XIV. on Chart No. III., and *sailed into it* from its Eastern side, and this apparently, as shewn by the log of another vessel, on the first day of its formation, sailing along with it and reaching the centre, where she was obliged to lay to. The track and rate of travelling of the storm, as also the position and run of the ship, are well ascertained, so that we can say with tolerable certainty that she found the central space to be about thirty miles in diameter, and the zone of hurricane around it not more than thirty-five miles in breadth! and either on account of this wide space,† or because the hurricane was not fully formed (though it *was* violent enough to oblige them to cut away the sails they had set after the

---

* Journal As. Soc. Beng. Vol. XIV.

† Proportionately speaking. Thirty miles would have probably been a moderate central space for a storm of 500 miles in diameter, and a small one for one of 800 or 1000.

shift), they found at this wide centre only a heavy "swell," and not the *sea* which is usually felt there, and which was felt by other ships as the Cyclone progressed.

255. Mr. THOM's view is (p. 201) speaking of the great hurricanes of the Southern Indian Ocean, that " in the early stages it is probable the calm is very extensive, and embraces several vortices, which gradually merge into one."

256. ELECTRICITY AND ELECTRO-MAGNETISM. ELECTRICITY. We can merely speculate on the part which electricity may play in Cyclones, and in so doing be careful that we do not mistake mere effects which are visible, for *causes* which are hidden.

It is remarkable that, at times, in some of these Cyclones electricity, in its common form of thunder and lightning, seems to be most abundantly developed, while in others, and this by far the greater number of those at sea, it seems not to have gone beyond common lightning, which sailors often omit to notice unless it is very severe. Mr. THOM makes this remark, and further says, that at the Mauritius (where he resided for some time) thunder and lightning are so rare during their hurricanes, that some affirm that it is never present! With respect to thunder, however, we may remark once for all, that only the very loudest would have any chance of being heard during the height of a hurricane Cyclone.

257. In an excessively severe Cyclone, 24th and 25th Nov. 1815, (track q on our Chart) which came in from the middle of the Bay of Bengal, and ravaged the Northern part of Ceylon from Point Pedro to Manar, it is expressly stated in the Colombo Gazette that at Point Pedro, several shocks of an earthquake were felt, but that "there was no thunder or lightning, a circumstance uncommon in this country."

In the Cyclone of June, 1842, of which the centre passed over Calcutta,* and which was of excessive violence there was certainly no lightning, nor any thunder heard, even at night, and mention is rarely made of it in the accounts of the Cyclones of the Bay of Bengal or China seas. On the other hand, in the Barbadoes hurricane of 1831, as quoted by Colonel REID, p. 30, we find that the development of electricity was awful as to its extent and appearance, but there

* Seventh Memoir, Jour. As. Soc. of Beng. Vol. XI.

seems to have been no special mischief occasioned by it; the whole atmosphere of the place appears to have been enveloped in an electric cloud while constant discharges were going on. A remarkable light and a "darkness" is also mentioned in some accounts* of these tempests,† and in the whirlwind of the *Paquebot des Mers du Sud*, detailed in my Fifth Memoir, and already alluded to p. 38, it appears that the onset of her tornado-Cyclone was accompanied with some electric explosion.

258. In many instances the passage of the centre, or the shift of wind which indicates it when there is no intervening calm, appears to be marked by electric discharges, and what is very remarkable, by a *single heavy flash* or two of lightning! This also sometimes occurs towards the close of the storm, and when the wind is a little abated, as if there was a peculiar zone or quadrant of the Cyclone in which the electric action was going on.

In a letter from Captain COMPTON of the *Northumberland*, to his agents in Calcutta, he describes this kind of lightning; and as he does not allude to lightning before, but only to a "lurid strange appearance," which may have been an electric light, I presume there was none of any remarkable intensity. He says—

"Between 3 and 4 A.M., two immense flashes of lightning took place. In half an hour afterwards the wind abated, and the frightful lurid appearance ceased."

Another account printed in the Calcutta newspapers, says that—

"At 2 A.M., a flash of lightning shewed them the loss of the foremast (which no one heard going over the side in the uproar of the elements);" and then, that "at 4 o'clock, *after another flash of lightning*, the wind suddenly stopt."

It would appear that only two notable flashes were seen.

In the log of the *Eliza*, Capt. MCCARTHY, which ship was dismasted, and near foundering in the Pooree and Cuttack Cyclone of October, 1842, which is the subject of my Ninth Memoir, (Journal As. Soc. Beng. Vol. XII.,) no lightning is adverted to until the calm

---

* *Judith and Esther.* REID, p. 76. *A great light* was observed during the Cyclone of December, 1845, at Baticolo, on the East Coast of Ceylon, PID. XIV. Memoir, Jour. As. Soc. Beng. Vol. XIV.

† Dr. PEYSONNEL (*Phil. Trans.* for 1756, p. 628), says that the lightning appears and the thunder is heard *at the close* of the West India hurricanes.

centre reached the vessel, where "much lightning" is noted, and immediately afterwards it is stated, that—

"The wind shifted in a flash of lightning suddenly to the S. S. E. from N. N. E., and blew instantly nearly as violently as before."

The ship *Thomas Grenville*, Captain THORNHILL, experienced a severe Cyclone in Lat. 25° South, Long. 62° East, (Track p on Chart No. II.) No lightning seems to have occurred till just at the approach of the centre, when it is said—

"At 2 P.M. Bar. 28.90 (from 29.80) one very vivid flash of lightning, and a loud clap of thunder now occurred. At 1-15 P.M., it moderated to a strong breeze for ten minutes, when the hurricane came on with (if possible) redoubled violence: at 4 P.M. Bar. 28.70."

In the Mauritius Cyclones of 1786 and 1789, as described by M. PERON, he says, of the first of them, that thunder and lightning were "nearly incessant throughout the whole of this terrible storm," and that a meteor was seen resembling a globe of fire, following the direction of the wind, then from the N. W., and which disappeared behind the mountains of Moka. It was considerably elevated in the atmosphere, and seemed nearly half the size of the moon. In the second of these Cyclones, thunder and lightning are not mentioned, but it was in this that the remarkable phænomena of flashes of light in the vacuum of the Barometer Tube, (see Part V.) were seen. This again would seem to indicate, that there are two kinds of Cyclones. That of Barbadoes also, as described in Col. REID's work, differs in this respect from the usual accounts. In a Chinese work quoted by Dr. MORRISON, which will be subsequently quoted, he notes that "they say if it thunders the gale (Tyfoon) breaks up."

259. In the great West India Hurricane of 1772, already alluded to, it is said, that at St. Croix (Santa Cruz) where its greatest intensity seems to have been felt, and the storm-wave to have been of tremendous height, so that the centre must have passed there, there was a "tenfold darkness made visible only by the meteors which like balls of fire skimmed along the hills." And in Luke Howard's Climate of London, p. 219 of the 3rd edition, the author gives from an anonymous manuscript in his possession the following passage relative to these electric phænomena: the paper is dated from Westend of Santa Cruz (Antigua), and signed M. SMITH.

" I must mention how dreadful every thing looked in this, in itself, horrible and dark night; there being so many fiery meteors in the air which I and others who were in the same situation were spectators of. Towards the East the face of the Heavens presented to our view a number of *fiery rods* (electrical brushes) which were through the whole night shooting and darting in all directions; likewise fiery balls which flew up and down here and there, and burst into a number of small pieces, and flew to and fro like torches of straw, and came very near where we lay in the road. This was over the town. In other parts another sort of fiery balls flew through the air with great rapidity; and notwithstanding all these phænomena common thunder and lightning were abundantly great."

In the Antigua Cyclone of 1848 it is stated, that "at midnight the wind raged furiously, lightning and thunder were incessant, accompanied by floods of rain."

In the Tobago Cyclone of Oct. 1847 (see p. 28) it is said "the lightning was vivid in the extreme and fearful in its brilliancy."

260. In the October Cyclone of the Bay of Bengal forming the subject of my 18th Memoir (Jour. As. Soc. of Bengal, Vol. XVIII.) in which many ships were involved, it appears from careful investigation, that on the advancing portion of the Cyclone, and even on each side of it, the thunder and lightning were, according to the general expression used in reply to my inquiries, "nothing to speak of," but that on its rear portion, and where as I have shewn by diagrams and numerous logs the Cyclone was "lifting up,"* there were heavy electrical discharges. This agrees with the Chinese and Mauritius popular view that when it thunders and lightens the "gale" ceases, and with that current in the West Indies according to Dr. PEYSONNEL as just quoted, p. 214. From a passage in a letter in the Nautical Magazine for April, 1853, it would seem that the distinction is, that *forked* lightning rarely, if ever, occurs in Cyclones. The writer states, that on the coast of British Yucatan, "some Indians, who were in much consternation at the threatening appearance of the sky, jumped for joy on hearing thunder, saying, 'no fear of wind now!'"

261. There is no doubt that during Cyclones immense electric action must be going on in the atmosphere, for we know that every change of state of all bodies in nature, as water into vapour, or gas (steam), or the moisture of the air again condensed into water and

---

* This term will be subsequently explained.

falling as rain, gives rise to it. Hence while prodigious torrents of rain are falling for days together over hundreds of square miles in these great Cyclones, a vast amount of electricity must be generated, but it is possible that at the same time it is carried off, *i.e.* conducted to the earth or sea, by the rain itself, as fast as it is formed. The fires of St. Elmo, as they are called, which remain for hours together at yard-arms and mast-heads of ships in blowing weather, are instances of permanent electric action existing between the sea and the atmosphere through the ship as a conductor.

262. Mr. THOM thinks, and gives data to shew, that lightning is much oftener developed on the equatorial sides of Cyclones than on the opposite sides, but he is on the whole inclined to think it an *effect* rather than a *cause* of rotatory storms.

The following is given by Mr. THOM from the log of the *Fairlie* when at the centre of her Cyclone, in Lat. 12° 2′ South, Long. 103° 41′ East.

"The wind now at S. S. W. and awful lightning flying about in all directions; large balls of fire or meteors were seen at every yard-arm, and mast-head, or boom-end: Appearance most awful. * * * * * *
From the flashes of lightning (no thunder) *a heavy cloud seemed to be down upon us.*"

This ship had, literally, every thing but the bare lower masts *blown* out of her, and narrowly escaped foundering.

263. The fact is that we have hitherto very few and very imperfect data on which to found any opinion, and for any practical purposes, our present knowledge is almost useless. Nevertheless careful seamen and all who are desirous of advancing human knowledge will, I trust, neglect no opportunity of noting all that may fall under their observation, how trifling or indifferent soever the appearance may be supposed as to its relations to the causes of the storm. In Part V., when speaking of water-spouts, we shall perhaps be obliged to go into some details of electric action in them which may suggest many points of inquiry; and in the section treating of the signs of approaching Cyclones, it will be seen that one kind of lightning, at least, is a remarkable precursor of them in some instances.

264. Although the following from KAEMTZ' Meteorology, p. 366, relates rather to what are called *storms* (thunder-storms) on shore,

than to our Cyclones, yet it is worth insertion, as shewing how much we have yet to learn even in these constantly recurring and most familiar of meteorological phænomena.

"ELECTRICITY OF STORMS. Notwithstanding the numerous researches that have been undertaken on this subject, it is still enveloped in great obscurity. Place yourself near an electrometer, and observe it during the whole course of a storm, and you will see how variable its indications are. The lightnings are actually very near without the most delicate instruments giving the least sign of electricity; suddenly the latter increases at the moment of a very powerful flash. Another day the storm arrives with all the signs of a very powerful electric tension, lightning plays in the clouds, the two straws of the electroscope collapse, and it is some time before they open again. At one time, the electric tension will vary for every clap of thunder, at another time it will remain the same for a quarter of an hour, although the lightnings rapidly succeed each other. In one storm the straws separate rapidly; a flash of lightning occurs, and they collapse; during another they fall together, and then diverge rapidly to approach slowly, until a fresh clap of thunder makes them diverge again. The electricity may be for a long time positive, its force also varies; but soon, while the rain, the clouds, the wind, and the lightnings, remain the same, the straws separate sometimes under the influence of positive, at other times under the influence of the fluid of the contrary sign.

"If we compare all that has been written upon storms, we do not hesitate to conclude that they are the most complicated phænomena of meteorology. I suspect that a long time will elapse before we can account for all the circumstances by which they are accompanied. First, a single observer is insufficient to collect all the data; we ought to note the electricity, the direction of the wind, the movements and the form of the clouds, the size of the drops of rain, and the direction in which they fall, the form and place of the lightnings, and the divergence of the straws of the electrometer; each of these phænomena requires all the attention of an observer, who also loses valuable time in writing his remarks. Several additional observers are necessary, who being dispersed over the whole surface where the storm is visible, should each notice all the indications in his station, and compare them with the rest.

"All the capricious indications of the electroscope are due to its being influenced by several strata of superposed clouds, which act and re-act on each other and on the earth, so that the electricities are developed and neutralised alternately. We are accustomed in storms to see the most powerful developments of electric tension, and it is difficult to conceive how lightnings and claps of thunder could occur without there being a very notable electric tension."

The probable solution of these anomalies is that where fire-balls and the like are seen, the electric discharge takes place *from* the earth into the atmosphere, and that where a flash or two more occur at or near the centre, it is occasioned by a "lifting up" of the disk of the Cyclone, like the separation of the plates of the electrophorus, and that this is also the cause of the thunder and lightning so frequent in the passage of the rereward of the Cyclone.

And again, speaking always of thunder-storms, he says, p. 364,—

"In all cases a rapid condensation of vapours is the essential condition for the formation of storms: if electricity is very powerfully developed, there is a storm, if not there are simply passing showers, accompanied by very marked signs of electricity. If we examine all the circumstances that accompany the development of electricity, we must consider the condensation of vapours as the cause of its production, and conclude that it is the storm that produces the electricity, and not the electric tension that produces the storm, as is the general opinion. Violent rains without thunder and lightning are distinguished from storms merely by a lesser development of electricity, whence proceeds the absence of lightning and thunder."

265. ELECTRO-MAGNETISM has been adverted to by Colonel REID as perhaps having some connection with the rotatory character of storms, and their opposite motions in the different hemispheres, and the lines of magnetic intensity with the occurrence of storms, but every thing on this part of the subject is so speculative, that we can do no more than simply allude to it. I may add, that Mr. THOM adduces some arguments and data in opposition to this theory.

266. ARCHED SQUALLS AND TORNADOS, &c. It has been presumed, and with some show of probability, that some Tornados, meaning by this word one kind of those which occur on the coast of Africa, have a relation to Cyclones; that is to say, they are in fact sometimes *supposed* to be truly miniature ones, as to their motions, extent, and duration. Their violence we well know to be excessive. I shall first make some remarks on the Arched squalls.

267. The most remarkable of the Arched squalls, perhaps in the world, as to the regularity of their formation, frequent occurrence, similarity of appearance, and excessive violence, are those of the Straits of Malacca. The North-westers of Bengal during the hot season, and one class of the Tornados of the coast of Africa may come next, and then the *Pamperos* of the Rio de la Plata, which seem, at least sometimes, to be of this kind. Those of the Straits of Malacca most usually occur at night or late in the afternoon, and very rarely if ever in the morning or before 4 P.M. which is another peculiarity. They are most

common when completely within the Strait, or between the latitude of 5° North, and the Straits of Singapore, and are most frequent about the Middle or opposite to Malacca, where the Strait is completely shut in by the Island of Sumatra.* They may be said (that is the heaviest of the wind invariably) to come from some point between N. N. W., and W. N. W., North West being the most usual.

"They rise," says Horsburgh, "with a black cloudy arch, rising rapidly from the horizon to the zenith, and scarcely allowing time to reduce sail." Perhaps they are better described (and I have seen many of them) by saying that a mass of black clouds collects and rapidly rises, forming a vast and magnificent arch, beneath which is always observed, even in the darkest night, a dull, gloomy, phosphoric light, like that transmitted through oiled paper by a candle, which at times becomes stronger, particularly on the approach of the arch to the zenith. Flashes of very pale sheet lightning are often observed crossing this space, which sometimes extends over 10 or 12 points of the horizon; the low grumbling of thunder, the falling of the rain, and even the distant roar of the wind, may, I think, be distinctly heard as the arch rises. They are sometimes, but not always, accompanied by heavy thunder and lightning, in which many ships have been struck, but the danger is from the wind, the first burst of which is always tremendous, and sufficient to dismast or upset the finest frigate, should she venture to meet it under any but storm-sail, and many vessels have been lost by sleepy or rash officers allowing themselves to be caught in them. Towards the end of the squall the wind veers a little, but there is nothing that I ever heard of, to induce the supposition that the blast is other than one blowing in a direct line, but I advert to these squalls, and to kinds subsequently mentioned, in reference to Mr. Hopkins' theory of their being descending winds or gusts; or, as he expresses it, parts of "a vertical wheel," (vortex.)

268. The North-westers of Bengal are at times excessively violent, and sometimes equal those of the Straits of Malacca in their regula-

* The heavy squalls, or short gales, of sometimes 6 or 8 hours' duration, which come from the S. Westward are called Sumatras; they more usually arise like common squalls, and have not the constant tendency to arch, which so strikingly distinguishes the others.

PART IV. § 270.]   *Pamperos of the Rio de la Plata.*   221

rity of arching. I have known them preceded a few minutes before their onset by a whirlwind, which had some appearance of a waterspout in the elongation of the cloud and the effects produced on the water of the Hooghly and on large boats upon it. On land it was strong enough to unroof a strong-built bungalow.*

269. The following from Mr. HOPKINS' work, p. 71, is the extract given by him from the voyage of H. M. S. *Beagle*, describing a " *Pampero*" in the Rio de la Plata.

"On the 30th January, 1829, the *Beagle* was standing in from sea towards the harbour of Maldonado. Before mid-day the breeze was fresh from the N. N. W., but after noon it became moderate, and there was a gloominess and a close sultry feeling, which seemed to presage thunder and rain. During three preceding nights banks of clouds had been noticed near the S. W. horizon, over which there was a frequent reflection of very distant lightning. The barometer had been falling since the 25th slowly but steadily, and on the 30th at noon it was 29·4 inches and the thermometer at 78°. At about 3 o'clock the wind was light, and veering about from the N. W. to N. E. There was a heavy bank of clouds in the S. W., and occasional lightning was visible even in day-light. There were gusts of heated wind. At 4 the breeze freshened up from N. N. W. and obliged us to take in all light sails. Soon after 5 it became so dark towards the S. W., and the lightning increased so much, that we shortened sail to the reefed top-sails and fore-sail ; shortly before 6 the upper clouds in the S. W. quarter assumed a singular hard and rolled or tufted appearance, like great bales of black cotton, and altered their forms so rapidly that I ordered sail to be shortened, and the top-sails to be furled, leaving set only a small new fore-sail. Gusts of hot wind came off the nearest land at intervals of about a minute. The wind changed quickly, and blew so heavily from the S. W., that the fore-sail split to ribbons, and the ship was thrown almost on her beam ends ! The main top-sail was instantly blown out of the men's hands, and the vessel was apparently capsizing, when top masts and jib-boom went close to the caps, and she righted considerably. Two men were lost. The starboard boat was stove by the force of the wind, and the other was washed away, and so loud was the sound of the tempest that I did not hear the masts break, though holding by the mizen rigging. Never before nor since have I witnessed such strength, or I may say weight of wind ; thunder, lightning, hail, and rain came with it, but they were hardly noticed in the presence of such a formidable accompaniment ! After 7 the clouds had almost all passed away ; the wind settled into a S. W. gale, with a clear sky."

270. In PURDY's Atlantic Memoir, Part II., p. 186, Captain PETER HEYWOOD, R. N., who has given excellent sailing directions

* The Indian term for a thatched, but European fashioned house.

for the Rio de la Plata, the results of three years' experience on that station, speaks of the *Pamperos* as S. West gales only, and does not allude to their veering at all.

271. In WEBSTER'S Voyage of the *Chanticleer*, the *Pamperos* of Buenos Ayres are thus described—

"The following indications of a Pampero have frequently fallen under my observation. The weather is getting sultry during a few days with a light breeze from the East or N. E. ending in a calm. A cool light wind then sets in from South or S. E., but confined entirely to the lower strata of the atmosphere, *while the clouds above it are moving in the opposite direction from N. W. to S. E.* The Northern horizon, as night advances, becomes dark, with heavy lowering clouds, accompanied with lightning from the East or N. E. The Southern wind now ceases, and is followed by variable winds from the Northward. Heavy clouds are thus brought over; and lightning, accompanied by thunder, follows in a most terrific manner. The wind veers gradually to the Westward in violent gusts, the lightning becomes more vivid and thunder more awful, a gale of wind follows from the S. W. more violent, but of short duration, and fine weather begins."

We may thus dismiss this class of the minor hurricanes* as certainly belonging to that of the right-lined and not circular winds.

272. TORNADOS OF THE WEST COAST OF AFRICA. Colonel REID, p. 512, says, that from explanations received from naval officers, as well as from some log-books, he should be convinced that the tornados on the West coast of Africa, as well as the *Pamperos* on the coast of South America, and also arched squalls, are phænomena altogether different from the whirlwind; but the evidence has not proved reconcileable, and he gives sundry logs of H. M. S. *Tartar*, from April 3rd to June 2nd, in which various *tornados* are noticed, one of which veered round the compass, while the others seem to have blown from the N.E., S. E. and Eastward, but he does not seem aware that seamen, and especially such as keep but brief logs like the one quoted, often mean to express by the words "came on a heavy tornado from the S. E." only that it *began* at that quarter, and not that it also ended there. Nevertheless, there is no doubt that many of these African tornados are merely squalls.

And it is to this point that I would direct the attention of the intelligent seaman, namely, that there may be really two distinct classes

---

* Using this word to express their violence only.

of these little tempests. The first, merely gusts of wind and rain in one direction, or straight-lined winds; and the second, true circular storms (Cyclones) in miniature.

273. In PURDY's Atlantic Memoir, Part I., pages 69, 70, the following description is given of these phænomena from M. GOLDSBERRY:

"Between Cape Verga and Cape Palmas, and during the months of May, June, July, August, Sept., and October, the countries near the sea are frequently exposed to hurricanes, which the Portuguese have denominated tornados, and which have obtained this name even amongst the negroes. During my stay in the river of Sierra Leone, I witnessed one of these tornados, but it was not one of the most violent. These meteors happen a few weeks before the rainy season, and continue till the month of November. The countries above described are, therefore, exposed to them for nearly six months, and these whirlwinds are more or less frequent, and of different degrees of violence, according to the state of the atmosphere.

"This part of Africa generally experiences ten or twelve of these hurricanes in a year, and it is easier to describe their effects than to discover their cause. They are characterised by circumstances which deserve all the attention of philosophers.

"The sky is clear, a perfect calm has prevailed for several hours, and the weight of the air is oppressive. Suddenly, in the most elevated region of the atmosphere, is perceived a little round and white cloud, the diameter of which does not appear to exceed 5 or 6 feet: this cloud, which seems to be fixed and perfectly motionless, is the indication of a tornado.

"By degrees, and at first very gradually, the air becomes agitated, and acquires a circular motion. The leaves and plants, with which the land is always covered, rise several feet from the soil; they keep incessantly moving and revolving around the same spot.

"The negroes, who pass their lives like children, amuse themselves with this rotatory motion; they follow the turn of the agitated leaves and plants, laugh at their innocent amusement, and announce the approach of the tornado.

"The cloud, which is the indicator of this phænomenon, has now increased in size: it continues to spread, and insensibly descends to the lower region of the atmosphere; at length, it grows thick and obscure, and covers a great part of the visible horizon.

"By this time the whirlwind has increased, the vessels in the bays double their cables, or drop anchor near the shore; the tornado becomes violent and terrible; the cables often break, and the violent agitation of the ship causes them to run foul of each other.

"Many negro huts are swept away, trees blown up by the roots, and, when these whirlwinds exert their full violence, they leave deplorable traces of their progress. These meteors happily last only a quarter of an hour, and terminate by a heavy rain.

"The maritime countries to the Northward, comprised between Cape Blanco and Cape Verga, are not subject to these phænomena; it is only to the South of the latter Cape, and as far as that of Palmas, that they are felt in their full violence; and they always occur at the same periods. Some topographical circumstances, peculiar to this part of Western Africa, are, doubtless, among the number of causes of these whirlwinds."

274. The following is abridged from Mr. HOPKINS' work, p. 74, describing, from LAIRD and OLDFIELD, an African tornado:—

"At the approach of a tornado, a dark mass of clouds collects on the Eastern horizon, accompanied by frequent, loud, but short noises, reminding one of the muttering and growling of some wild animal in a voice of thunder. This mass or bank of clouds gradually covers one side of the horizon, extending to it from the zenith; but generally before this a small and beautifully formed radiant arch, on the verge of the horizon, appears and gradually increases. Long before it reaches the vessel the roaring whistle of the whirlwind is heard, producing nearly as much noise as the peals of thunder that seem to rend the very clouds apart from each other. The course of the squall is distinctly marked by the line of foam it throws up, and I have stood on the taffrail of a vessel, and felt the first rush of the wind while her head sails were becalmed. The sensation of relief from the oppressive heat which the tornado produces afterwards is most cheering and delightful."

The following description of the African Tornado, I take from the Quarterly Journal of Science for 1827, p. 486, where it is copied from Jameson's Journal. I have marked a few passages in Italics.

"*Squalls of wind on the African Shores.* The following description is by Dr. MILNE of MILNEGRADEN from the relations of his father. The approach of the squall is generally foreboded by the appearance of jet black clouds over the land moving in a direction towards the sea, at the same time that a gentle breeze blows towards the shore. In those circumstances the precaution, which my father usually adopted, was to take in immediately all sail so as to leave the ship under bare poles, and send the whole of the crew below deck.

"As the tornado approaches nearer, the rain is observed to be gushing down in torrents and the lightning darting down from the clouds with such profusion as to resemble continued showers of electric matter. *When, however, the squall comes within the distance of half a mile from the ship, these electric appearances altogether cease;* the rain only continues in the same manner, As the tornado is passing over the ship, a loud crackling noise is distinctly heard among the rigging, occasioned by the electric matter streaming down the masts, whose points serve to attract it, and I think that I have been told that when this phænomenon takes place at night, a glimmering of light is observed over every part of the rigging. But when the squall has removed to about half a mile beyond the ship, exactly the same appearances return by which the squall was characterised in coming off the shore, and before reaching the same distance from the ship. The lightning is again seen to be descending in continued sheets, and in such abundance as even to resemble the torrents of rain themselves which accompany the squall. These squalls take place every day during a certain season of the year called the Harmattan season. The jet black clouds begin to appear, moving from the mountains about nine in the morning and reach the sea about two in the afternoon. Another very singular fact attending these tornados is, that, after they have moved out eight or nine leagues to sea, where they become apparently expended, the lightning is seen to rise up from the sea. The violence of the wind during the continuance of the storm is excessive."—*Jameson's Edinburgh Journal of Science*, 1834, p. 367.

In the following part, in the section in which I speak of the formation and breaking up of hurricanes (Cyclones), I shall perhaps advert again to some of the peculiar appearances and effects just described.

275. EARTHQUAKES. There appears no reason to doubt that at the approach of, and during *some* Cyclones, shocks of earthquakes have been felt; but it is impossible to say, as yet, if these have any relation as cause and effect, though earthquakes are supposed now to have more connection with the state of the atmosphere than was formerly allowed. No one indeed can have resided in tropical countries subject to these commotions without having remarked, that some connection certainly exists, by the peculiar state of the weather preceding the shocks.

276. We should bear in mind that at sea, unless the shock was one of a most peculiar kind and extraordinary force, it certainly would not be felt, and on shore also, unless during the calm, it would require to be much stronger than common shocks usually are, to be distinctly felt in the shaking and uproar of a storm. Hence we must allow that slight shocks may occur during Cyclones much oftener than is supposed. Nevertheless Colonel REID's remark that "it is very material to the success of the present investigation, that the phænomena of Cyclones and earthquakes should not be connected without proof," should not be forgotten. The earliest notice I can find of a storm and earthquake occurring together, is the following in the Gentleman's Magazine for 1738-39, which I copy from Dr. MARTIN's Medical Topography of Calcutta: the italics are mine. It is stated that—

"On the night between the 11th and 12th October, 1737, there happened a furious hurricane at the mouth of the Ganges, which reached sixty leagues up the river. *There was at the same time a violent earthquake* which threw down a great many houses along the river-side in Golgoto (i. e. Calcutta) alone, a port belonging to the English. Two hundred houses were thrown down, and the high and magnificent steeple of the English church sunk into the ground without breaking. It is computed that 20,000 ships, barks, sloops, boats, canoes, &c. have been cast away: of nine English ships then in the Ganges, eight were lost, and most of the crews drowned. Barks of sixty tons were blown two leagues up into land over the tops of high trees; of four Dutch ships in the river, three were lost with their men and cargoes; 300,000 souls are said to have perished. The water rose forty feet higher than usual in the Ganges."—*Gentleman's Magazine for* 1738-39.

It would seem, judging by what we know of the storms of the Bay of Bengal, that in all probability this must have been a true Cyclone, and if so it is a very exact case of the hurricane and earthquake occurring together.

277. The next is more modern, being the Ongole storm of October, 1800, Track *a.* on our Chart, No. III. In the Asiatic Annual Register, Vol. III. for 1801, it is stated in a report from Ongole, that—

"On the 19th instant (October), about 10 minutes after 4 A.M., the wind blew a hurricane, *when suddenly we felt a severe shock of an earthquake; which kept the earth in continual agitation for nearly a minute.* It shook down many houses, but I believe no person was killed in consequence."

The writer, evidently himself a partaker in the calamity, goes on to describe the sad distress occasioned by the storm, which was evidently one of tremendous intensity.

In the severe Cyclone of Nov. 1815, which ravaged the Northern part of the Island of Ceylon from Point Pedro to Manar, it is expressly stated in the Colonial Gazette, that at Point Pedro *several* shocks of an earthquake were felt during the hurricane, but no thunder and lightning. This Cyclone had travelled in, at least from the centre of the Bay of Bengal in about the same latitude, having thereabouts dismasted the ship *Cornwallis*.

Since the former editions of this work I am enabled, so far as relates to Bengal, to affirm upon excellent testimony that in *some* Cyclones at least shocks of earthquakes certainly occur. Mr. CRANK, who has been already alluded to (p. 209), is a gentleman who has had much experience of earthquakes in Persia and Assam, and therefore not liable to be mistaken in them. He had, moreover, during the utmost fury of the Cyclone, to remove his family from the house, which could no longer be safely inhabited, to a strong low built outhouse where there could be no vibration, and thus his account is of

PART IV. § 278.] *Earthquakes during Cyclones.* 227

the greatest authenticity. And he states distinctly in reply to my queries, that with the first shock, being then in the open air, he experienced a very unpleasant smell,* which he attributes to some gaseous exhalations, having noticed it on other similar occasions, that the shocks were oscillating ones, and repeated several times during the night. (See Journal of A. S. Bengal, vol. xxiv. pp. 425-428.)

278. In Dr. BLANE's account of the great hurricane at Barbadoes in 1780, he gives several notices to prove that there undoubtedly must have been an earthquake during its continuance. At St. Lucia it is said to have happened "some hours after the greatest severity of the gale." Dr. ARNOLD also, a member of the Wernerian Society, Edin. Phil. Journal, Vol. VII. p. 188, in an article on the climate of Port Antonio, states that during a hurricane on the 18th Oct. 1815, two shocks of an earthquake were felt all over the island of Jamaica, but he mentions it as being "remarkable," from which we may infer it was so considered by the inhabitants.

In the Antigua Hurricane of August, 1848, it is stated (Ann. Register, p. 110, for 1848) that "at midnight, the wind raged furiously; lightning and thunder were incessant, accompanied by floods of rain. *At this time a severe shock of an earthquake was felt, attended by very heavy gusts;* by half-past one the mercury had fallen four-tenths of an inch, and the storm at this time was dreadful. By two A. M. it had abated."

In the Tobago Hurricane of Oct. 1847, a severe earthquake is said to have preceded "the first out-break," which was at about ten at night; and that abundant proofs of it were apparent.—*Globe Newspaper.*

In Col. REID's work he has given the log of H. M. S. Packet *Spey*, Lt. JAMES, for the Antigua Hurricane of August, 1837, and at p. 62, on the remarks at St. Thomas', where the "hurricane appeared to have concentrated all its power, force and fury," we find it also stated, that "in the midst of the hurricane, shocks of earthquakes were felt."

In the Chronological list of the West India Hurricanes, already quoted, it is stated that in the great Martinique Hurricane of 13th

* Possibly Ozone.—H. P.

Q 2

August, 1766, a shock of an earthquake was experienced in the night of the Cyclone.*

279. In the log of the American ship *Unicorn* of Salem, forwarded to me by Mr. REDFIELD, under date of October 30th, 1843, the ship then lying at Manilla, (it is not said if at the Bar or at Cavite,) it is stated that at—

"7 P. M. thick and rainy, and has been so for 24 hours; *felt two heavy shocks of an earthquake;* blowing fresh. The barometer (which was at 29.90 A. M.) was now 29.84, and at 10 P. M. a severe hurricane (Cyclone) commenced." This is an instance of the earthquake preceding the onset of the storm.

280. The following is printed in the *Quarterly Journal of Science* for 1829, p. 436, from the *Bibliotheque Universelle* of March.

"*Coincidence of Storms and Earthquakes with a depression of the Barometer.* Feb. 21st, 1828, the Barometer at Geneva indicated 26 inches $\frac{1}{1}\frac{1}{2}$ of a line (F and equal to 28.69 E). The 19th, 20th, 21st and 23rd of the same month furious tempests raged throughout the South of Europe, and on the 23rd the shock of an earthquake was felt in the North of France, and in the Netherlands, a new example of the coincidence of these three phenomena."

281. It is a curious question, if we assume these Cyclones and earthquakes to be connected, since the storms last for hours, and the earthquakes but for minutes, to investigate the precise points both of place and time of the storm, at which, if for the present we may use the term, the earthquake "explodes." Colonel REID, from careful inquiry on the spot, is inclined to believe that there was not any shock during the Barbadoes hurricane. What has been stated, however, will I trust direct attention to the subject, and we shall no doubt obtain gradually a much more extended and correct view of our subject.

* See Savanna la Mar hurricane and earthquake, at p. 192.

PART V.

1. BAROMETER AND SIMPIESOMETER, CAUSES OF THEIR MOTIONS.
—2. RISE OF BAROMETER BEFORE CYCLONES.—3. OSCILLATIONS
OF THE BAROMETER.—4. BAROMETER AS A MEASURER OF THE
DISTANCE OF THE CENTRE.—5. HEIGHT OF CYCLONES ABOVE
THE SURFACE OF THE OCEAN.—6. BANKS OF CLOUDS.—7. SIGNS
OF APPROACHING CYCLONES.—8. SEASONS AT WHICH HURRI-
CANES OCCUR.—9. WHIRLWINDS AND WATER-SPOUTS.—10. FOR-
MATION AND BREAKING UP OF HURRICANES.

282. BAROMETER AND SIMPIESOMETER. I am aware that if I begin by telling the seaman that he *should* have a good barometer and simpiesometer on board, I may to some, who though fully aware of their value and uses cannot afford these instruments, be like the physicians who order change of air, nourishing food, and wine to poor patients who have not the means of paying for them. I will then address this to owners, underwriters, freighters, and above all to GOVERNMENTS, Home and Colonial, which have so often such large amounts of stores, and so many hundreds if not thousands of valuable lives,* costing them thousands of pounds in mere outlay, afloat. Owners

* In a memorandum submitted to the Government of Bengal in October, 1846, I shewed that within a few months, 50 ships and 10,000 men, coolies and troops, would be afloat from India, and that most of these would have to cross two if not three hurricane tracts on this side of the Cape, of which no one in England, judging by the instructions sent out, appeared to have had any sort of knowledge! In our wars thousands of men, and hundreds of thousands of pounds worth of stores are constantly afloat. While the Memorandum was before Government, the ship *Sophia Frazer*, from Amoy with Chinese emigrants, ran headlong into a furious tyfoon, and at its close it was found that nearly thirty of the unfortunate Chinese had been suffocated for want of air! all hatches being necessarily closed. The ships *Briton* and *Runnimede* (see pp. 56, 112) had a whole regiment on board. The *Golconda* (p. 97) had upwards of 300 Sepoys, and the ships *Collingwood* and *Camperdown*, with 831 persons of H.M. troops on board, ran into an extremely perilous position in the Cyclone of Oct. 1843, in the Bay of Bengal. (See eighteenth Memoir, Journ. A. S. Bengal, XVIII.)

and underwriters do not want to be told that the policy is vitiated, if the ship be not properly found in the usual sea-stores, ground-tackle, and the like, and yet how few of them dream of the enormous amount of mischief, for which some one, eventually, must pay, arising from the want of books, charts and instruments! I trust that those who may read this work will find in it so many additional motives for ships being provided with a barometer and simpiesometer, that they would no more think a ship or themselves, fairly dealt with if sent to sea without them, than if sent without pumps. Pumps indeed are but precautions against leaks *after* they occur; barometers, simpiesometers, and the Law of Storms duly studied, are precautions against that which for the most part gives rise to the leaks, and which may prevent their occurrence; and whether the owner wishes his ship to escape racking and straining, and losses of spars, sails, &c. of a hurricane, the underwriter to escape the payments which so much reduce his profits, or the Captain to acquire the credit of a fortunate if not of a clever master, they will all find that *the commander who is watching his barometer is watching his ship*, and that in the most efficient manner.

283. I find since my first edition a striking commentary on this passage, and on my remark at p. 163 § 186; a commentary not the less forcible that it is afforded by a foreign officer of high reputation and unquestionable talent, who has nobly borne signal testimony to the masterly skill with which the fleets of JERVIS, NELSON and COLLINGWOOD were managed, M. JURIEN DE LA GRAVIERE, in Part 6, of an admirable series of papers entitled LA DERNIÈRE GUERRE MARITIME, in the *Revue des Deux Mondes*, Vol. XVII. Number for 15th January, 1847. After describing NELSON's putting to sea in haste upon the news of VILLENEUVE's having left Toulon, and carrying his fleet at night through the passage between Biscie and the Sardinian coast, so as to pass to the eastward of Sardinia, he says—

"The weather was uncertain and threatening; the wind which had blown fresh in the channel was now rising and falling and variable. NELSON foresaw a gale, and before midnight the fleet was under handy sail with top-gallant masts housed. Studying attentively the minutest precursory indications of every atmospheric disturbance NELSON placed the highest confidence in barometrical observations, and his journal contains notes of the highest interest relative to them, which he entered daily with his own hand. Worthy it is of all remark that the fiery Admiral took more care of his yards and sails in ordinary circumstances than he did of his ship, or his whole fleet in decisive cases. He knew with what sudden violence the Mediterranean hurricanes came on, and expecting to meet the enemy he did not mean to do so with disabled ships."

The French fleet were, as NELSON expected, partly disabled, and a letter from VILLENEUVE to DECRÉS describing their distress, with another from NELSON in which he tells LORD MELVILLE that his fleet had defied these hurricanes for twenty-one months without losing a mast or a yard, are placed side by side by M. DE LA GRAVIERE—and this occurred in January, 1805. A copy of one page of NELSON's Barometrical Notes from a book in his own hand-writing is given by Sir H. NICOLAS in the LETTERS AND DESPATCHES OF LORD NELSON.*

The oldest Barometer Registry, however, which I have found as applied distinctly to the useful purpose of forewarning the Mariner of the approach of a Cyclone, is in the Log of Capt. W. STANLEY CLARKE, then commanding the Hon. E. I. Comp. Ship *True Briton*, on her passage from China, by the Pacific Ocean route, in July, 1797. And so accurately is the gale registered, that the distance of the centre can be very fairly estimated by it. The fleet, which was large, was separated into two divisions, and by the positions of these, the situation of the centre can be very closely calculated, all these Logs being in my possession. (See 17th Memoir, J. A. S. Beng. Vol. XVIII.)

284. CAUSE OF THE MOTIONS OF THE BAROMETER AND SIMPIESOMETER. The atmosphere of the globe in which we live, may be best likened, for sailors, to a thin coating of air all round the earth. It is thin in comparison with the globe of the earth, though supposed to extend to 40 or 50 miles in height, (thickness;) but then this is round a ball of 8,000 miles in diameter, so that (using always round numbers, it is only $\frac{1}{200}$th part of the diameter or $\frac{1}{100}$th part of the semi-diameter of the earth. If we dipped a small globe of eight inches in diameter into a varnish, or a solution of gum Arabic, till a coat of $\frac{1}{25}$ of an inch thick was formed over it, this would represent the atmosphere as to height, but then we must recollect that the air is a highly moveable and elastic fluid, and is constantly changing also its heat and the quantity of moisture it contains. Moreover, it is not, like the varnish, of the same density (consistence) throughout, but

* [See also below, p. 368.]

varying from the surface of the earth, where it is most dense, to the extreme limits of the atmosphere, where it is much lighter than any thing we can express.

The effects of winds or other causes also are to *heap up*, as it were, the air over one part of the earth like a wave, and this *seems* to form a corresponding hollow in other parts. These waves and depressions may be of course straight, or crossing each other, and circular, or of any other form, just as in other fluids.

285. It is evident that if there is more air in one part and less in another, or if the air in one part is lighter or more dense (as from containing more vapour or being heaped up) than in another, a column of it extending from the earth's surface to the limits of the atmosphere in that part, will be heavier or lighter than in other parts; or than it usually is in that spot. Now the barometer* and simpiesometer *measure the weight* of the atmosphere, or the weight of the column of air *over the spot in which they are placed, but only there.*

How they are demonstrated to do this will be found in any treatise on Natural Philosophy. The *average* weight, or *pressure*, as it is usually called, of the atmosphere, at the level of the sea, is a little more or less equal to that degree of weight, or pressure, of it which would support a column of mercury of 30 inches in height; and in the common wheel barometer and simpiesometer, it may be seen that the tube instead of being inserted into a cistern of mercury, is curved up only for a few inches at the end, and that these few inches of mercury or oil, *with* the pressure or weight of the atmosphere, support (or weigh as much as) the 30 and some inches of mercury, or the whole column of it in the long tube on the other side, so that if the air is heavy, or there is more of it, the barometer rises, and when lighter it falls.†

* The name Barometer signifies "*measurer of weight.*" Simpiesometer "*a measurer together with,*" because it measures *with* the barometer. It had better have been called a Proterometer "*a measurer before,*" for it always shews the changes before the barometer.

† The upper closed part of the tube, which in good barometers is a perfect vacuum, is left so, to allow of the free rise and fall of the mercury, which is thus measured. In the simpiesometer this part is filled with hydrogen gas, which acts like a spring against the column of oil, and as its elasticity varies with every change of temperature, the moveable thermometer scale compensates for this.

As I have before observed, heat makes the air lighter, but enables it to absorb more vapour or steam, and this vapour itself affects the weight of the atmosphere; but at present, and for common practical purposes to which I now come, the seaman has little or nothing to do with this. For him, and for our science of storms, the considerations are simply, the rates and causes of the fluctuations as far as we can discover them, and what amount of danger or safety they indicate to us, and how they do so.

286. To Mr. REDFIELD is due the honour of the theory which appears to explain completely the extraordinary fall of the barometer in Cyclones, and moreover the rise which often precedes their onset; and I do but use his mode of illustration in the following explanation of it, though I add a little to it, to make it more distinct for the plain seaman, who, if he will make the little experiments which I now describe, will perfectly understand the fall of the barometer in these tempests; and moreover, will, whenever he pleases, be able to make a miniature Cyclone for his amusement or instruction, or that of others.

a. Take a common, and a plain flat-bottomed beer tumbler, such as are usually carried on board ships, and mark on a sheet of paper a circle of the size of the bottom of it.

b. Make a few marks or figures for ships, one about the centre of this circle, one or two more at different distances from it, and one outside of the circle, all in the same line.

c. Fill the tumbler a little more than half-full of water, and place it over the mark.

d. Stir this water round smartly with a tea-spoon or paper-knife, and then look at the tumbler *sideways;* you will see that the water forms a cone at the top, and that if we call the column (depth) of water two inches at the edges, it is only about one and a half inches at the centre.

e. Now consider your paper as the ocean, the circle as a storm circle of 500 miles in diameter, if you like, and the water to be air representing the various heights, and consequently different *weights* of the atmosphere over different parts of the storm-circle, and upon the ships (and consequently on their barometers) which we have marked in it. It is clear that the ship at the centre has the *least* pressure on her barometer, which is therefore very low, and that those at different distances have more and more, (their barometers higher) while the one outside has it about the mean height, or even a little higher, if we suppose the atmospheric wave rolling off or driven up above the usual height, as will be afterwards explained.

f. The seaman may very instructively vary this experiment by throwing a few grains of powdered biscuit, black lead or coal (which is better) into the tumbler, stirring it well as before, and *looking down* from above at the little ship-marks. He will see the grains passing over the ships, and shewing the course of the wind-arrows in a Cyclone, and if he marks a track and moves the tumbler along it, he will see how a Cyclone successively strikes, passes over, and leaves ships; lowering and raising their barometers accordingly; and how the circular currents of winds veer differently on different sides of the track! He has thus a complete miniature of the Cyclone in his hands, and he may *make one,* for either hemisphere, by stirring the water with or against the hands of a watch.

287. Before I conclude this section, I would impress on seamen the importance of obtaining good barometers and simpiesometers; and these cost but little more than bad ones; and the utility of comparing them with any known good ones at any ports they may arrive at, as for instance with those at public establishments, &c. It is not requisite that this should be done with all the accuracy of comparisons for altitudes with standard instruments, but it is highly useful to know what the error of a barometer is when we come to compare with other ship's barometers, and when we wish to use it in assisting our judgment in forming an estimate of our distance from the centre of a Cyclone.

The following note is from my former work on this subject.

"I have little doubt that the oil in the common simpiesometers is affected by light, and becomes viscid when exposed to it. Messrs. TROUGHTON and SIMS, at my suggestion, have manufactured a 'Tropical Tempest Simpiesometer,' of which the two principal improvements are, a door to keep the light from the oil, except when observing, and a tube of such a length, that it will allow of the great depression (to 26 or 27 inches) which sometimes occurs, with a temperature of 75 or 80° in tropical hurricanes, without any risk of the gas escaping. See Jour. As. Soc. Beng. Vol. XII."

Learning after this note was published, that the house of ADIE and Co. of Edinburgh still existed, (I had been told that Mr. ADIE who

PART V. § 288.] *Rise of Barometer before Cyclones.*   235

first invented the simpiesometer was no more,) I wrote to them, and in their reply they inform me, that they *had* found oil to thicken, and now use an acid; as also that they had made their tubes long enough to act safely at a temperature of 120° and 27 inches, but should now make them long enough for a pressure of 26 inches. The improvement in both sorts of the instruments will be thus accomplished.

288. RISE OF THE BAROMETER BEFORE CYCLONES. Colonel REID, p. 518, explains this as follows: and from what we have just said of atmospheric waves, it may be easily supposed that something of the kind suggested does take place.

"A progressive whirlwind of great extent, might have the effect of arresting the usual atmospheric current, and of heaping it up to a sufficient extent on one side of the storm, so as to affect the barometer, by increasing the atmospheric pressure; whilst on the opposite side of the same whirlwind, the atmospheric pressure, beyond the limit of the storm, might be found to be somewhat less than ordinary."

"The following diagram is intended to render this explanation more easy. The circle is intended to represent an extended storm in high latitudes; and the parallel lines the prevailing westerly atmospheric current."

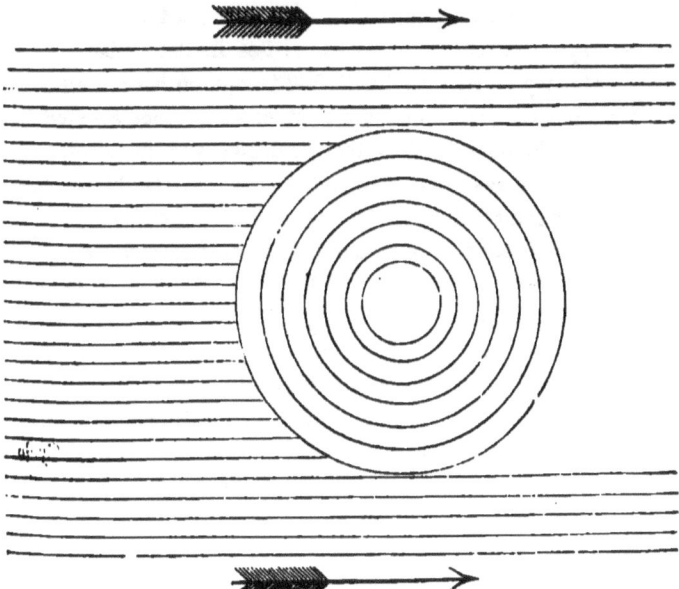

"The same figure may also serve to explain why progressive revolving storms are often preceded by calms; and why a rise in the barometer may sometimes precede the setting in of a storm."

In his new work Colonel REID gives an interesting instance of this and of the precursor swell, being that of the *Medway* Steamer in which he was a passenger, which vessel experienced fine weather on the eastern side of the Atlantic, but a long low swell coming from the N. W. prevailed, being at first westerly, then increasing and coming nearly from the North, on the 1st October, in Lat. 46° N., Long. 14° W. This swell was proceeding from a Cyclone raging on the western side of the Atlantic.

"And two barometers on board the *Medway* rose half an inch above their usual level, the weather at the place of the Steamer being fine, affording an additional proof that the atmospheric pressure is augmented just beyond the limit of whirlwind storms."

289. OSCILLATIONS OF THE BAROMETER AND SIMPIESOMETER. Another peculiarity of the barometer and simpiesometer which undoubtedly often occurs, is their "Oscillation" (rising and falling within a short space of time) before and during a Cyclone. In one case in a Tyfoon-Cyclone, in the China Sea in September, 1840, in which the transport *Golconda*, with 300 troops on board foundered,[*] the simpiesometer was observed in another ship, which avoided it, to oscillate for 24 hours before the Cyclone; and in several other instances also it has been observed, as well as the barometer, to fluctuate very remarkably on the approach of a Cyclone.

290. The vibration of the mercury in the Barometer *during* Cyclones is, I think, first noticed in 1703 by Mr. FRAS. HAUKSBEE, who mentions it (Phil. Trans. Vol. XXIV. p. 1629) as occurring in the Great Storm of that year, (see p. 184) and some experiments were performed at Gresham College in support of a theory to account for it.

291. In a report from the office of the *Captaine du Port* (Master Attendant) of Pondicherry, giving an account of the severe Cyclone of October, 1842, (Eighth Memoir, Jour. As. Soc. of Bengal, Vol. XII.) of which the centre passed over that place, and five or six ships foundered near the coast, it is said that "from 2 to 5 in the after-

---

[*] Fourth Memoir, Jour. As. Soc. Beng. Vol. X.

noon, in the most violent part of the storm, *the oscillations of the mercury in the barometer were so apparent, that it rose and fell instantaneously two or three lines, as though somebody had shaken the barometer.*" The calm took place between 5 and 6 P.M.

The fluctuations of the water barometer at the Royal Society's Rooms are described by Professors DANIELL and BARLOW as resembling, on the approach of and during bad weather, "the breathings of some huge animal." They amounted in one instance to 0.28, while those of the mercurial barometer at the same time were only 0.02.

292. KAEMTZ (*Meteorology*, p. 316) says, speaking of the barometer during tempests,

"These changes (of air) are rarely brought about without agitation: the air moves with velocity, and tempests are the result. The barometer *oscillates* and falls rapidly and rises again in the same manner. These characteristic oscillations are made at short intervals, they are irregular and should be regarded as a consequence of the inequality of pressure that gives rise to the tempest. What we have said concerning winds, fully confirms this opinion; continued tempests (for I am not speaking of those which last merely for a few minutes) *are almost always preceded by great barometric oscillations which as it were announce their approach.*"

293. Colonel REID, p. 447, speaking of the Barbadoes Cyclone of 1839, says—

"During the hardest part of the gale several persons observed remarkable oscillations of the mercury in the tubes of their barometers."

In the log of the *Sophia Reid*, p. 89 of Colonel REID's work, is noted, at the approach of a Cyclone in the Gulf Stream,

"Mercury much agitated, and inclined to fall."

294. Mr. THOM, p. 184, says—

"It has been observed that the mercury in the tube is *in certain parts of the storm* subject to sudden oscillations within the space of a few minutes, or even a continued motion upward and downward is said to have taken place. As this occurs near the margin of the calm, or where the atmosphere begins to ascend in sudden and spiral gusts into the focal current at the point of minimum pressure, it is not at variance with the principle of a regular gradation in the descent and rise of the barometer."*

* We may note here that the up-current, as it is termed by Mr. ESPY, is in this passage assumed as a fact; but no evidence has as yet been produced to shew that it exists; and Mr. REDFIELD's theory accounts equally well for the fall of the barometer. See pp. 18, 20, and 233.

295. In my Third Memoir, (Jour. As. Soc. Beng. Vol. IX.) in the log of the brig *Freak*, Captain SMOULT, which vessel was near foundering in a Cyclone in the Bay of Bengal in April-May, 1840, it is noted as the vessel approached the storm, and when the wind is marked as "strong breeze and threatening weather," "*barometer vibrating very much.*" The vessel was at this time about 240 miles from the centre of the Cyclone, and I have estimated that the verge of the circle may have extended to within about 130 miles of her position at 1 P.M., when this remark was made. The entire fall of Captain SMOULT's Barometer was from 29.30 to 27.25.

In my Seventh Memoir (Jour. As. Soc. Beng. Vol. XI.) in the Calcutta hurricane of June, 1842, Mr. WILLIS' note from Garden Reach, 5 miles W.S.W. of Calcutta, speaking of the simpiesometer, says, and this at the approach of the centre : Simpiesometer, "*breathing, as it were, or fluctuating with the blasts.*"

In the October Cyclone of 1848, in the Bay of Bengal, this vibration of both the Barometer and Simpiesometer, was particularly remarked on board the *Barham*. The Barometer (a first-rate instrument) was affected for about two days, on one occasion it amounted to 0.4, four tenths, in three hours, and at 7 A.M. they shook out reefs, for the Barometer had risen to 29.65, but at noon it was again at 29.20, and then rose again before finally falling.

296. In an extract from the log of the ship *Tigris*, which vessel was caught in one of the small Tornado-like Cyclones of the " Hurricane Tract" (see p. 49 § 58) in the Indian Ocean, being in Lat. 9° S. Long. 83° East, (Track *ee* upon our Chart), Captain ROBINSON remarks,

"Wind blowing in tremendous squalls from W.S.W. to West, the barometer rising and falling the tenth of an inch before and after them."

Captain SHAW, of the ship *Kilblain*, describing a hurricane in the Southern Indian Ocean in 12° S. and 80° East, says, while lying to in the severest part of the gale, " Simpiesometer *dancing* as it were from 29.10 to 29.20, the oscillations being so quick that it is very remarkable ;" and he observes, that it shortly afterwards became more steady, though there was a heavy head sea on, " shewing that it was not the ship's motion that made the glass vibrate."

297. In M. Peron's relation of the voyage of *Le Naturaliste* and *Le Geographe*, speaking of the winds and hurricanes at the Isle of France, he says that the two most remarkable ones of modern years, (he was writing in 1801) were those of December, 1786, and December, 1789, in which the fall of the barometer exceeded an inch (English); and in the last storm it is noted that the oscillations extended through the space of two lines, (0.178 inches English) and that "*flashes of a pale light were evolved from the surface of the mercury that fills the whole vacuum of the tube.*"*

298. There is, then, it will be seen, no sort of doubt of the existence of this oscillation, and we shall, I trust, in future have it carefully observed and noted, and the seaman will readily understand how this may occur, and that it may be, and probably is, occasioned by the passage of successive waves of air, circular or straight, over the place of the instrument, and that these waves may be more or less abrupt, and high, and gradual, according to a hundred modifications of the storm above. For him the practical fact is, the important truth that the barometer and simpiesometer if "uneasy" announce bad weather in the vicinity. The water barometer before alluded to shewed the effects of every successive gust of wind, which was in fact a wave of air passing over the spot.†

* M. Peron was a first-rate naturalist and observer, and his name stands so high that we may suppose he did not set down this very remarkable phænomenon without fully satisfying himself by careful inquiry, that it had been observed and registered at the time by more than one person. In De Luc's "*Recherches sur les Modifications de l'Atmosphére,*" Vol. I. a considerable section is devoted to the consideration of luminous Barometers, which scientific men now agree with him as considering to be an electric phænomenon. But this term is applied to those Barometers in which a pale phosphoric light is produced by suddenly tilting them, so as to fill up the vacuum by the mercury; but it is evident that the light described by Peron was not so produced, unless indeed we suppose the vibrations to be far more |extensive and violent than he describes them.

† If a syphon with a barometrical vacuum was fixed in the side of a diving-bell, the outer branch being left open and the bend filled with mercury, the rise of the mercury in the inner branch would measure, proportionally, the depth to which the diving-bell sunk, and the pressure (*i. e.* weight) of water upon it. If a wave passed over the spot the mercury would rise for a moment and sink again, and so on successively as each wave passed. This is the oscillation of our barometers and sim-

299. Another sign of an approaching fall in the barometer, if not of a Cyclone, which I have noted, is that *it does not rise* at the usual barometric tide hours, but remains stationary. When this occurs, there can be no doubt that some disturbance is going on somewhere, and the barometer and simpiesometer should be carefully watched.

This indication has in many instances since it was first published by me, in 1844, been found of notable advantage as a warning sign, affording thus some hours more for preparation against a Cyclone.

300. *During a storm*, the most careless seaman, I presume, looks often at his barometer, and few if any require to be told that while it continues low, more bad weather must be looked for, *and this even if they have gone through all the phases of one Cyclone*, and their barometer has fallen at its onset, remained very low during its fury and the calm of the centre, and has gradually risen again a *little;* still, low square sail* should not be made, for it is possible that there is another Cyclone close upon the one which has past.

301. This case of successive Cyclones following one upon the other, has occurred in repeated instances in both the East and West Indies, and off the coast of North America; or as in the case of the *Culloden's* Cyclone of 1809, as traced by Colonel REID, the track being a curved one, the ships—in that case a fleet of them—after falling out of the storm circle on one side, ran again into it in a day or two afterwards, when it had curved back and again crossed their route homeward; but their barometers were always low in the interval: see p. 96 to 100, for what has been said on these various kinds of Cyclones. In all such cases it is clear, that while the barometer and simpiesometer have not risen to their average height, and all other appearances have not become favourable, the seaman should still be on his guard. I say here "*all other appearances*," because we may suppose a case in which if two Cyclones approach each other at a wide angle, as was certainly the case in the *Golconda's* Cyclone before quoted, there might be a position in which for a time, and before the borders of the two were in contact, the barometer might be *driven* up, *i. e.* the two atmospheric

piesometers, which measures the depth and waves of air in the *ocean of air* in which we live, as the fish do in theirs of water.

* And, if not absolutely necessary, no square sail should be set.

waves might produce a considerable rise, though in such a case the weather would, as far as we can judge, be still gloomy and threatening enough.* We have as yet no exact account of any vessel's barometer and simpiesometer between two following or parallel Cyclones, though such cases must doubtless frequently have occurred. A good account of this kind, with careful barometrical observations, would be very valuable.

302. The ship *Eliza*, Captain McCarthy, whose able report is given at length in my Ninth Memoir, (Jour. As. Soc. Beng. Vol. XII.) appears† to have been a little after midnight on the 2nd and 3rd Oct. 1842, off the Sand-Heads, not far from the point where two parallel storms were passing to the North and South of her, and one of them was a furious and almost destructive storm to her. Her log states that about this time "barometer fell very suddenly since midnight from 29.30 to 28.30! the simpiesometer (which before is marked at 29.22) to 28.22!"

303. RISE OF THE BAROMETER AND SIMPIESOMETER BEFORE THE STRENGTH OF THE CYCLONE IS OVER. Seamen are well aware of, and anxiously look for this beautiful indication, which is often, and truly to them a rainbow of hope in the depth of their distress. Mr. REDFIELD accounts for it by supposing that, as the lower part of the whirling cylinder (disk?) of the Cyclone, meets with resistance at the surface of the earth, it cannot advance so fast as the upper part, and thus the cylinder is not a truly perpendicular one, but more or less inclined in the direction of the track, just as we often see the cloud of a water-spout move on above, and the spout itself remain seemingly stationary below, so as to become sloping or curved. By means of the little experiment with the glass tumbler, described at p. 233, we can easily understand this, if we suppose the glass a little inclined over,

---

* As will be subsequently shewn, this case of a high Barometer with gloomy threatening weather on the approach of a Cyclone may also occur where the track of the Cyclone crosses the Monsoon or Trade wind.

† I speak with caution here, because though the *Eliza's* place is well ascertained, the position of the centres of the parallel Cyclones is not so exactly determined at this time as to enable me to speak with certainty. Nevertheless, this instance is the nearest example we have, and I will not therefore omit it.

and that the vortex was to remain at the centres above and below; for that part of it above will then have a less column (of water in the glass, and of air in the case of the Cyclone), between it and the paper, than that below, which is really the centre of the supposed Cyclone, as raging at the surface of the ocean, and the augmented column is an addition of *weight*, which consequently causes the rise of the mercury.

304. In the Nautical Magazine for 1841, p. 726, in remarks on the winds and weather off the West Coast of Australia, already quoted, (p. 50) by Commander WICKHAM, of H. M. S. *Beagle*, it is remarked that, in two Cyclones which occurred between St. Paul's and Amsterdam, and Swan River, though of equal strength and duration, and both attended with heavy rain, the barometer in that "from the N. W." (it was N. N. W., veering to the Westward), fell nearly 6-10ths of an inch, while in that, "from the South Eastward" (which appears not to have veered at all, as nothing is said of it), it only fell 0.25, and this appeared very unaccountable.* The solution of the difficulty will appear simple if the Horn Card be moved over the spot which, (approximately only, for I have not been able to obtain the exact position of these Cyclones,) I have marked on Chart II.; for it will at once be seen that, if the card be moved so as to take the N. N. W., and gradually the West wind-point over the spot, this places the ship *before* the track of the Cyclone, while the centre passes close to the South of her; but with the South Eastern gale, she is *behind* it, and thus the fall is far less considerable. I have little doubt that many of the barometric anomalies of gales from different quarters, causing different degrees of fall, will eventually be explained in this way.

305. THE BAROMETER AS A MEASURER OF THE DISTANCE OF SHIPS FROM THE CENTRES. This question is not an easy one; nor do I flatter myself we shall completely resolve it. Nevertheless, its mere examination may throw much light on the mechanism of storms, and any guide, under due caution, is highly desirable to the mariner. I am therefore desirous, not so much of setting my own views forward, as of directing the attention of sailors *and of residents on shore, whose*

---

* The gales are quoted to shew their different effects on the barometer being from "opposite quarters," and such they seemed to be at the time.

observations are the most valuable for the solution of it,* to this infinitely important problem in our science.

This question, briefly stated, is this—"*Assuming the nearly circular figure; the rotatory and progressive motions of Cyclones; and that some of them are, at times, stationary; and that there is a point of veering or a calm space, which we may call their centre—and all this is tolerably certain; can the fall of the barometer be made an approximate measure† of the distance of that centre from us? or if its rate of approach to a ship or a station on shore? and can it be used as such, both within and without the tropics?*

306. I have suggested this question since 1845, (in the first Edition of the Horn Book of Storms for the Indian and China Sea, p. 11), and it had indeed engaged my attention since 1842, when I exhibited to the Asiatic Society of Bengal, at one of its meetings, a large Barometrical Chart of the curve formed in the storm of June of that year, which is the subject of my VII. Memoir in its Journal. Mr. THOM, whose work was published in 1845, had also taken up this question without having seen my work; and he has given a plate of the range of the barometer during hurricanes, and a diagram on his chart of the approximate distances from the centre, by barometer, for the Cyclones of the Southern Indian Ocean. The fall in a given *time* he has not adverted to in the text, considering, probably, that time, and distance from the centre, must vary according to the rate of the Cyclone's travelling and distance. In the following remarks, I shall endeavour always to distinguish his views from my own, should they meet. I think the difference between our theories is, that he inclines to suppose the absolute (entire) fall as the best guide, and I prefer the *rate* of fall, though I would not neglect the entire amount, particularly where sudden.

307. And first, it is evident that there are very great differences in the fall of the barometer in various storms,‡ though they may all

---

* On account of their observing in fixed positions, and much more at leisure than sailors can be at sea; at least until we have vessels sent out for the purpose of investigating storms.

† For the plain seaman: "*approximate* measure," is a measure *approaching to* and usually not far from a correct one.

‡ In one case (Sixth Memoir, "Storms of the China Sea," Jour. As. Soc. Bang.

be true Cyclones, and this in the same seas; so that the absolute (total) fall of the barometer, say of one, two, or even more tenths may greatly mislead us, if we trust to that alone; for we may be supposing that "as the barometer has only fallen three-quarters, or one-tenth, that is not so very much (from a height say of 29.80)" taking here no account of *the time in which this fall has occurred;* whereas with a fall of one or two-tenths in a short time, we may be really close upon the dangerous centre, or a hurricane may be rapidly coming close down upon us. The seaman will clearly understand this by the consideration of the plate, and what will be said of it.

308. Let us first consider our data, and the amount of their certainty or uncertainty, with reference to our problem. I allow that the barometer is one which, if not correctly 'set,'* as it is phrased by sailors, is tolerably sensitive and acting well. The questions then are—

I. AS TO THE CYCLONES THEMSELVES.

First. Cyclones being in themselves of different intensities (strengths), *i. e.* causing the barometer to fall more or less at their centres, as seen in the plate; one Cyclone causing a total fall of one inch, and another of two and a half, or even more.

Secondly. Cyclones being of the same or of different *extent*, as one being 100 and another 400 miles in diameter, but both occasioning the same total amount of fall at the centre—say of one inch.

Vol. XI.) there is a remarkable instance of a furious Tyfoon-Cyclone, in which the barometer of the ship *Ariel* gave no warning, being at 30.10, falling to 29.80, and rising to 30.25, with Tyfoon blowing as hard as ever, and at noon it was 30.90. I ascertained from the then chief officer of the ship, who subsequently commanded her, that the barometer was a good one, and had always acted well. I traced, however, two storms (Track XXV. on Chart No. IV.), one coming from the N. E., and the other from E. S. E., and a heavy Monsoon from N. E. blowing in from the Pacific at the same time, and it is to this that I attribute this remarkable instance. In the *Buccleugh's* Cyclone also (Track *n* on Chart II.) the barometer fell only 0.24, but the *simpiesometer* 0.82. The *Ariel* had no simpiesometer. Mr. THOM (p. 177), who had not seen my Memoir, thought that "there is no well authenticated case of this kind on record," but I consider the *Ariel's* as one, and it is highly instructive.

* *Set, i. e.* the Zero point, or level of the mercury in the cistern or bag, too high or too low.

PART V. § 310.]   *Conditions of the Problem.*   245

Thirdly. Cyclones of the same extent and intensity, but travelling with greater or less rapidity.

II. AS TO OUR OBSERVATIONS.

First. The previous height of the barometer.

Secondly. The *time* it takes for a given fall.

Thirdly. The change of the observer's position if at sea, and especially if running or scudding.

We have thus in the storm itself to consider—1, Intensity; 2, Size; and 3, Rate of travelling.

And in our observations—1, Amount of the fall; 2, Time in which the fall occurs; and 3, Change of position.

It is clear that all these may replace each other, and that therefore they must all be taken into account. On shore, as there is no change of position, that element does not interfere with our estimate.

309. Let us now consider the first of these conditions, the different intensities of the Cyclones themselves—of course as shewn by the greater total depressions (fall) of the barometer, which we suppose have always occurred where the wind has been of the greatest strength. In truth, we have no sort of *measurement* of the wind's strength in tropical hurricanes, and only judge that the most violent wind has occurred with the greatest fall of the barometer, from its extraordinary and often incredible effects. It would appear, moreover, that where these great falls occur, they usually take place in remarkably short intervals of time.

310. In this question of the intensity of Cyclones, as indicated by the fall of the barometer in a given time, it is clear that we must take as standards only such observations as have been made on shore, and again of these only such as have been made when the centre has passed over the spot are really efficient ones to our purpose; because in all others, the *distance* at which the centre passed from the place of observation is in some degree uncertain, and thus such cases can only be used to test the accuracy of any rule which we may deduce. The observations made by ships at sea, however valuable in themselves and for their own purposes, are also out of the question for this part of our research, except also as testing the value of our rules or estimates; for even where the centre has reached the ship, all the obser-

vations must, it is clear, be heights of the barometer in *different places* in the area of the storm.

311. The instances in which the barometer has been accurately registered in places where the centre of the Cyclone has passed over the spot—and as just shewn, it is only these which will give us trustworthy data—are very few. They are laid down in the annexed Barometrical Chart, and I shall presently describe the principle upon which it is constructed.

The Plate thus comprises,—   Authority.

I.	Madras Hurricane	. .	1836,	Observatory records.
II.	Mauritius	. . .	1836,	Cols. LLOYD, LEWIS, REID.
III.	Calcutta	. . .	1842,	H. PIDDINGTON, 7th Memoir, Jour. As. Soc. Beng. Vol. XI.
IV.	St. Thomas', West Indies,		1837,	Professor DOVE, in Part X. of Scientific Memoirs.
V.	Duke of York, Kedgeree, mouth of the Hooghly,		1833,	Mr. JAMES PRINSEP, Jour. As. Soc. Beng. Vol. IV., REID, p. 291.
VI.	Havannah	. . .	1846,	Bermuda Royal Gazette.
VII.	Madras .	. . .	1841,	Observatory Report, 5th Memoir, Jour. As. Soc. Beng. Vol. XI.

312. The principle on which the diagram is constructed is, to take, without reference to the hour of the day or night at which it occurs, the lowest point of the barometrical depression in a Cyclone, as the CENTRE or axis of that Storm. This is placed on the double, or axis, lines in the middle, and the fall and rise, and the time in which these occur, are shewn on a scale of hours below, and of inches to the right and left. We have thus the Cyclones brought together and *placed upon each other*, as it were, for comparison, under exactly equal conditions *as to time*, and as to the fall of the mercury in that time. The scale of miles above will be explained.

313. And we are immediately struck with the fact that there are evidently two distinct classes of Cyclones, in one of which the fall and rise are more or less gradual, forming an easy curve, while in the

others it forms not so much a curve, but almost an angle,* or rather the figures called by opticians caustic curves, and in these last Cyclones the fall has been excessive and the fury of the tempest far beyond the average of such visitations. We may thus divide the storms into a first and second class; the first class being those of the greatest (excessive) and sudden falls near the centre.

There is also evidently another peculiarity, *i. e.* that all the rapid part of the fall seems to begin at from three to six hours before the passage of the centre, and that before that time, the fall even of the violent Cyclones is comparatively gradual, and in fact approaches closely to the second class.

314. The following table explains itself by its title. It is calculated to the nearest third decimal from the authorities given, and arranged like the plate, of which it is in fact the measurement of the different hours.

As an hundredth of an inch may be worth notice in these calculations and estimates, and as this is easily missed in the measurement of the drawing, the copying of the plate, or the shrinking of the paper after the impression, I have extracted and calculated this table from the registers of the storms, calculating to the nearest decimal where the distance in time from the centre does not fall exactly at the hour of observation; and where the calm centre has lasted for any marked epoch of time, as we see it notably has in the plate, taking the middle of that interval of time as the central instant. The heads 2*h.*, 3*h.*, 4*h.*, 6*h.* and 12*h.* signifying the *total* fall in that space of time *before* the passage of the centre; and the column of rate, that of the rate of fall per hour, *in the interval given at the heading only*, and not the total mean fall.

* The two kinds of curves are compared by Professor Dove, of whose Memoir I have only access to brief abstracts, to deep ravines with precipitous sides, and to extensive valleys with gentle declivities.

Table of the fall of the Barometer, and of the Rate of fall per hour at various distances (in time), from 12 hours before, up to the passage of the Centre; to accompany the Barometrical Chart.

No. of Plate	Storm and Authority	Mean falls between 12 h. and 6 h. from Centre.	Rate of fall per hour.	Mean falls at between 9 h. and 6 h. from Centre.	Rate of fall per hour.	Mean falls between 6 h. and 3 h. from Centre.	Mean rate of fall per hour.	Mean falls between 4 h. and 2 h. from Centre.	Mean rate of fall per hour.	Mean falls between 3 h. and Centre.	Mean rate of fall per hour.	Mean falls between 2 h. and Centre.	Mean rate of fall per hour.	Remarks.
		In.	In.	In.	In.	In.	In.	In.	In.	In.	In.	In.	In.	
I.	Madras, Oct. 1836. Observatory Records	0.157	0.026	0.319	0.106	0.454	0.151	0.422	0.211	0.715	0.238	0.473	0.236	
II.	Mauritius, March 1836. Col. Reid, p. 175.	0.500	0.083	0.345	0.115	0.445	0.148	0.225	0.112	0.100	0.033	0.015	0.007	Lowest depression before the calm.
III.	Calcutta, June 1842. H.P., J. A. S., Vol. XI.	0.120	0.020	0.048	0.016	0.434	0.145	0.392	0.196	0.286	0.095	0.083	0.041	
IV.	Madras, May, 1841. Observatory Records	....	....	....	....	0.135	0.044	0.098	0.049	0.301	0.100	0.234	0.117	No registry, 12 h. to 6 h.
	Means of Nos. I. to IV.	0.259	0.043	0.237	0.079	0.467	0.147	0.284	0.142	0.350	0.116	0.201	0.100	
V.	St. Thomas and Porto Rico, August, 1837. Professor Dove, Part X. Scientific Memoirs	....	....	....	....	0.265	0.088	0.467	0.233	1.481	0.494	1.237	0.618	No registry, 12 h. to 6 h.
VI.	Duke of York's Hurricane, Redgeree, 1833. Mr. Prinsep, as quoted by Col. Reid and J.A.S.	....	....	....	....	....	....	....	....	3.600	1.200	2.370	1.180	No other registry.
VII.	Havannah Storm of Oct., 1846. Bermuda Royal Gazette, Oct. 20th, 1846	0.440	0.077	0.190	0.063	0.310	0.103	0.580	0.290	1.190	0.397	0.980	0.490	
	Means of Nos. V. to VII.	0.440	0.077	0.190	0.063	0.287	0.095	0.523	0.261	2.090	0.697	1.529	0.763	

315. If we look at the mean fall for the *two* hours before the centre, (or from || to 2h. towards the left hand, in the Plate) in the Cyclones of the first class, V., VI. and VII., we shall find it to be as much as 1.529 for the whole two hours, or 0.763 per hour, while those of the second class give but 0.201 for these two hours, or 0.100 per hour!

Again, at 3 hours from the centre we have,

	Total fall.	Mean rate per hour.
First Class,	2.090	0.697
Second Class,	0.350	0.116

And at from two to four hours from the centre it is,

	Total fall.	Mean rate per hour.
First Class,	0.523	0.261
Second Class,	0.284	0.142

It is evident then that, taking into account also the differences of rates of travelling,* which might occur in Cyclones of exactly the same intensity, these two classes are, at least towards their centre, so different, that no common rule can apply to them.

316. The examples of the plate and tables are, as I have already noticed, necessarily from registries of Cyclones on shore, but there is also no doubt in my mind that this distinction, of the sudden and excessive falls near the centres, which constitutes our two classes, exists also at sea, where in those Cyclones from which ships have but barely escaped from the centre, we usually find this kind of fall. As instances of this, I may quote the following:

The H. C. War Steamer *Pluto*, in the China Sea—nearly foundered (see pp. 57, 106, 157)—Bar. fell in all from 29.90 to 27.55 or 2.35 ins. and in the last three hours about 1.25 of this fall occurred while steaming and drifting into the centre.

The transport *Briton* in the Cyclone in the Andaman Sea, in which with the *Runnimede*, she was lost on the Andaman Islands, (see pp.

---

* We do not know what the rates of travelling of any of these Cyclones were, except No. III. It is possible that the more rapid falls of some may have been occasioned by their quick approach, but the two classes seem essentially different in so many respects that I treat them as such, at least up to four hours, counting from the centre.

56,110) appears to have had a fall of about 1·0 inch or more of her simpiesometer within the three hours of her being at the centre, and of about 2.30 ins. in all; and in this storm the force of the wind was so terrific that it was tearing up the front of the poop in both ships!

The ship *London* in the Bay of Bengal, quoted by Colonel REID, p. 291, from the Journal of the Asiatic Society of Bengal, had a fall from 29.70 to 27.80 or 1.90, of which one inch took place within the last four hours.

More instances may be adduced, but these will be quite sufficient to shew that the peculiarity of an excessive fall in about three hours, (of time or distance,) preceding the passage of the centre, takes place at sea, as on shore, with the Cyclones of one peculiar class.

317. But we may also say that, generally speaking, the seaman, fortunately, does not require, or very rarely so, to know what we are now discussing; for when once so far involved in the Cyclone as to be within four hours (of time) of its centre reaching him; or at from 50 to 60 miles, allowing it an average rate, he has probably but little choice of manœuvreing left him.

What he wants is, a rule to serve him as some sort of a guide at the approach and beginning of the storm, when his plan of management, as explained at p. 126, may depend on his estimate of the distance of the centre from him. We shall now see that, at greater distances, the fall for both classes of Cyclones is so nearly the same, that we may, *so far as our present knowledge extends*, deduce some useful hints at least from the barometer in this respect; if we cannot fairly and safely trust entirely to it or call it *a measurer* of the distance from the centre.

318. For it is evident both from the plate and table, that, at from $0h.$ to $3h.$ from the centre, we have a mean fall,

    For the First class of, ... 0.095 per hour.
    For the Second class of, . . . 0.147

    The mean of these two being . . 0.121

and with the exception of the Madras storm of 1841, No. IV., this mean is near enough to each separate instance, to guide us, and to admit of its adoption for our purpose.

Part V. § 320.]   *Fall in time converted to distance.*   251

At from 6h. to 9h. distance from the centre, the means are—
First Class, ... 0.063   One instance only.
Second Class, ... 0.079

Mean of this, ... 0.071

At from 6h. to 12h. distance from the centre, the means are—
First Class, ... 0.077 per hour.
Second Class, ... 0.043   „

And the mean of these is, . . 0.060

But as we have only one registry for the storms of the first class, we shall perhaps be safe in taking 0.050 as a mean fall for this interval.

319. So far then as to the mean fall *in time;* but the sailor requires this in distance for his purposes, and we have seen (p. 245) that time and distance can replace each other according to the rate of the Cyclone's travelling. We can only then affix an arbitrary distance,* and then try how far our rules would have shewn correct results in actual cases, where we have the rate of fall for any fair average of hours, and can calculate nearly enough for all practical purposes the actual distance of the centre from the ship. The seaman will clearly understand, I hope, that what I mean to express is, that while it may be at least worth his masts to him to know if he is at 200, 150, or 100 miles distance from the centre, it is of little or no use, generally, for him to know if he is at 50 or 80 miles from it; for he has probably room, time, and weather to manœuvre in the first case, and in the latter has little or none of these, and must therefore confine his care to getting out of the way of the centre, or allowing it to pass with the fewest chances of damage to his ship. When the centre is within 50 miles (or say three hours) of him, the Cyclone is pretty nearly master. What human skill can accomplish must be done before that time.

320. The scale then which, after much consideration, I have found the nearest to the probable truth, is that marked in miles on the upper part of the plate, though this is to be considered as by no means strictly a limit; for as regards limits, I repeat that it may possibly be found in

* I do not do this by mere guess, but after much consideration, and all the calculation which the imperfect elements afford of *about* what the distances have really been at the times, as will be subsequently shewn in a table.

the end that there *are* no strict ones at all; and that even the various extremes may be wider apart than the following table will indicate—

An average fall of the Barometer per hour of		Shews the distance of the Centre from the ship to be in miles		Remarks.
FROM	TO	FROM	TO	
0.020	0.060	250	150	
0.060	0.080	150	100	The third decimal of the Barometer heights is replaced by a cipher.*
0.080	0.120	100	80	
0.120	0.150	80	50	

I have not set down anything for the centre division of our table, *i. e.* from the centre to 3*h*. before its passage, for it will be seen that the rate of fall per hour doubles after the Cyclone has fairly begun and lasted six hours; and that then (from 3*h* to || or from nine hours after the commencement up to the centre), it may either continue to fall at the same rate of *about* 0.1 per hour or a little more, or that its rate of fall per hour may be, if it should be a Cyclone of the first class, as 100 to 400, when compared with that of the former three hours; or in other words, that it will now begin to fall *four* times as fast, or 0.40 per hour!† We have plenty of instances of this, and even of a fall of more than 0.5 or 0.75 (half or three-quarters of an inch) in the hour! I doubt not that this peculiarity will fully account to the seaman for, and I hope put him well on his guard against, cases of sudden falls; which if they occur at the beginning of Cyclones, as they sometimes do, are warning enough of course; but which may also advise him of his too near approach to danger of such imminence, that we may at least say that no ship can hope to escape from it with her masts

---

* For we must always bear in mind errors of observers and instruments, and except perhaps in Observatory registers, allow something for weather in which doors, windows and roofs are shaking hard to fly out of their frames. I trust also that both the seaman and the scientific reader will make large allowances for the exceeding scantiness of our materials. If we had *seventy* sets of averages in lieu of only seven, we might see our way clearly at once; and indeed, I have at times felt doubtful if it was not yet too soon to publish this investigation, but have been led to do so, by the hope that the sailor may be doubly served by it; for it may perhaps assist him, and it will no doubt obtain for us more materials—and perhaps better workmen with them than myself!

† The plain seaman must be very careful in reading all this to pay attention to the values of the decimals, and not confound the tenths with the hundreds.

PART V. § 321.] *Application of the rule; Precautions.* 253

standing; and he should in such cases have the axes upon deck: a precaution too often neglected by young commanders and officers, who are apt to suppose that precaution indicates fear—and they *are* sometimes afraid of being *thought afraid* of the storm.

It will be remembered also, that it is quite impossible by any previous rates of fall to estimate, when so near the centre, which of the *classes* of storms we have to deal with; and I repeat that what we have to do with our ship must all be done before this time.

321. But before applying this rule we must recollect that there are many circumstances to which attention is to be paid, and the best precepts I can at present give for the application of it are the following. If I make these minute, and encumber them as some may think with "considerations," it is because I will not set down any thing as positive which is not so, or on which the smallest doubt may remain; and because I should much regret that any one should be led to suppose, for want of explanation, that he had an infallible *Law*, when he has only an empirical rule.*

1. The barometer, at the approach of bad weather, should be observed carefully every hour, and especially at night. If every *half*-hour so much the better: and it should be entered immediately on the log.

2. At the end of every two or three hours, *the rate of fall per hour* should be carefully averaged to the nearest decimal.

3. It should then be considered, if the barometer has *not risen* at the usual times of its tides,† or if the interval of the falling tide, or any period of it has formed a part of the time elapsed, or if it should in this interval have risen. In both cases an allowance should be made for this at the onset of a storm; for or against, as may be. Thus:

From 11 P.M. to 2 A.M., are 3 hours.
Total fall of barometer, ... 0.18

Which is per hour, ... 0.06
But as this is within the falling barometer tide-time, and the whole fall is not yet considerable, we may call the fall, about, ... 0.05

---

* A rule founded partly on facts and partly on assumptions.

† I need not explain to any sailor, I hope, the *tides* of the barometer, and the importance of noting if anything affects these.

and this sort of allowances is exactly what the sailor makes when working up his log in a tide's way or for a heavy sea.

4. As the ship's run and the probable bearing of the centre are known, and as, if it is one, the average track of the Cyclone may also be known in many places, it should be considered if the ship has been approaching it or increasing her distance from it, as this would also affect the fall.

5. The application of our rule depends not only on the fall, for the last hour for instance, but on the *rate* of falling. Nevertheless, the seaman will easily understand, after he has made his best estimate, that if latterly the barometer has fallen at an increasing rate, he may be farther on (to the right hand or towards the || of the centres) upon the curve than he supposes ;* and the careful man will always keep on the safe side, and make that sort of allowance which he does for excessive tides or currents in the neighbourhood of dangerous passages.

6. As will be seen in the Notes to the Table, the vicinity of land certainly affects the indications of the barometer, whether the storm be supposed to be advancing toward, or coming from, or passing by it; and this should always be borne in mind.

7. The trades and heavy monsoons also appear to affect the state of the barometer, at least on that side of the storm circle on which they blow, whether with or against its track; but we require more data before we can say how and under what circumstances this occurs.

322. I proceed to shew in the following table, by actual examples carefully calculated, what the results of the foregoing rule would have given, if it had been used in the storms and on board the ships they refer to.†

The columns of this table require some explanation.

---

* A common remark in Log Books we know is: "The barometer falling very fast towards Noon"—or towards Midnight, &c.

† These examples are not selected partially, but I have taken every one I could meet with which affords the elements necessary, viz., good barometrical observations, with the place of the observer and that of the centres of the storms for two days, as accurately determined as could reasonably be expected. But it is rare to find all these present, and the want of any one of them debars us from using the instance as a fair test.

PART V. § 323.]   *Explanation of the Table of Examples.*   255

I., II., III., and IV. explain themselves, as being Number, Date of Storm cited, Ship or station, and Authority from which the data are taken.

V. Is the Average *rate* of fall of the barometer per hour.

VI. Shews the Number of hours for which the average has been taken.

VII. Shews the Distance which would be given by the rule *at the beginning and end of the time for which the average is taken :* Thus, if we refer to the table at p. 248, we shall see that a fall of 0.1 (one-tenth) per hour, gives a distance of 135 miles to 45 miles; that is, that it would indicate at the first hour 135 miles of distance, in two hours 105, and in three 75, and so on, reading the table backwards as it were, or diminishing the estimate of the distance as the fall has lasted a longer or shorter time, so as to obtain an approximation to the actual distance at the time of making the last observation and the calculation for it.

VIII. Mean distance by rule: is the distance of the centre of the storm at the *mean* time between the observations quoted. Thus, suppose observations from Noon to 4 P.M. give an *average* fall of 0.080, this will give a distance of 125 miles *at* 2 *p.m.*

This column is necessary, because the fall, for instance, from 3 to 4 may have been more rapid, or was not observed. If it was observed, this would shew that the centre and ship were approaching rapidly, and the next hour's average *with* this one, or that from 3 to 5, will demonstrate this. If it was *not* observed, it is clear that it *may* have been more rapid (since, for example, the weather looks more threatening and the squalls are heavier); and the careful seaman will then suspect directly, that though at 2 P.M. he *was* at 125 miles distance, or thereabouts from the centre, yet it may be coming fast upon him, and the next hour, as before, will shew a much greater fall, and consequently a greatly reduced distance: in a word, the whole time of the approach of a Cyclone should be a contest between the vigilance of the seaman and the vagaries of the winds and waves.

323. We should also in fairness remark that the want of regular horary observations in the instances at sea, place us in an unfair position for testing the accuracy of our rule; for we must at present

take intervals of from 2 to 24 hours, and assume the mean as the time of the mean fall, which for 2 hours is well enough; but if we come to 4 or 6 hours, it may much mislead us. Thus we may say that—

From 8 to 9	the fall is	0.01 or 0.02
,, 9 to 10	,,	0.02 or 0.03
,, 10 to 11	,,	0.03 or 0.04
,, 11 to 12	,,	0.04 or 0.05

or something of this sort.

Now, as above said, if we have only observed at 8 and 12, we shall say that the mean fall was 0.025 or 0.038 at 10 A.M.; but if we look at that between 10 and 12, we shall call it 0.035 or 0.045; and if we went on thus to 2 or 4 P.M., we should still be undervaluing the rate of falling, and consequently under-estimating our distances. Hence we may, especially where the true interval is large, suppose that the rule might give a nearer approximation if we had had more observations.

324. In considering this table also, the seaman will upon reflection find it *much nearer to correctness in its result than it appears to be!* For he must recollect that what he usually does is, probably, to look at his barometer, and take the average fall for say 4 hours, and then, as I have done,* estimate that at 2 hours (the mean time) it was at a mean distance.

But he should bear in mind, that either because the time of the usual rise has intervened in the latter part of his observations, or from any other cause, the fall has been *less* than it probably would be; or that the ship in scudding one way and the storm the other to meet her, have approached each other with excessive rapidity, and that he will in the next—as he may have had in the last hour or two—have a much greater fall, though this has not been yet observed; so that really the highest distance given by the limits was nearest to the true distance of the centre in the early part of the averaged hours. Thus if from 12 to 4 A.M. we say that the fall has been 0.050 per hour, this gives between 250 and 150, or an average of 200 miles at 2 A.M.; but possibly, had the barometer been observed, it would have given

* For want of hourly observations.

[Part V. § 326.]   *Explanation of the Table of Examples.*   257

this average also between 12 and 1, or say *at* 1 A.M. and that from 3 to 4 the fall was such as to place the limits of distance between 100 and 150.

Hence, as above said, our table *may* be more correct than it at first sight appears, or at all events, with careful observations the rule may serve the mariner for all that he requires better than it promises to do.

325. I should remark finally on the following table, in reference to the care taken to construct it with fairness, that it has been difficult to use the rule in some cases in which readers who are acquainted with my Memoirs would naturally look for an example: thus, No. III. on the plate is the curve of the Calcutta Cyclone of 1842, and yet I have not given any example from that storm. This is because, as will be seen in the Memoir, though we have a very good series of barometrical observations at Calcutta, which are moreover my own, yet we have only the log of one vessel to settle the position of its centre at noon the preceding day—and this may therefore be very uncertain; added to which the Cyclone itself was perhaps not fully formed when it passed her.

326. Another remark that I should make is, that I have mostly taken the instances at the time the sailor requires them, or that represented by the interval between 12h. and 6h. on our plate, and the reason of this will be evident, by referring to § 318, p. 250, and to the fact that (§ 316, p. 249) the rule becomes none at all when the centre is too nearly approached. On this last account, also, I have often omitted what are at first sight good examples, but in which, upon consideration, it will be seen that though the one height of the barometer is given at a proper time for our purpose, and the distance of the vessel or station is perfectly well ascertained at both, yet the next observation for the average is so close to the centre (and sometimes in it), that it would give no fair criterion of the average fall in the interval, for the reasons already given.

Table shewing the results given by the Rule at page 252 for estimating the distance of the centre by the average fall of the Barometer, when applied to actual cases at Sea.

I. Number.	II. Date.	III. Ship or Station.	IV. Authority.	V. Average fall of Barometer per hour.	VI. For how many hours average taken.	VII. Distances from centre given by the Rule.	VIII. Mean of distances by the Rule.	IX. Actual distances from the centre, measured by Chart.	X. Mean Actual distance.	XI. Rule. $+ \mid -$ of true distance.	Remarks.
				Ins.	*h.*	*Miles.*					
1	28th and 29th Sept. 1840.	*Flowers of Ugie,* Bay of Bengal.	H. Piddington,IIId Memoir, ……	0.13	7.00	100 to 50	75	180 to 20	100	— 25	Cyclone and ship approaching each other.
2	16th May, 1841.	Madras Observatory.	H.P. Vth Memoir, ……	0.054	1.00	100 to 150	125	90	90	— 35 ?	Cyclone not violent, no Ships lost, No. 3 too near the centre.
3	23rd Oct. 1842.	*Sarah,* Bay of Bengal.	H. P. VIIth Memoir, ……	0.080	2.00	100 to 80	90	90 to 70	80	— 10	Ship on the verge of the Cyclone.
4	……	Madras Observatory.	……	0.020	5.00	250 to 150	200	170 to 145	157	+ 43	See Remarks (1).
5	……	Ryacottah Observatory.	……	0.030	2.00	250 to 150	200	80 to 60	70	+130*	An Inland Report from Surveyor General of S. India.
6	21st May, 1843.	Barque *Teazer,* Bay of Bengal.	H. P. XthMemoir, ……	0.018	15.30	250 to 150	200	198 to 65	131	+ 69	
7	21st & 22d May, 1843.	Barque *Candeshar,* Bay of Bengal.	……	0.075	4.00	150 to 100	125	152 to 88	120	+ 5	Ship and Cyclone approaching each other very rapidly.
8	……	Masulipatam, ..	……	0.030	24.00	250 to 150	200	100 to 280	190	+ 10	
9	……			0.082	2.30	80 to 100	90	82 to 104	93	—	
10	……	Ship *Genl. Kydd*								— 3	A shore observation, but locality flat.

				Fas.	A.	Miles.					
11	27th November, 1843,	Ship *Tyne Dale*, H., P. XIth S. Indian Ocean, H. Indian Memoir, ...		0.054	4.50	250 to 150	200	87 to 70	79	121		Position uncertain,‡ See Remarks(2)
12	28th and 29th Nov. 1843,	*John Fleming*, S. Indian Ocean,		0.033	24.00	250 to 150	200	37 to 83	60	140		
13	Sept. 3rd, 1842,	Barque *Wm. Eng*, Gulf of Florida,	Mrs. Redfield. Am. Jour. of Science, Jan. 1846, p. 8,	0.050	8.00	250 to 150	200	....	200 †			Place of centre, not very well ascertained, but nearly at this distance.
14	Oct 5th, 1844,	Key West, Gulf of Florida,	Mr. W. C. Redfield. Am. Jr. of Science, 1846, p. 342, Cuba Hurricane, W. C. R.	0.048 0.079 0.082 0.04	13.00 4.00 4.00 4.0	250 to 150 150 to 100 100 to 60 250 to 150	200 25 90 200	....	300 180 158 928		100 53 63 28	Cyclone crossing the Island of Cuba (3).
15 16	Oct. 6th, 1844, Oct. 7th, 1844,	*U. S. B. Pioneer,* H. M. S. *Illustrious,* Halifax Harbour,	p. 346,	0.05 0.06	4.0 4.0	250 to 150 250 to 150	200 200	350 to 288 345 to 240	319 292		119 22	See Remarks (4).
17 18		H. M. S. *Pique*, Coast of Cape Breton,	p. 356,	0.05 0.07	4.0 3.0	250 to 150 150 to 100	200 125	290 to 190	240° 120°	5	40	See Remarks (5). Mr. Redfield's estimate of Ship running for the Gut of Canso. But he states, p. 165, that the centre of 7th should be placed 30' further in advance, which would make these Nos. 210 and 215.
19			p. 357,									
20	4th and 5th April, 1843,	Barque *Blanche*, South Indian Ocean,	Mr. Thom, p. 305 and 37,	0.01	24.0	250 to 150	200	180 to 130	155	45		Ship to S. E. and S. of Cyclone in the S. E. Trade.

NOTES AND REMARKS ON THE FOREGOING TABLE.

327. Some of the instances here given appear to contradict the rates we have taken as guides. They are however as authentic as other instances, and I have set them down as what are called, in researches on questions of Natural Philosophy, *outstanding instances* :*
or instances which for the present do not agree with the rule given, but which should be registered if thought correct, as they often lead to new discoveries and laws. I note, in the order in which they are numbered, these exceptional cases.

(1.) No. 5. Madras Observatory. The average fall here would give 200 miles of distance, while the centre of the Cyclone was really at about 70 miles only. The track of the storm is one coming in directly towards the shore, and the outer verge of it may have been so far influenced by the Pulicat Hills, behind Madras, as to have diminished the fall of the barometer; and this appears the more probable, that in the same Cyclone the result shewn by the *Sarah*, No. 4, in the open ocean, though on the extreme verge of the storm, is a very fair one. That by the Ryacottah Observatory, No. 6, though completely inland and subject there to many irregularities, between the great chains of the Eastern and Western ghats, is also as fair a result as can be expected.†

(2.) These two instances, *Futtle Rozack* and *John Fleming*, Nos. 11 and 12, are wholly outstanding ones, but with the exception of the Note at No. 12, explaining that the *Fleming's* position was uncertain on that day, I have no reason to doubt that the instance, at least in the case of the *Futtle Rozack*, is a correct one as to data, and tolerably so as to that vessel's position.

But this Cyclone was a very remarkable one, for while it was raging to the South of the Equator, another, which was probably forming

---

* Herschell's Discourse on the Study of Natural Philosophy.

† The same anomaly occurs in applying our rule to a case on the coast of Ceylon, in the log of the P. and O. Co.'s Steamer *Hindostan*, Captain Moresby,(XIV. Memoir J. A. S. Bengal, Vol. XIV.) in which that vessel steamed through the centre of a Cyclone, coming up from the E. b. S. and striking the (North and South) shores of Ceylon. By the average fall of the *Hindostan's* Barometer (within 10 or 15 miles of the shore), the distance of the centre would have been at 4 P.M. 1st December at 200 miles distance, but it really was not at more than 70 miles from her.

on the 27th, was blowing on the North side of it in about the same latitude North, and not far from the same meridian; and between the two, along the Equator, a heavy North Westerly and Westerly monsoon was blowing. All these causes together may have influenced the barometer so far as to occasion a considerable diminution of its usual fall. The Cyclone also was a very slow moving one, and we do not yet know if these almost stationary storms are subject to exactly the same laws in this respect as those which have moved onwards from the first, or gradually acquired their usual progressive rates.

(3.) These three instances are very instructive, for this Cyclone at the time of the observations was forcing its way over the lofty mountains of Cuba, as its centre crossed that island, to travel up between the Bahamas and Florida, and the observations were taken at Key West about 118 miles to the left (Westward) of the track. Its centre at noon of the 5th,—and our latest averages extend only to 6 A.M. of that day,—had but just cleared the North coast of Cuba. The great differences are evidently owing to the irregularities of the pressure, and this view derives much corroboration from the fact, that as the front of the storm was clearing the island and reaching the ocean, from 6 A. M. to 11 A. M., there were evidently, though the gale was then in full force, very marked barometrical waves, as follows, viz. —

6 A.M.	29.402	
		− .067
7 "	.335	
		+ .083
8 "	.418	
		+ .118
9 "	.536	
		− .205
10 "	.331	
		+ .059
11 "	.272	

with, then, a gradual fall to 2 P. M., when the depression was 29.134 the lowest. The rise was not subject to these waves, but I do not in any case calculate from it. This instance is a full proof, that, as has been said, the vicinity of land influences the effect of the storm on the barometer. The next is also a proof of it.

(4.) Nos. 16 and 17, H. M. S. *Illustrious*, lying in Halifax harbour, during the passage of the great Cuba Cyclone of October, 1844.

The centre of this hurricane passed within 130 miles to the East of the harbour, at noon on the 7th, on a N. E. track, but the left hand portions of it were affected doubtless by the hills of the Nova Scotia coast, for the barometer of the U. S. brig *Pioneer*, No. 15, gave, as will be seen, correct approximations on the seaward side of the same storm circle.

(5.) These two instances are also very notable, for between the storm and the ship was at first interposed the high lands of Cape Breton and Nova Scotia, the ship being at the Northern entrance of the Gut of Canso, and in the second instance she was running in for it. Accordingly we find the following barometric waves marked between noon of the 6th and 8 A. M. of the 7th, when the regular fall takes place to 3 P. M. of the 7th (29.19) which was the minimum hour.

$$\begin{array}{lll} \text{Noon 6 A. M.} & 29.63 & \\ & & +\ 12 \\ \text{3 P. M.} & .75 & \\ & & +\ 15 \\ \text{8 P. M.} & .90 & \\ & & -\ 27 \\ \text{7 A. M.} & .63 & \end{array}$$

We may remark also of this, as of so many other instances, that had it been a case in which a ship was lying in an open roadstead, or bay, and the Captain had desired to put to sea, he would have had by the wind and average fall of his barometer, a sufficiently correct notice of the bearing and distance of the centre to guide his course in safety, so as to do the best he could to avoid the centre, or profit by the gale; which is practically all that we desire. Mr. REDFIELD's corrected review of the position of the centre, I have noticed in the remarks. Our result would by it approach much nearer to exactness.

(6.) I have noted here that "the ship was to the S. E. and S. of the Cyclone in the S. E. trade," because I think it probable that a stormy trade or monsoon, rising as they often do to the strength of a gale, may influence the barometer so far as, on that side of the Cyclone towards which they blow, to diminish the total amount of fall, and consequently the average *rate* of fall; so that in such cases we shall find the distance given by the rule to be always too large. This instance, moreover, is one giving us only a 24 hours' average with a total fall of 0.23, which for anything we know may have occurred with the last six or four hours. The rates of fall of the following days are however exceedingly small for a vessel so close in with the centre. Mr. THOM has given, at p. 181, a table of the barometer of the *Vellore*, which is altogether an exceptional instance from our rule, and which, if there be no misprint, is very remarkable; for it would seem that she had, from the 1st to the 4th of the month, a regular fall of 0.10 in the twenty-four hours while running before the gale, and this would give but a rate of fall of 0.004 per hour, though from the diagrams she seems to have been at from 120 to 160 miles of distance only. This anomaly probably arises in some way from the effect of the trade wind, but it is impossible in the present state of our knowledge to do more than conjecture vaguely as to the causes of this remarkable difference from the result of the test by the log of the *Blanche*. We must leave this notice for further investigation: observing, by the way, that the *Vellore* was during part of the time close in with the Island of Rodriguez, though one would not, *a priori*, suppose that this mere speck in the ocean could have influenced her barometer. We have, unfortunately, no other registers from the numerous logs given by Mr. THOM to aid us in forming any sort of judgment. This anomaly occurs again in a very remarkable instance, that of the *Buccleugh's* hurricane, in which it is remarked, in an extract from her log now before me,

"It is surprising that previous to so severe a gale a greater fall of the barometer had not taken place, having not been lower than 29.76 inches. It may be accounted for from the wind blowing from the Southward. The Simpiesometer had been for the last week about 38 decimals (0.38 is meant) lower than the barometer, but on the morning of the gale it fell 82 decimals lower than the latter, therefore the indications of this sensitive instrument ought to be attended to."

In this instance also, as in that of the *Blanche* and *Vellore*, the barometer of the *Asia*, quoted by Mr. THOM, seems to have indicated by a fall of 1.25 in 48 hours, or .018 an hour, the vicinity of the storm with tolerable correctness, but we are too uncertain of the position of the centre, and the barometer is given at too great intervals of time, to admit of our using this instance as a test of our rule.

Table of instances of excessive falls of Barometer.

Storm and Locality.	Fall of Barometer. From.	Fall of Barometer. To.	Total Fall.	Authority and Remarks
H. C. Ship *Duke of York*, Kedgeree, 1833,	29.00	26.30	2.70	Col. REID and Journal As. Soc. Beng.; may have fallen more.
Brig *Gazelle*, China Sea, 1849,	29.80	27.00	2.80	Overland Mail, Sept. 1849.
H.M.S. *Pluto*, China Sea,	29.90	27.55	2.35	Log, from Government of India.
Am. Ship *Hosgue*, Timor Sea, 1848,	29.80	27.60	2.20	Log, from Captain LOVETT.
Ship *John O' Gaunt*, China Sea, 1846,	29.65	27.50	2.15	Log, from Mr. ELLIOTT, R.N. Master H.M.S. *Agincourt*.
Brig *Freak*, Bay of Bengal, 1840,	29.30	27.25	2.05	Log, Captain SMOULT.
Ship *Exmouth*, S. Indian Ocean, 1846,	29.00	27.00	2.00	Mr. THOM and H.P. copy of Log, mercury disappeared, exact fall uncertain.
The Havannah, 1846,	29.68	27.74	1.94	Bermuda paper.
Ship *London*, Bay of Bengal, 1832,	29.70	27.80	1.90	Col. REID, p. 291.
Ship *Ann*, China Sea, 1846,	29.90	28.05	1.85	Mean of 2 Barometers, Log from Mr. ELLIOTT, R.N., Barometer fell 0.3 in 50 minutes.
Mauritius, 1824,	29.93	28.23	1.70	Col. REID, p. 167.
St. Thomas and Porto Rico, 1837,	29.76	28.07	1.69	Professor DOVE in TAYLOR'S Scient. Mem. Part X.: reduced to English measure.
H. C. S. *Neptune* and *Scaleby Castle*, 1809, China Sea,	29.85	28.30	1.55	HOMBURGH, p. 289, H. P. VI. Memoir; Logs from E. I. Company; H. C. S. *True Briton* foundered.
Port Louis, Mauritius, 1819,	29.50	28.00	1.50	Col. REID, p. 161.
Brig *Mary*, W. Indies; Gulf Stream, 1837,	29.60	28.10	1.50	Col. REID, p. 96.
Barque *Nimrod*, off New Caledonia,	29.70	28.20	1.50	Captain ESPINASSE, letter to H.P.

328. EXCESSIVE FALLS OF THE BAROMETER. In reference to what has been said of excessive falls of the barometer, the preceding table collects instances of them where they have amounted to an inch and a half or more.* Falls of an inch, and of an inch and a quarter are by no means so uncommon as might be supposed; and it is not I think improbable that, as our knowledge progresses, this remarkable distinction of Cyclones of the same intensity (so far as we can judge from descriptions, and from the damage sustained by well managed and well found ships) occurring sometimes with moderate and sometimes with excessive falls of the barometer, may lead us to some new views, arising not only from the great extent of the fall of the mercury, but also from the incredibly short space of time in which some of these falls occur. Thus the ship *Lady Feversham*, dismasted in the Bay of Bengal, in a severe Cyclone, on the 22nd, 23rd October, 1842, (see Eighth Memoir, Jour. As. Soc. of Bengal, Vol. XII. p. 341), had the Barometer at 29.70 at 2 P.M. on the 22nd, and at 11 P.M. or 9 hours afterwards, at 29.40. But from 11 P.M. on the 22nd, to 1 A.M. of the 23rd, or in two hours only, it fell to 28.40 or *an inch* in two hours! Shortly after this, at 1.45 A.M., the calm of the centre reached her, when the Barometer fell to 28.30, at which it seems to have continued till it rose. Many more instances might be adduced, but this will suffice to warn the careful seaman that he must look to every sign in the progress of these his battles with the winds and the wild waves, if once engaged in them.

329. HEIGHT OF CYCLONES ABOVE THE SURFACE OF THE OCEAN. The height to which storms extend is always a question of interest. We do not, as yet, know that if we were exactly acquainted with this element of our science, we could apply it to any useful purpose, but it is possible that it may turn to account some day or other.

Meteorologists are much divided as to the height to which clouds ascend, and M. PELTIER,† whose work I shall presently allude to more

---

* I have omitted amongst these that of the barometer of Admiral KRUSENSTERN on the Coast of Japan in 1804, because the account in the Chinese Repository for 1839, quoted in my Notes on the Law of Storms, written for the use of the China Expedition, does not give the necessary data for ascertaining the amount of fall; which however was probably more than 2 inches.

† PELTIER sur les Trombes; Paris, 1840.

in detail, shews satisfactorily, that there are two kinds of clouds; the common fog-clouds (as they might be called) and *transparent* clouds! or bodies of air loaded with moisture at a different temperature from the surrounding atmosphere, which are not yet condensed into vapour (fog) so as to be visible to us as clouds, but which nevertheless may be subject to, and produce all the phænomena of visible clouds, though in a different electrical state.

The warm blast which proceeds from the mouth of a glass-blower's or iron-founder's furnace chimneys, long after the fires are allowed to "blow out" as it is called, forms in fact a small transparent cloud of heated and dry air, which would, if we looked at objects on the other side of it, be seen to create a *mirage* like that of the Desert from its heated sands, and if this blast met with a little fog cloud it might dissolve it, and carry its moisture up till it found an equal temperature by gradually losing its own in the surrounding air on its passage.

What I speak of here, however,—and till our science has made much more progress, we must always, for practical ends, speak thus— is to be understood as applying only to the height of the cloud-portion of the Cyclone, or what one might call the height of the storm-bank of clouds (see the next section) above the sea.

330. KAEMTZ (pp. 365 and 366, English translation) says that *thunder* storms have been seen to pass above the summit of Mount Blanc (15,680 feet,) and he continues, speaking always of course of thunder storms only, to say,

"It is sometimes possible to determine approximately the height of a storm. When lightnings pursue a horizontal course we measure the interval separating the thunder and the lightning; now, as sound travels 333 metres (1,092 feet English) in a second,* we have only to multiply by 333 the number of seconds that have elapsed, in order to estimate the distance of the lightning from the observer. If at the same time, we measure the angular height of the lightning, we can hence deduce its vertical height. Thus in 1834, when there were several very elevated storms at Halle, I found, on the 5th of June, that the lightnings were at a height varying from 1,900 to 3,100 metres, (6,233 to 10,171 feet English) on the 21st of July, the minimum of certain lightnings traversing the zenith, was 1,300 metres (4,265 feet English)."

"When storms also are not very elevated, we must admit that the clouds we see are formed after the more elevated strata, that principally constitute the storm.

* 1130 is, however, perhaps a better rate as determined by the French Academicians.

Part V. § 332.]   *Thickness of Cyclone Disks.*   267

The rapidity with which the lower clouds are condensed gives rise to a strong electric tension that is manifested by repeated discharges; this is due to the inductive action of the higher masses acting on the lower."

331. Mr. Espy pronounces that some clouds, which he supposes to be condensed vapour carried up by vortices, rose to the heights of ten and fourteen miles ! the writer in the North American Review, No. CXXIII. April, 1844, alluded to at p. 4, reduces these heights to four miles and two miles and a half.

332. Mr. Redfield says, (and I will not do him the injustice to attempt any abridgment) in relation to this, (American Journal of Science and Arts, p. 184),

"Vertical height of the storm wind.—What is the general height or thickness of the storm, and by what means can this be approximately determined ? These questions and their solution, are doubtless of some importance in their bearing on meteorological theories, and seem to deserve our attention.

"In nearly all great storms which are accompanied with rain, there appear two distinct classes of clouds, one of which, comprising the storm-scuds in the active portion of the gale, has already been noticed—above this, is an extended stratum of stratus cloud, which is found moving with the general or local current of the lower atmosphere which overlies the storm. It covers not only the area of rain but often extends greatly beyond this limit, over a part of the dry portion of the storm, but partly in a broken or detached state. This stratus cloud is often concealed from view by the nimbus and scud clouds, in the rainy portion of the storm, but by careful observations, may be sufficiently noticed to determine the general uniformity of its specific course, and approximately, its general elevation.

"The more usual course of this extended cloud stratum, in the United States, is from some point in the horizon between S.S.W. and W.S.W. Its course and velocity do not appear influenced in any perceptible degree by the activity or direction of the storm-wind which prevails beneath it. On the posterior, or dry side of the gale, it often disappears, before the arrival of the newly condensed cumuli and cumulo-stratus, which not unfrequently float in the colder winds on this side of the gale.

"It appears, therefore, that the proper storm-wind revolves entirely below the great stratus cloud, which covers so large a portion of the storm; and we may infer also, that the production of the accompanying rain and the depressing effect of the storm's rotation on the barometer, are chiefly confined within the same vertical limit. In regard to rain this result is in accordance with observations on the quantity which falls at different elevations above the earth's surface; and in the case of the barometer, a like accordance is shewn in the diminished range of the mercury in storms which is found in ascending from the ocean level.

"The general height of the great stratus cloud which covers a storm, in those parts of the United States which are near the Atlantic, cannot differ greatly from one mile; and perhaps is oftener below than above this elevation. This estimate, which is founded on much observation and comparison, appears to comprise, at the least, the limit or thickness of the proper storm-wind, which constitutes the revolving gale.*

"It is not supposed, however, that this disk-like stratum of revolving wind is of equal height or thickness throughout its extent, nor that it always reaches near to the main canopy of stratus cloud. It is probably higher in the more central portions of the gale than near its borders, in the low latitudes than in the higher, and may thin out entirely at the extremes, except in those directions where it coincides with an ordinary current. Moreover, in large portions of its area there may be, and often is, more than one storm-wind overlying another, and severally pertaining to contiguous storms. In the present case, we see from the observations of Professor SNELL and Mr. HERRICK at Amherst, Mass., and at Hamden, Me., (115 and 135*b* of Mr. REDFIELD's Memoir) that the true storm-wind, at those places, was superimposed on another wind; and various facts and observations may be adduced to shew that brisk winds, of great horizontal extent, are often limited, vertically, to a very thin sheet or stratum."

333. A more ready means, however, of placing this clearly before the mind of the plain seaman is the following —

If he will take any Mercator's Chart of which the scale is one inch to a degree of longitude, and lay close down upon it the thinnest of the two Horn Cards with this book, the storm-circle marked upon the card will *about*† represent a Cyclone disk which reaches one mile in

---

* *Note from the American Journal of Science, Vol. XXXI.* p. 127-128. If a disk be cut from the thin paper of Chart IV. (Mr. REDFIELD's) of a size which will represent one thousand miles in diameter, it will be found to have a thickness which represents more than a vertical mile, by the scale of the Chart. A disk of the same size, but on a scale representing a storm of but 400 miles diameter, if cut from the paper of this Journal, will also represent more than a mile of vertical thickness in the storm. These and other analogous considerations, deserve the attention of those who may think that winds are mainly induced, and supported by movements or influences of a vertical character or tendency. It might be useful for those holding such views, to attempt to draw out the supposed paths of vertical induction and geographical progression in the winds, on an accurate and uniform linear and vertical scale, for the purpose of attaining a more precise standard for estimating the supposed vertical action or influence.

† *About.* We do not require exact accuracy in these things, but the calculation is in round numbers: that on a scale of one inch to a degree the 360° would require a globe of 360 inches in circumference, or 120 inches in diameter, and 8,000 miles

height, and yet is 180 miles in diameter, so that he might suppose the minutest microscopical atom or grain of dust below the Horn Card to be a ship, and if he whirled the card round, he would represent the action of a rotatory storm, and its relation as to height to the surface of the ocean.

334. He will then have to bear in mind for the future, that in a Cyclone he cannot be so much said to be enveloped in a moving *column* of whirling winds, as to be caught in a thin flat disk or circle of such winds, which may be from 100 to 1000 miles diameter, but only from one to three or four miles, or perhaps at most five miles high; and that from such an elevation as the Peak of Teneriffe or Mouna Roa, he might perhaps look down upon a stratum of storm-clouds beneath which his ship was cutting away her masts or foundering in a hurricane: just as travellers in the Alps have looked down upon thunder or hail-storms which were devastating the corn-fields and vineyards of the valleys below. In the section on banks of clouds, we shall shew that hurricanes (Cyclones) *have* been, pretty nearly, so seen from elevations!

335. To assist the reader further in forming a correct idea of what a disk of storm of moderate extent and height is, I have also traced below the framing line of the Barometrical Chart, dotted lines representing the disks of Cyclones of 300 miles in diameter, and of ten, seven and three miles in height, with a supposed vortex *v.* at the centre which has a calm of ten miles at its base. He may from this estimate what it would be if five or even fifteen miles high, and how fallacious all our notions are apt to become when we consider these storms as whirling *columns*, and insensibly go on to liken them to water-spouts as to height, which it is evident they cannot at all resemble, since their size (diameter) may be said to have been in many cases estimated to a few miles with tolerable correctness; and in frequent instances the next stratum of clouds above the storm, either at rest or moving altogether differently, has also been clearly distinguished and noted: so that we may boldly affirm that at most the height (thickness is the

(diameter of the earth) : 120 inches :: 1 mile : to .015 inches. Now 100 average horns measure about 1.5 inches in thickness when close together, so that each averages about .015.

more correct word) of the disk never exceeds ten miles, and usually falls far short of it. Mr. REDFIELD, as just quoted, thinks it oftener below than above *one* mile.

336. Indeed, I am inclined to believe that the Cyclone-disk is sometimes so *thin* that at, or near the centre, whether calm or not, it has often been *seen through*; and the following I take to be instances in which this has occurred. Colonel REID remarks that the Spaniards call the clear space seen at the centres of Cyclones the "Eye of the Storm."

In Professor FARRAR's Memoir, quoted at p. 3, he says,

"A clear sky was visible in many places during the utmost violence of the tempest, and clouds were seen flying with great rapidity in the direction of the wind."

In my Eighth Memoir (Jour. As. Soc. Beng. Vol. XII.) Dr. MALCOLMSON, Surgeon, Political Residency, Aden, in giving an account of a Cyclone in the Arabian Sea, in which the ship *Seaton* of Bombay was dismasted and in great danger, says,

"During the height of the storm the rain fell in torrents, the lightning darted in awful vividness from the intensely dark masses of clouds that pressed down, as it were, on the troubled sea." *In the zenith there was visibly an obscure circle of imperfect light of ten or twelve degrees.*

In the whirlwind of the *Pacquebot des Mers du Sud*, forming the second part of my Fifth Memoir (Jour. As. Soc. Beng. Vol. XI.) which was a true Cyclone by its veering, and a tornado as to duration and violence, is the following passage—

"A very remarkable fact is, that while all around the horizon was a thick dark bank of clouds, the sky above was so perfectly clear that the stars were seen, and one star shone with such peculiar brilliancy above the head of the foremast that it was remarked by every one on board."*

In the Cyclone of October, 1848, in the Bay of Bengal, (see Eighteenth Memoir, Journal Asiatic Society of Bengal, Vol. XVIII. p.

---

* We are forcibly reminded here, and there may be more than a mere poetical figure in it, of the beautiful invocation to the Virgin by the Mediterranean mariners—

"In mare irato, in subita procella,
Invoco te MARIA *nostra benigna stella*."

"In raging seas and in the tempest's war,
We hail thee, MARY! as our guardian star."

PART V. § 336.]   *Cyclone Disks seen through.*   271

904) the superintendent of the Light-house at False Point Palmiras distinctly states that at the time of the passage of the centre, or for about two hours of calm, the stars were seen very clear over head with a thick bank of haze all round.

In other instances, though the Cyclone is not distinctly seen through, yet the appearances are such as to leave no manner of doubt that its thickness must be very insignificant.

In the Cyclone of October, 1848, already referred to, several ships speak of a circle of light, or of its being much clearer over head at the centre, and this is exactly the appearance which should occur to an observer situated at the centre of a thin disk, as well as to one in the focus of a thick vortex. Both would allow of more light reaching him, and if the altitude of the surrounding bank was definite enough to be measured the thickness of the disk might be calculated. Thus, apart from the earth's curvature, the appearance of a bank or wall of cloud of $15°$ in altitude to a spectator at the centre of a calm space of 20 miles in diameter, would indicate that the Cyclone was $2\frac{3}{4}$ miles high, one of $20°$ that it was $3\frac{3}{4}$ miles, and one of $30°$ that it was six miles high (thick), but the difficulties here are always to know the extent of the calm, and the want of any defined edge to the bank of cloud. Nevertheless this may be approximately taken, and if noted, would be of great interest.

The ship *Tigris*, Captain ROBINSON, experienced a short but severe tornado-Cyclone in April, 1840, in Lat. $37°$ to $38°$ S., Long $68°$ to $75°$ East, track *e e* on Chart No. II., and in the midst of it while lying to, "the clouds broke away and the sun shone out, the whole surface of the water as white as snow with foam, and coloured like the rainbow in all directions. At 11 the wind blew with such fury that the three top-gallant masts were blown away, the spencer split to pieces and the furled sails blown to shreds from the yards." We have no express note here, it is true, of the surrounding bank of clouds, but no doubt there was one, for the ship had still to lay to for three or four hours under bare poles.

Colonel REID states, 529, that

"During a gale in the North Atlantic, about Lat. $40°$, in a ship hove to, after the clouds broke sufficiently to see through the lower ones, the upper light clouds appeared in a quiescent state, as if the storm was confined to an altitude little above the surface of the globe."

We have many instances on record, indeed, in which, at the onset or towards the close of storms, the upper clouds have been seen moving from 8 to 16 points differently from those in the storm below, and the weather above appeared altogether fine. And the whole tendency of this is to shew that not only a Cyclone has no resemblance to a column, but that it may be at the centre and towards its edges, nothing more than a very thin disk.

337. BANKS OF CLOUDS. These are at times very remarkable on the passage, or approach of, and during Cyclones; and there is no doubt that the clouds and other celestial objects, carefully watched, may often afford the vigilant seaman far more advice and warning than is usually supposed. We are perhaps too much accustomed to rely on our instruments now-a-days, and we neglect these signs, which must after all have been the barometers and simpiesometers of DRAKE, CAVENDISH, DAMPIER, and all our daring band of naval and commercial navigators up to the end of the last century, and still are so for our hardy fishermen and coasters. Some of these signs may at least serve as corroborative indications, and the careful observation of them (to which indeed I am desirous of directing attention) lead to farther and perhaps useful knowledge.

338. It seems certain that Cyclones have been frequently *seen* as forming thick banks or walls of cloud at a greater or less distance from ships which have escaped from, or have afterwards been caught in, or by the common and too frequent error have *sailed into* them. The following are instances.

Colonel REID, p. 47, 2nd Ed. gives a letter from Mr. MONDEL, Master of the *Castries*, West Indiaman, who saw a bank of clouds so thick and close, that in broad day-light at 3-30 P.M. it was taken for land by all on board, though at the time the ship was 350 miles from St. Lucia, at which island a severe Cyclone was felt on the following day.

The Master of the ship *Shakespear*, in a Cyclone in the Atlantic in 32° N. 79° West, in October, 1848, which I shall again refer to, in speaking of the red sky, describes the appearance of the clouds "as the ship was running out of the storm, like to a thick dark fog blowing away to the North and East."

Dr. PEYSSONNEL (Phil. Trans. for 1756, p. 629) says, that being upon Grande Terre, and viewing thence the whole island of Guadaloupe during a Cyclone,

"I observed that the storm, which had affected us in the night, was now very violent upon the island of Guadaloupe; it was a frightful thick, black cloud, and seemed on fire,* and gravitating towards the earth: it occupied a space of about five or six leagues in front, and above it the air was almost clear, there appearing only a kind of mist."

The distance from the centre of Grande Terre to the centre of Guadaloupe is about 20 miles. We cannot judge what the horizontal or vertical angles were from what is here said, but the description remarkably accords with the following from Balasore and Point Palmiras.

"At Balasore, Mr. BOND, Master Attendant, informs me that the gale of June, 1839, (see First Memoir, Jour. As. Soc. Beng. Vol. VIII.) was foretold by the blackness of the heavens to the Eastward."

"In the *Duke of York's* hurricane in 1833, though it did not reach to Balasore, 75 miles to the S.S.W. of Kedgeree where that vessel was lost, the bank to the Eastward in the heavens so plainly indicated a gale that every person barred up his doors and nailed them. We had only a good top-gallant breeze.

"Mr. RICHARDSON, Branch Pilot in the H. C. Service, also informed me, that during this hurricane, while he was driving about with all his anchors gone, some passengers whom he had previously landed at the Black Pagoda were upon the top of it, and felt no excessively violent wind, though they *saw* the horizon very black, and the sea dreadfully agitated to the N. East."

From the Black Pagoda to Kedgeree, and the centre of the Cyclone was to the South of that place, the course and distance is about 135 miles, and I have placed the line of the track of this Cyclone (*j*. on our Chart No. III.) 130 miles to the N. E. of the Black Pagoda.

In the China Sea, the brig *Virginie*, which narrowly escaped running into a tyfoon by being on that quadrant of it where the wind was foul for her, states in her log, abridged in my VIth Memoir on the storms of that sea, that they "saw a heavy bank of clouds to the E.S.E." Mariners so often see this appearance of a bank of clouds, that when they mention it we must take it to be of a remarkably threatening kind. I have many such instances on record where there is no doubt,

* From flashes of lightning, I presume is meant, or the effect of the sun's rays.

to my mind at least, that ships have seen the banks of clouds forming the outer verges of storms; much resembling, as we may suppose, those which surround the upper extremity of a water-spout.

839. The cases also in which ships have seen remarkable banks or walls of cloud surrounding them wholly or in part, or as just quoted at p. 270, where the clouds have been so dispersed in the zenith during the utmost fury of the Cyclones, as to allow the clear sky or stars to be seen overhead, are by no means uncommon. Colonel REID gives, p. 116, the log of the *Duke of Manchester*, the master of which ship in a Cyclone in 32° North, 77° West, in the month of August, says—

"A most extraordinary phænomenon presented itself to windward, almost in an instant, resembling a solid black perpendicular wall, about fifteen or twenty degrees above the horizon, and disappeared almost in a moment; then in the same time made its appearance, and in five seconds was broken, and as far as the eye could see: from this time to midnight, blowing a most violent hurricane with a most awful cross-sea breaking constantly on board fore and aft."

In LUKE HOWARD'S Climate of London, Vol. III. p. 151, is a letter from an officer of H. M. S. *Tartarus*, giving a capital account of a Cyclone (Sept. 30th, 1811) off the coast of North America. Barometer fell till mercury was out of sight (to 28 ins. probably). The scenery of the sky it is impossible to describe. *No horizon appeared, but only a something resembling an immense wall within ten yards of the ship.*

Capt. G. J. O. SMITH, of the American Clipper Barque *Mermaid*, homeward bound from the Straits of Sunda, describing a very remarkable Cyclone in 25° S., Long. 66° East, of which the track seems much to have resembled that of the *Futtle Rozack's*, marked VIII. on Chart No. II., says—

"At daylight the lower stratum of clouds cleared away, and an awful sight appeared from N.W. to N.E. being a heavy solid mass of black matter about 30 degrees high. We were able to see it but for a few minutes as it commenced to rain again."

These black masses were no doubt the dangerous central portions of the Cyclones, in which so many fine ships have disappeared.

I have already alluded, p. 270, to the whirlwind of the *Paquebot des Mers du Sud*, which is certainly an instance of the bank of clouds surrounding a vessel in a Cyclone.

Captain MILLER, of the *Lady Clifford*, whose good management I have adverted to at p. 166, observes, as the Cyclone was approaching, and before it was travelling past to the N.E. of him, and while his barometer had as yet fallen but very little, being at 30.05 at noon, and at 29.91 at midnight, that—

"Towards evening a thick cloud or bank gathered in the N. E., and a long swell set in from that quarter.* At 10 P.M. the whole sky was overcast and the barometer began to fall."

I have quoted at length, at p. 148, the capital instance of the *Earl of Hardwicke*, Captain WELLER, who undoubtedly saw the body of the Cyclone which he avoided, in the bank or arch of clouds to the North and N.W. which he describes.

In the Nautical Magazine for January, 1847, is an excellent account of a typhoon-Cyclone in the China sea in September, 1842, by Captain HALL of the *Black Nymph*, who also saw distinctly the body of the storm approaching. He says—

"Towards evening I observed a bank in the S.E. Night closed in and water continuing smooth, but the sky looked wildish, the scud coming from N. E., the wind about North. I was much interested in watching for the commencement of the gale which I now felt sure was coming, considering that Colonel REID's theory being correct, it would point out my position with respect to its centre. That bank in the S. E. must have been the meteor approaching, and the N. E. scud the outer N. W. portion of it, and when at midnight a strong gale came on about North to N.N.W., I felt certain we were then on its Western and Southerly verge."

In the letter of the Commander of the *Judith and Esther* to Colonel REID, p. 75 of his work, it is stated at the conclusion of the hurricane (Cyclone),

"At 6 P.M. the gale abated and the sea fell fast, the appearance of the sky at this time was very remarkable, being of a deep red colour to the North, and looking very dark to the West, as if the gale was moving in that direction."

The same kind of appearance after a severe Cyclone in the Arabian sea is noted by Dr. MALCOLMSON in his account of the *Seaton's* disasters, quoted at p. 270.

Colonel REID, p. 27, speaking of the Barbadoes hurricane of 1831, which also devastated St. Vincent, says—

---

* See page 178 for what is said there of distant swells.

"A gentleman of the name of SIMONS, who had resided for forty years in St. Vincent, had ridden out at day-light, and was about a mile from his house, when he observed a cloud to the North of him so threatening in its appearance, that he had never seen any thing so alarming during his long residence in the tropics: he described it as appearing of an olive-green colour. In expectation of terrific weather, he hastened home to nail up his doors and windows, and to his precaution attributed the safety of his house."

Colonel REID quotes also the log of the *Rawlins*, in which, after a Cyclone, is noted "A dismal appearance to the N.W.," and Mr. REDFIELD and Dr. MITCHELL of New York, as stating that the labouring people in N. York had learned to prognosticate from what quarter the wind would set in, in their storms, by noting where "the haze or *cirrus* which appearing at sunset indicates its approach," first appeared.

I have found and could quote many more instances from recent logs sent to me since this class of observations was first urged on the attention of mariners in the first edition of this work, but for brevity's sake I will cite but one more, differing a little from the foregoing. It occurs in the paper of Commander FISHBOURNE of H. M. S. *Hermes*, already alluded to at pp. 38, 39. Speaking of a Cyclone, which he allowed to cross ahead of him when off the South Coast of Africa, he says—

"While steering along the land with the wind from the N.W. which changed to W.S.W. and West, and increased in force as we approached the centre, it crossed ahead of us. About 6 P.M. the position of the centre abeam of us, as we conceived, was distinctly visible; dense masses of cloud in a state of agitation, with a blueish space in the centre moving away to the E.S.E. *Immediately* after the passage of these masses of clouds abaft our beam the Barometer rose rapidly."

The "blueish space" here described might very possibly be an electric light or the reflection of the sun's rays in the central space, or the disk of the Cyclone (see the following section) might have been lifted up, so as to allow the position of the vortex to be discerned. The fact is always an important one and of high value from such good authority.

340. SIGNS OF APPROACHING CYCLONES. I have remarked in the section on the banks of clouds, that it may be well worth the careful seaman's while, to observe watchfully the atmospheric and other signs which were the barometers of our forefathers. As there

Part V. § 340.]   *Signs of approaching Cyclones.*   277

are many dispersed notices of these indications, and these would require much detail to explain them, I have thought that, both to save space and to enable the mariner to refer to them at a glance, the form of the following tables would be the shortest and best in which to set them out for our purpose. I have divided them into Celestial and Terrestrial signs for greater convenience, and followed them by a few remarks on some of the phænomena. It is not intended by anything that may be said, or cited from others, that the seaman should consider any one of these signs as a *certain* prognostic of a coming Cyclone, but he will allow readily that in certain seas, and at certain seasons of the year, his attention *and that of his officers* cannot be kept too much awake; and any one or more of our signs may, by inducing a closer look-out on the barometer and simpiesometer, give him from an hour to six hours more TIME for his precautions, whether at sea or in harbour. Of time no class of men should better know the value than sailors, and none should more anxiously ponder upon what " the little cloud no bigger than a man's hand, which riseth out of the sea"* may portend to them, and to the lives and property with which they are entrusted.

The barometer and simpiesometer too, it should be borne in mind, are not always perfectly faithful advisers, for any complication, such as the approach of double Cyclones, or a Cyclone crossing or travelling against a strong trade or heavy monsoon may affect their indications. "Instrumental observations," says Dr. Buist of Bombay, "are not the essence of Meteorology, if unaccompanied by due discerning of the face of the sky."

* 1 Kings xviii. 44.

Tabular View of the Celestial and Atmospheric Signs stated by various Authors to indicate the approaches of Tyfoons, Hurricanes, &c.

Authority.	Place of Observation.	Sun.	Moon.	Stars.	Clouds & Lightning.	Atmosphere.	Notes.
1. Captain Langford, Phil. Trans, 1698, p. 407; from the Carib. Account of European settlements in America in 1766.	West Indies.	More red. Often with a burr about it.	Occur at full, change, or quarters. A great burr about it.	Look big with burrs about them.	N. W. Sky very black and foul, and turbulent.	A great calm, sometimes for an hour or two, strong Westerly wind, mists, and clouds.	These Signs often at change or quarter of Moon, if a Hurricane is to be expected at full, and at full if it is to be expected at change.
2. Dr. Peyssonnel, Phil. Trans, 1756, Part II.	West Indies.	Setting sun blood-red.			Little clouds, moving with great rapidity.	Calms and frequent shiftings of breezes from all points.	
3. Dr. Morrison, Chinese Repository, Sept. 1839, p. 230. C. B. Greenlaw, Esq., Homburgh, Captain Shire, various logs, M. La Place, Dampier. U. S. S. Plymouth and Saratoga.	China Sea and Coasts of Heinnam, and Pacific Ocean.	Capt. Nisbet of the H. C. S. Essex, had observed in five Tyfoons in the China Sea and one on the Malabar Coast, the sun to set of an excessive fiery red colour.		Are as distinctly seen rising and setting as the sun and moon Horizon being peculiarly light and clear.	Irregular masses moving against or across the usual monsoon.† Fiery in thick masses, frequent rainbows broken.*	Remarkable clear weather, distant land very distinct.† Slight noises heard at intervals a few days before, wheeling round and stopping quick and also thick muddy atmosphere.* Peculiar moaning sound in the atmosphere as of wind rising and falling; H. P. Capt. Shire.	† Many European accounts. * The descriptions of Chinese fishermen, and possibly an imperfect translation.
4. Derrotero de las Antillas, as translated in Purdy's Atlantic Memoir; and Lt. Evans, R.N.	Coasts of Mexico for the Nortes.				Lightning on the N. W. and N. E. horizon; at sunset Cumuli, changing to dark nimbus of a deep purple.	Thick fog or low scud flying fast to the south, or (Evans) extraordinary white haze or mist distinct from common fog, principally to the North.	Weather warm, hot, & oppressive, sometimes affecting the breathing.

#	Source	Location					Remarks
5.	Mr. GITTINS, of Barbadoes, quoted by Col. REID, P. 36, and Bermuda Gazette.	West Indies. Barbadoes.	Of a blue appearance; the same where shining in a room. Appearance of the sun at day like that of full moon.	…	…	During forward in divided portions and with a fleet irregular motion, not borne by the wind, but driven, as it were, before it.	Distant roar of the elements, as of winds rushing through a hollow vault. White objects of a decided light blue colour.
6.	Capt. RUNDLE, quoted by me XI. Memoir J. A. S. Vol. XIV. Capt. JENKINS H. C. S. *City of London*, MSS. from E. I. C.	Southern Indian Ocean.	…	…	Captains RUNDLE & RITSO dancing, sickly appearance.	A remarkable kind of lightning, shooting up in stalks or columns in the horizon or with a dull glare, and remaining for a short time.	Captain RUNDLE, H.P. Captain SHIRE, and Cyclone Storm at Baticolo; a moaning noise, rising and falling with the wind.
7.	Mr. VAILE, Ship *Barham*.	Bay of Bengal.	…	…	…	Lightning, like flashes from a gun, and sparks from a flint and steel.	
	Capt. STEWART, *Rajasthan*; XIV. Memoir J. A. S. Vol. XIV., and other logs and accounts.	Arabian Sea.	…	…	…	Light clouds moving with great rapidity in different directions. Thin, low damp vapours flying rapidly across the upper clouds.	
8.	Popular and well-known at the Mauritius. THOM, p. 92. *Exmouth's* log; Newspaper account of *Northumberland's* Cyclone; Capt. BIDEN and others.	Southern Indian Ocean and Mauritius.	…	…	…	Remarkable red colour of the clouds, and all objects appearing tinged with red colours; described as lurid red to bright *crimson* and brick-dust haze.	Brick-dust haze in the horizon. This is well known and considered as an almost infallible sign at the Mauritius and Bourbon. With the ship *Manchester* in 22° S. 86° East, it also occurred during the height of the Cyclone.

[PART V. § 340.

CELESTIAL and ATMOSPHERIC *Signs of Cyclones continued.*

Authority.	Place of Observation.	Sun.	Moon.	Stars.	Clouds & Lightning.	Atmosphere.	Notes.
9. PIDDINGTON. Various Memoirs, brig *Freak*, Captain SMOULT, ship *Albion*, Capt. MCLEOD, and *Barham*, Capt. VAILE, Capt. WILLIAMS, in ROMME's Work, Vol. II. p. 149.	Bay of Bengal	. .	Remarkably bright halo round her.	Remarkably bright & twinkling and seen at very low altitudes	Red sky and light even at night. Rising rapidly, appearing ragged and black with white-feathering edges, and stretching out with long tails towards the Zenith. Thick banks over the place of the Cyclone, growing longer and darker, webs of light gauzy cloud or of thin smoky scud flying low. Dense clouds surrounding the horizon to 10° or 15° in height; upper edge tinged with red, and red light reflected from them on all objects. Thin portions flying rapidly, at very low altitudes.	. .	Seeremarks & 18th Memoir Jour. As. Soc. Beng.
10. LYNN, Star Tables H. C. S. *Buccleugh*'s Typhoons.	Northern Pacific Ocean.	. .	. .	. .		Remarkably clear, murmuring noise in the rigging.	

Tabular View of the TERRESTRIAL *and* OCEANIC *Signs, stated by various Authors to indicate the approach of Cyclones.*

Authority.	Place of Observation.	The Sea.	Sea Shore.	Land Generally.	Notes.
1. Captain LANGFORD, and account of America.	West Indies.	Smells stronger than at other times, and rises in vast waves without wind.	. . . .	Hills clear of clouds and fogs. Noises in caverns or wells like a storm.	Hollow rumbling sounds in the clefts of the earth, wells, &c.
2. Dr. PEYSSONNEL, Captain ANDREW.	West Indies.	Said to increase its temperature.	Sea birds come to land.		
3. Dr. MORRISON, U.S.S. *Plymouth*, 1849, Chinese fishermen, &c.	China Sea.	Gives a bellowing sound, & boils up, dashing loose rocks against each other. Sea weeds plentiful, nets come up covered with much muddy sea.	Water fowl fly about.		
4. Captain ANDREW.	. . . .	Temperature of the sea said to increase.			
5. Colonel REID.	Bermuda.	Sea assumes a muddy or brown colour.	Swell rolls in, though hurricane is at 600 miles distance.		
6. ULLOA.	Western Coast S. America, Coast of Peru and Chili.	Birds called by the Spaniards Quebrantohuessos (probably the *Falco ossifragus*) are seen to appear in great numbers on the shore and ships.			
7. *Derrotero de las Antillas*, as translated in PURDY's Atlantic Memoir.	Coast of Mexico, for the *Nortes*.	Sparkles.	. . . .	Moisture on walls and pavements. Mountains clearly defined; some with thick clouds like a sheet, distant hills remarkably clear.	

Tabular view of the TERRESTRIAL *and* OCEANIC *Signs, stated by various Authors to indicate the approach of Cyclones.*

Authority.	Place of Observation.	The Sea.	Sea Shore.	Land Generally.	Notes.
8. Mr. GITTENS, of Barbadoes, quoted by Col. REID, p. 36, and Bermuda Gazette.	West Indies, Barbadoes.	. . . .	. . . .	The motions of the branches of trees; not bent forward as by a stream of air, but constantly whirled about.	
9. Mr. COCKS, Annual Register, 1771.	West Indies.	Remarkably clear, so that the lead is seen at great depths. Boils, as it were, before the wind comes on.	. . . .	. . . .	
10. Capt. MCLEOD, ship *Albion*, Captain WILLIAMS, in ROMME, Vol. II. p. 149.	Bay of Bengal.	Vast numbers of Turtle floating in the calm preceding the Cyclone, *apparently in a state of stupor.* Gossamer webs appearing in the rigging.			
11. Capt. WICKHAM, H.M.S. *Beagle*, and *H.P.*	W. Coast of Australia.	Great increase or entire change of usual currents and sets of tides.	. . . .	. . . .	
	Bay of Bengal Sand Heads.				
12. Mr. BOND.	Balasore.	Tides earlier (or later) than usual.	. . . .	. . . .	
13. Dr. MACGOWAN, Jour. A. S. Vol. XXV. p. 368.	Ningpo.	. . . .	. . . .	Water rising in wells and ponds.	Probably brought from the shore by the approaching Cyclone.

PART V. § 842.]   *Remarks on Table of Signs.*   283

REMARKS ON THE FOREGOING TABLE OF SIGNS.

341. RED SUN, SKY, AND LIGHT.—In this table we find that the redness of the sun, is a commonly known sign from the Mauritius to China and the Pacific. It is also noted by Virgil for the coasts of Italy.* We shall examine this with the remarkable red colour of the clouds and of all objects, which is so well known at the Mauritius as a precursor of the Cyclones there.†

It is very certain that this phænomenon of red sky, red clouds, and red light does occur as described, and moreover that it must therefore occur under unusual circumstances; for were the conditions which occasion it usual ones, it would frequently be seen, and be no sign at all. It is its rarity, its appearing only at one season, generally, we may suppose, closely preceding the Cyclones, which has caused it to be specially noted as a sign.

It is very difficult to give the plain seaman an idea of the causes which may produce this apparent excess of the red rays of light over the others; and indeed meteorologists would not altogether agree in their views were they to undertake the explanation, not only of the appearance itself, but of why it should be seen at particular times; and it is not the object of this work to explain *causes*, but rather to deal simply with effects, and to point out here and there the researches which may lead to the discovery of causes. Hence we shall only briefly say, that *probably* the red colour of the sun and sky on this occasion is due to the absorption of the greater part of the blue rays of the sun's light,‡ which thus will leave only the red and yellow rays to form red, orange, and yellow lights of various shades and intensities, the greens and violets being so faint, from the want of blue, as to lose all power of acting their part in producing pure white light.

342. How this absorption and refraction of the rays (for both probably occur) take place, we cannot readily say. We know by the

---

* L Georg. v. 453.

† On a recent occasion it occurred in Calcutta; and the general remark amongst all persons who knew the Mauritius, and with Creoles of the island residing in Calcutta, was immediately, that a hurricane would have been expected at that island.

‡ Every seaman, I hope, knows that the pure *white* light of the sun is a compound of the seven prismatic colours, which are themselves compounded from the red, yellow, and blue rays.

experiments of HASSENFRATZ and others, that the light of the sun passing through dense strata of the atmosphere, or as at sunset, through a much greater extent of it, loses a large portion of its blue rays, and we may suppose *theoretically*, when the redness occurs at noon-day, that not only the light has to force its way through a dense atmosphere, but also through clouds of at least three different kinds, the common fog-clouds, the invisible clouds (266,) and perhaps snow-clouds.*

It is also not improbable since, as will be presently shewn, it occurs by moonlight also, that it may be really an electric phænomenon, or due to the polarization of light by the disk of the Cyclone.

The attentive mariner, then, will be satisfied to know that this red light *must* be produced, like the Cyclone, of which it is in certain parts of the globe the warning sign, by some peculiar state of the atmosphere, in which causes are at work that he is ignorant of, but which will produce effects against which he must be on his guard; and that it is not a mere chance occurrence, which has been superstitiously made a *sign* of by the ignorant. The blue light mentioned by Colonel REID, p. 36, as having occurred at Barbadoes and Bermuda is very remarkable.

343. For such of my readers as may not have seen, and may therefore be desirous of having some description of this red appearance, I insert the following:

Mr. BARNETT, passenger in the ship *Exmouth*, in the Cyclone of May, 1840, of which the log is printed by Mr. THOM, Track *s.* on Chart II. says in a letter printed in the Calcutta *Englishman*, which very graphically and minutely describes their distress—

"On the morning of the 30th, we witnessed a most extraordinary phænomenon; the day appearing to break full an hour before its time, though there was no appa-

---

* Clouds which are well-known to aeronauts and to travellers in high mountains, and are in fact the frozen fog-cloud. Then each of these clouds may be, and probably is, differently electrified, which again may vary their properties of reflecting or refracting the light which reaches the earth. The fact that even strata of air (invisible clouds?) are in opposite states of electricity during a calm, has been distinctly shewn by M. M. PELTIER and BECQUEREL. See *Peltier sur les Trombes.* Introduction, p. VII. to IX.

rent break in the heavens, from which it could be said light broke; all was seen through the medium of bright crimson. Sails, men, sea, and even gray clouds, appeared as acted on by a Claude Lorraine glass; it gradually decreased till sun-rise.*
The 1st of May was one of the most lovely days I ever remember to have witnessed, and on it, and the 2nd, we made the first two fair runs we had had, but the sea continued still unaccountably high, and was running against the wind."

The *Exmouth* was dismasted on the night of the 3d and morning of the 4th. Captain BIDEN, in the log of the *Princess Charlotte of Wales*, Track *k.* on Chart No. II. says—

"At sunset on the 26th, though dark and cloudy, the sea was completely tinged with a red colour."

In a newspaper account of the *Northumberland's* Cyclone, Track *q.* on Chart No. II.,

"The approaching storm was indicated in the morning of the 6th by a fall in the barometer of one-half *inch* (?) and a peculiar brick-dust, hazy appearance."

In the log of the ship *Sulimany*, in a heavy Cyclone in 10° S. 85° East in April, 1848, in which she was involved in the centre, occurs the following remark, after the shift of wind when it was " impossible for it to blow harder :" " From 8 to 10 P. M. vivid lightning with a remarkable red appearance to the S. E. *throughout the night.*" No lightning is noted before, and this it will be observed is at night and (the wind being at S. W. and West) in the direction of the centre of the Cyclone. See also, at p. 149, Captain WELLER's notes in the ship *Earl of Hardwicke*, in which the red lurid haze and red tint of every thing on board, is noticed.

In the newspaper account of the Mauritius Cyclone of January, 1844, in which five vessels were lost on one spot only, and much other damage was done, it is remarked, that " In the course of the day the wind had shifted from N. E. to N. W. *and the sun went down in the midst of frowning clouds of a lurid red colour,* which but too truly foretold the approaching bad weather."

The following is from an old work, reprinted in the Nautical Magazine, Vol. for 1841, p. 666, under the title of "Eolian Researches," and I extract it entire as shewing that it is probably sometimes an

---

* Mr. THOM, p. 92, gives the Log, which says that the appearance lasted for five minutes.

atmospheric appearance, and not owing to reflections or refraction of clouds:

"Sometimes there appears first, like a flaming cloud in the horizon, from whence proceeds the fiery tempest, in a most astonishing manner, and some of these hurricanes and whirlwinds have seem'd so very terrible, as if there had happened one entire conflagration of the air and seas. I was inform'd by Captain PROWD of Stepney, a person of great experience and integrity, that in one of his voyages to the East Indies, about the 17th degree of South latitude, he met with a tempest of this nature, towards the coast of India; of which I had some particulars extracted from his Journal: First, contrary to the course of the winds, which they expected to be at South-east, or between the South and East, they found them between the East and North, the sea extremely troubl'd, and, which was most remarkable and dreadful, in the N.N.W., North and N.N.E. parts of the horizon, the sky became wonderfully red and inflam'd, the sun being then upon the meridian. These were thought omens of stormy weather, which afterwards happen'd according to their suspicions; and as the darkness of the night increas'd, so did the violence of the wind, till it ended in an extreme hurricane; which an hour after midnight, came to such an height that no canvas or sayles would hold; and seven men could scarce govern the helme. But that which I mention as most considerable to our purpose, was, that the whole hemisphere, both the heavens and raging seas, appeared but as one entire flame of fire; and those who are acquainted with the reputation of this grave person will find no just reason to distrust the truth of the relation."

In the Northern Pacific, in the H. C. S. *Buccleugh's* Log, Tracks *bb, cc, dd*, on Chart No. IV. Mr. LYNN says—

"At sunset the clouds predicted another severe typhoon. This appearance was that of remarkable dense and large clouds surrounding the horizon at an altitude of about 10° or 15°, having thin edges tinged with a deep crimson border, as if bound with a ribbon of that colour, and reflecting an awful redness upon the sails, *which appearance had also preceded the former gales*, and which I shall ever conceive are certain indications of their approach."

344. In DAMPIER's description of a China Sea Typhoon, in the second chapter of his Voyage to Acheen and Tonquin, many of these signs, and particularly the red bank of clouds as above described by Mr. LYNN, and our banks of clouds of the preceding section,* may be recognized. He says—

"The typhoons are a sort of violent whirlwinds which reign on the coast of Ton-

---

* In this and other quotations from DAMPIER, I am either quoting at second-hand or translating, (mostly from the French), having been unable to procure the work in Calcutta! The Italics are mine.

kin in the months of July, August, and September; they happen generally when the moon changes or becomes full, and are almost always preceded by fine, clear, and serene weather, accompanied by light and moderate winds. These light breezes vary from the usual wind of this time of the year, which is more from the S. W. and become North and N. E.

"*Before these whirlwinds come on, there appears a heavy cloud to the N.E. which is very black near the horizon, but towards the upper part it is of a deep dull reddish colour, higher up it is more brilliant*, and then to its extremity it is pale and of a whitish colour which dazzles the eyes. This cloud is frightful and alarming, *it is sometimes seen twelve hours before the whirlwind comes.*[*] When it begins to move with rapidity, you may be sure that the wind will blow fresh. It comes on with violence, and blows for twelve hours more or less from the N.E. It is also accompanied with terrible claps of thunder, with sharp and frequent lightning, and rain of excessive violence. When the wind begins to abate, the rain ceases all at once and a calm succeeds. This lasts thus an hour more or less, then the wind coming from about S. W., blows with as much violence from that quarter, and as long as it has blown from the N. E."

345. In the West Indies and Atlantic, we do not find in the published works this appearance of a red sky and red light sufficiently marked to assume it as a frequent, if not almost a constant sign of an approaching Cyclone. At pp. 75 and 88 of Colonel REID's work, we find it casually noticed only, but not in such connection as to authorize us to call it a premonitory sign there. I am, however, informed by Mr. J. PALMER, Chief Officer of the ship *Charles Kerr*, that before the Barbadoes Hurricane of the 26th July, 1832 or 33, in which he was wrecked in Bridgetown harbour in the ship *Pacific*, the red sky was distinctly seen at sunset and during the night preceding the hurricane. It was like a cloud dark above and red below, which hung all over the anchorage. This cloud came in from the S. W. and the Hurricane began at ½ p. 9 A. M. next day.

In a chronological list of hurricanes in the West Indies, mostly from the Annual Register, published in the *Nautical Magazine* for Sept. 1848,—it is said, speaking of the great hurricane of 1780 at Barbadoes, that "The evening of the 9th, preceding the storm, was remarkably calm, but the sky was *surprisingly red and fiery* ; during the night much rain fell, and at 10 A. M. on the 10th, the storm com-

[*] This is exactly a bank of coloured clouds indicating a Cyclone coming down from the N. E. or E. N. E. like many of our tracks thereabouts.

menced. In a detailed notice of the recent Antigua Hurricane of 21st August, 1848, the afternoon and evening of the day on which the Cyclone occurred are thus described—

"During the afternoon and latter part of Monday last, heavy masses of clouds gathering imperceptibly from all directions, and hanging motionless, together with a sensation of oppressive heat and closeness, as if a vacuum existed in the atmosphere, occasioned some comments, but as the barometer did not indicate unusual severity of weather, it was conjectured by the prophetic, that a dash of rain, accompanied, perhaps, by a few electrical discharges, would be the only result ; *the redness of the sky* and the sudden bursts of occasional eddies of wind at sunset, however, produced some anxiety, though not sufficient to induce any very precautionary measures of security. As night came on, these gusts of wind increased in severity, and serious fears of an approaching storm began to be entertained. At 11 there was every appearance of an approaching tempest, though the mercury had only fallen one-tenth. By 11, the wind was raging furiously, with incessant vivid flashes of lightning, thunder, and floods of rain. A few minutes before 1 o'clock the mercury fell two-tenths in an incredibly short space of time, and by half-past one it had fallen two-tenths more, being then in several barometers at 29.40. By 2 o'clock it had risen three-tenths." This extract probably gives a very fair average account of the onset of a West India Cyclone. In this case, it was passing to the Southward of the Island ; the veering being from N. and N. E. to East, S. E. and S. S. E."

In the Atlantic Ocean, by a report forwarded to me by Captain HUTCHINSON, of the Bark *Mandane* of Liverpool, of the Log of the British ship *Shakespear* in a Cyclone of which the calm centre passed over her in about Lat. $32°$ N., Long. $79°$ W. on the 18th Oct. 1848, Barometer stated to have fallen to 28·00, it is stated that—

"For three days before they had the gales." " There was about an hour, or near before sunset, a blood red sky to the Westward, and also to the Eastward. The three strata of clouds were shaded with red which seemed to be reflected from those to the Westward, the whole having a peculiar appearance which had never been seen before."

346. Since the first edition of this work, I have found two undoubted and highly remarkable instances of this red light, occurring in the Bay of Bengal *and at night*. The first is from Captain NORMAN McLEOD of the ship *John McViccar*, who was in the ship *Albion*, in the *London's* Cyclone of October, 1832, track *i.* on Chart No. III. His letter says, after describing five days of oppressive calms and other signs, which I shall elsewhere allude to, that on the evening of the 5th, in Lat. about $14°$. $50'$ N. Long. $89\frac{1}{2}$ East,—

"At sunset the sea and sky became all on a sudden of a bright scarlet colour, I do not remember ever seeing it so red before, even to the very zenith, and all round the horizon was of this colour. The sea appeared an ocean of cochineal, and the ship and every thing on board looked as if it were dyed with that colour; the sky kept this appearance till nearly midnight, and it only diminished as it came on to rain. No sooner was this phænomenon over, than the sea became as if it were all on fire with phosphoric matter. We took up several buckets of water, but even with the microscope few or no animalcules were detected. Having lost my log, I cannot give you the temperature of the water."

It will be observed that here the red light was prevailing from sunset till nearly midnight, or for 4 or 5 hours, and the moon was full on the 10th, so that she was then 10 days old, and must have given a good light. The second instance is in the recent Cyclone of October, 1848, in the Bay of Bengal, forming the subject of my 18th Memoir Jour. As. Soc. Beng. Vol. XVIII. It was observed by more than one of the many ships which felt the Cyclone, but by none so carefully and remarkably as by Mr. VAILE, Chief Officer and then in command of the *Barham*, who stated to me very carefully every particular in addition to his capital log. In the sequel to the abstract of the Log I have given the substance of this statement as follows:—

"In this case too, we have the singular, and for scientific purposes the very valuable peculiarity, that the red sky occurred at night, viz. from 2 to 4 A.M., and at a time when the moon was shining as brightly as it could for the clouds, it being the day before the full moon when she had at that time an altitude of 40° or 50°.

Mr. VAILE states that at this time, the whole sky was clouded with dense, heavy looking clouds, some of which were opposite to the ship on the side of the moon. The red colour extended over all, but was in patches, deeper in some parts than in others, and that some clouds facing the moon, were of a very deep orange red, and that occurring at night it was more particularly remarked."*

With respect to the instances we have cited, we should also

---

* Captain SHAW in a letter describing the hurricane referred to above, at p. 238, on the 14th Dec. 1849, says, (on the evening of the 13th) "observed that *peculiar and true sign the dark blood red sky*, extending all along the N. eastern quadrant, and observed to his officers to 'remember,' and see if this was not the forerunner of a breeze not far off."

remark that this light must sometimes have been *reflected* and at others transmitted, and therefore *refracted* light. It is to be hoped that future observers will not omit to notice this phenomenon more in detail, so as to inform us of the state of the whole sky at the time, position of the sun or moon, and every other particular.

Captain MILLER, whom I have quoted at p. 275, informs me that in a Cyclone encountered by the Screw Steamer *Lady Jocelyn* in the Southern Indian Ocean, of which an account is given in the *Naut. Mag.* for 1854, and which occurred in the storm Tract in March in that year, the red light, which was very remarkable, did not appear till the Cyclone was moderating.

The following phænomenon, observed by myself at Calcutta, may perhaps contribute to explain this of the Red Light.

Being, on the 14th March, 1851, at sunrise, on a broad road (the old course) which divides the Esplanade of Fort William into two parts, I observed that I was in the centre of a low fog bank, perhaps not more than 40 or 50 feet, at most, in height from the ground, and which was distributed over the Esplanade in unequal wreaths like the layers of a stratus cloud. On the metalled road on which I was, there was little or no fog, but it extended all over the grass of the Esplanade, and the breadth of the fog bank to the East was as nearly as possible $\frac{1}{4}$ of a mile. I was thus in the middle of a broad, flat, disk or stratus cloud, much as we may suppose a Cyclone to form upon a large scale, or when upon its descent it brings down colder air from above and condenses that nearer the earth's surface through which it passes. The sun was at this time rising, and as soon as its disk was clear of the houses to the East (Chowringhee Road), which might require from 5° to 7° of altitude, a very remarkable phænomenon took place, which would seem to *assist* towards the explanation of this appearance. There was seen below the sun a complete semicircle, of a glowing crimson red, but hazy, light; the diameter of the semicircle passed in a horizontal line through the sun's centre and extended about 4° or 5° on each side of him, and at its greatest intensity the line of the diameter about coincided with the upper level of the fog; so that the effect was that of a section of half a complete cone of red rays passing through a low, flat fog bank, and refracted to the eye. As the sun rose the whole became indistinct, the red changing to yellow, and the form of the halo* changed from a semi-circle to a hazy yellow oval.

347. APPEARANCE OF THE STARS. It is probable that these bodies may often afford to the attentive observer a warning, in corroboration at least of other signs. Captain LANGFORD, in our table notes only a " burr" about them, which must be common in all hazy states of the atmosphere; but Captain RUNDLE mentions a " dancing sickly appearance" of them.† In commenting upon this in the Memoir in which his log is published, I have remarked that it may be owing to wreaths of vapour, and the intervals between them; and I have also adverted to the tremors of objects seen through telescopes in the mornings, as occasioned by the rarefaction of different strata of it, an appearance and cause usually well known to intelligent seamen. Meteorologists are also well agreed as to the usual causes of the scintillation of the stars, which I take to be the "dancing" alluded to; only that it was so excessive that it gave an idea of motion rather than of mere scintillation (sparkling); but they add also, what we should in future note, that apparent changes of *colour* as well as of brightness and place occur in the stars, especially at the times when they are most scintillating; or the changes of colour may occur with the planets also, though they rarely sparkle. Thus a blue or red star may change to the opposite colours, or become remarkably silvery and white for a short time, and then return back to its usual colour. All these are matters to be noted and carefully registered; they may be the first indices of the actions going on in the atmospheric strata above, and, by their frequency and constancy in certain circumstances, may be far more useful as premonitory signs than we at present suppose.

In three recent instances, all noted by good observers, this remarkable brightness and twinkling of the stars with an atmosphere so clear, that the rising and setting of the stars could be almost seen like that of the sun and moon, has been noticed; the first by Captain SHIRE in the China Sea, preceding a Tyfoon off the coast of Luconia (Seventeenth Memoir, Jour. As. Soc. Beng. Vol. XVIII.); in the second by Mr. VAILE in the *Barham* in the October Cyclone of 1848

---

* Or rather " glory," if *halo* is taken to mean a narrow circle.

† Which was carefully observed; for the log says, " I thought at first my eyes deceived me, but my mates observed the same; I suppose occasioned by some dense vapour."

in the Bay of Bengal, and in the third instance by myself at Calcutta at the time of a severe and apparently stationary Cyclone at Chittagong, 120 miles from Calcutta, in May, 1849. Captain SHIRE states that at Singapore it is perfectly well recognized as one of the warnings of the approach of a Tyfoon in the China Sea.

In a letter received from Captain OSGOOD, of the ship *Horsburgh*, he states that this remarkable clearness of the night atmosphere enabling the stars to be seen close to the horizon was also observed off the Cape of Good Hope previous to the onset of a Cyclone in Lat. 37° 20′ S., Long. 12° 49′ West.

Captain HUDSON, Master Attendant of Vizagapatam, states in his report of the May Cyclone of 1851, (see 21st Memoir Jour. As. Soc. Beng. Vol. XXI.) that this remarkably clear atmosphere, smooth deep blue sea, bright stars at night, and distant hills of wonderful distinctness are from his experience invariably the forerunners of a storm on that coast.

At Patchung San, Northern Pacific, on the day preceding the U.S.S. *Saratoga's* Tyfoon (see p. 278), it is remarked that the atmosphere was "very clear, and distant objects seen with great distinctness."

GREEN LIGHT OR SKY. Captain Duncan, of the ship *Duke of Wellington*, in describing the weather previous to the onset of a Cyclone gale in from Lat. 33° to 37° South, Long. 80° to 83° East, says,

"At sunset we had a beautiful sky to the Westward; light hazy clouds shaded from deep crimson to the lightest pink, *with streaks* of green between them; near the horizon the green was of a very deep colour. My passengers were all admiring it, I told them that old sailors said that green in the sky betokened no good, and so it proved with us."

Captain JONES, of the barque *West*, describing an approaching Cyclone in the South Indian Ocean, says,

"Towards sunrise, blowing very hard, &c. A heavy black bank of clouds at N.E. which had wooden looking tops, and between the breaks in the bank appeared tinged with a dull orange colour. *They were very green.*"

The light shade of green at twilight is a well known optical phænomenon (see *Kaemtz's* Meteorology, p. 499) and may be frequently observed in clear weather forming part of the colours then illuminating the sky, and generally in succession to the yellow, but when it becomes so strong as to be remarked "in streaks," as here described, it is doubtless an indication of some derangement in the usual state of the atmosphere, to which the sailor who wishes to have time at his command cannot too closely attend.

348. CLOUDS. Amongst the cloud signs, light divided portions of them carried rapidly on with irregular motions, seem to be one of the commonest signs, after the banks of clouds and red light of which we have just treated in detail.*

349. LIGHTNING. I have noticed in a preceding page the peculiar kind of Aurora Borealis-like lightning noticed by Captain RUNDLE, and the following is his remark at length, with my note upon it, XI. Memoir. Jour. As. Soc. Beng. Vol. IV. :

"I observed those modifications of lightning more like the Aurora Borealis, which I have seen in the north sea, or rather more like the Aurora Australis, which I have seen off Van Diemen's Land and New Zealand. I have never seen it in low latitudes, but as a precursor of strong weather. It gradually lightens up the Western horizon with a sudden dark red glare, and thus flickers about for a few seconds and gradually disappears.

"Again visible to the W. S. Westward, the sullen red glare and flickering lightning; midnight squally, sea presenting flashes of phosphoric light in all directions."

My note is as follows :

"I have found while correcting this page for the press, a single instance in which this remarkable kind of lightning is described. It occurs in one of the replies to a circular, addressed, at my suggestion, by the Hon'ble the Court of Directors, E.I.C., to their retired officers, requesting information on storms in the Indian Ocean and China Seas, by Captain JENKINS, then commanding the H.C. ship *City of London*: who says, speaking of an approaching hurricane in March, 1816, in Lat. 12° to 16° South, Long. 78° to 76° East, for which, warned by his barometer, he was preparing : ' At 7, the appearance of the atmosphere altered, constant vivid lightning resembling in the distance the Northern lights, with frequent hard gusts of wind,' &c. We are not to suppose from its being so unfrequently noticed that it is therefore of unusual occurrence ; seamen are so accustomed to lightning, that they rarely take the trouble to describe it."

The following note is by Captain STEWART, of the ship *Rajasthan*, printed in my Fourteenth Memoir. It refers to the approach of a small but severe Cylone, experienced by him in the Arabian Sea—

* *Double* arched squalls too have been noticed in the Bay of Bengal and Southern Indian Ocean, when in close proximity to Cyclones. By double arched squalls I mean squalls rising with two distinct and sharply defined black arches, one within and below the other at some 10° to 20° distance, and usually rising with great rapidity and blowing very sharply while they last.

"On the evening of the 4th December,* I observed a remarkable kind of lightning to the North Westward, shooting up perpendicularly from the horizon in stalks or columns of two and three at short distances; it was not at all bright, but rather of a dullish glare."

In the October Cyclone of 1848, in the Bay of Bengal, just referred to, Mr. VAILE also notes in the *Barham's* Log, at the approach or rather shortly before the *settling down* of the Cyclone, that "the lightning has a very peculiar appearance similar to the flash of a gun," and upon farther conference with that gentleman he compared it both to the flash from a gun, and at times to sparks as if from a flint and steel, and altogether a most remarkable kind of lightning.

350. If we consider that the bank of clouds seen at the approach of a Cyclone is the edge or side of a disk, such as that shewn in our Barometrical Chart, p. 246, and admit that extensive electric action is going on towards the centre, while the outer part is a dense ring of cloud, we shall understand how this Aurora Borealis-kind of lightning may be the reflection of a series of continued discharges seen above the wall of clouds, just as we see the distant red sky of a fire, behind masses of houses or trees, or the reflected flashes of the discharges of a volcano behind a range of mountains. Continuous electric discharges (such as, by the way, no European electrician has ever dreamed of), are by no means uncommon in tropical climates, and especially amongst the Eastern islands. In the Java sea, off the South Coast of Borneo, it is no exaggeration to say, that the lightning sometimes *pours* down in cascades or columns from the clouds, and this in four or five places at once! and in Madras roads the discharges of sheet lightning behind, or rather above, thin transparent stratiform and cirrhous clouds, covering the whole sky, are sometimes so incessant for hours together that small print may be read by them; they resemble, in fact, the corruscations of the glow-worm or fire-fly. Mr. FARADAY, in the Philosophical Magazine, Vol. XIX. 1841, notices a "distant illumination of the clouds," which he thinks, if I recollect rightly, is simply a reflection of distant lightning on the edges of clouds, but no European lightning can give any idea of the terrific

---

* His Cyclone occurred on the 5th, and was then travelling up to him from the S. E. *if it was formed* (of which we have no evidence), but there *was* another, and this is very remarkable, raging to the N. N. W. of him at this time with the ship *Monarch*, at about 2° of distance.

magnificence of that of the Eastern Archipelago and Eastern Seas in general.

Amongst the terrestrial signs we find, at p. 281, that Captain LANG-FORD, in 1698, notes that the Caribs (Indians) had informed him amongst their signs the sea "had a stronger smell than usual." In a letter from Captain SPROULE of the ship *Magellan*, describing the August Tyfoon of 1850, in the China Sea, he says, "the sea-birds also flew high and wild, the sea also was agitated, *and had a very strong unpleasant odour.*" This was in Lat. 15° North and 112° 53′ East, or about the middle of the China Sea. It might really be a strong odour evolved from the sea or a peculiar sensitive state of the nervous system; but this might perhaps be settled if a few bottles of the water were filled at the time and carefully corked, and as many more a few days after, when the Cyclone was over and fine weather had returned. Care should be taken that the bottles are clean and the corks new, and the whole should be placed in the hands of a first-rate chemist for examination.

851. SEASONS AT WHICH CYCLONES OCCUR. I have placed here what we *as yet* know of the times at which Cyclones have been known to occur in various parts of the world, from which the mariner may deduce, *with some general probability*, the chances of one on his voyage. On this our information at present is very imperfect (except perhaps for the West Indies), but nevertheless it may be found useful now and then to look at the following table, in which the number of Cyclones is marked for the years of which we have any record. The use of it will be seen at once to be, that in running the eye along the line, of the West Indies for example, we see that in certain months more or fewer Cyclones occur; and in some none at all are recorded during the long period of years shewn by the first column. It is by no means to be supposed that we have any accurate registry of the meteors for any part of the world, but the numbers given *may* approach nearer to correctness than we should at first imagine, because if a Cyclone was to occur out of the usual months or seasons it would be much remarked.

The seaman also should not fail to keep in mind, what I have said of the difference between Monsoon-Gales and Cyclones; and that his monsoon or trade may rise to the strength of a very strong gale, but it will still be steady, and his barometer without depression.

352. *Table of the average number of Cyclones in different months of the year, and in various parts of the World.*

For what No. of years ascertained.	Locality.	Authority.	Jany.	Feby.	March.	April.	May.	June.	July.	August.	Sept.	October.	Novr.	Decr.
123	West Indies, ...	Naut. Magazine,.....	..	..	..	..	..	1	2	13	10	7	..	..
59	....	U.S. Journ. 1843, p. 3.	..	..	..	..	..	1	5	13	13	9	..	..
300	....	Roy. Geograph. Soc.	5	7	11	6	5	10	42	96	80	69	17	7
39	Southern Indian Ocean, 1809 to 1843	Reid, Thom, H. Piddington, ...	9	13	10	8	1	..	..	1	1	4	3	
24	Mauritius, 1820 to 1844........	M. A. Labutte, Trs. Roy. Soc. of Mauritius, 1849	9	15	15	8	..	..	..	..	..	6		
25	Bombay	....	1	1	1	5	9	2	4	5	8	12	9	5
46	Bay of Bengal, 1800 to 1840 ..	H. P.	1	..	1	1	7	3	..	1	..	7	6	3
64	China Sea, 1780 to 1845	H. P., Capt. Kirsopp ..	..	..	1	2	5	5	18	10	6	..		

In the former editions of this work I have stated that no Cyclones have been known to occur in the month of May in the China Sea; but from a letter from Captain E. T. F. KIRSOPP, commanding the Steamer *Juno*, at Manilla, I learn that a severe Cyclone was experienced in the Bay of Manilla and in the adjacent China Sea as early as the 4th May, 1850. Capt. KIRSOPP forwards me brief extracts from the logs of three ships, besides that of his own vessel, regarding this Cyclone, which like that of the *Easurain* (Track Z in Chart IV.) appears to have had a track out from the Bay of Manilla to the northward; but not having the detailed logs of the ships and their position being often deficient also, nothing very positive can be affirmed. The essential matter however for us at present is the fact, that severe Cyclones may occur in May, in the China Sea, and thus upon the appearance of doubtful weather or an uneasy Barometer, the careful seaman will take due precaution.

353. WHIRLWINDS AND WATER-SPOUTS. There seems, as far at least as effects are concerned, so complete a gradation from the trifling and harmless dust-whirlwind, to the larger and mischievous-ones of the same kind on shore, *all of which become water-spouts when they reach or cross water,** to the mischievous fine weather whirlwinds which have dismasted ships at sea, and from these again up to the great water-spouts and the smaller tornados or tornado-Cyclones, that a book on Storms would be incomplete that did not advert to them. I shall perhaps do so at some length, but I desire to say here, that I do it without especially advocating any particular theory, but rather as desirous of indicating from various facts what appears highly worth inquiring into.

354. The seaman cannot be too often told that a theory, whether directly proposed or hinted at by a writer, or framed almost unconsciously by himself, is not a *rule*, but a torch to assist in guiding him, and to be exchanged for another as soon as he finds a better one; and that it often occurs, that by the very destruction of the wrong theory by well observed facts, we get hold of the right one. I proceed first to describe these phænomena by my own knowledge, or by extracts in the order in which I have named them.

355. The simplest forms of whirlwinds are such as are not unfrequently seen in Europe, but in tropical climates they are far more frequent, and indeed in some countries and districts, during the dry season they occur daily, and often in numbers. On the plains of India they are seen most frequently in the mornings, when one or more slender whirling columns of dust, leaves, &c., are suddenly seen to rise and to move about, at first slowly, and then to start off as it were on some line of direction, whirling light bodies about as they pass them, and when they meet with obstacles, as trees, houses, or the like, sometimes passing over them, and at others being broken up and lost amongst them. In open spaces they become fainter and disappear.

356. No perceptible change of temperature accompanies these little whirlwinds, and I am not aware that any electric observations have been made on them. They appear in India, by the evidence of a friend, Mr. J. BRIDGMAN of the Goruckpore district, to turn indifferently either to the right or left, *i.e.* either ◯ or ◯. This gentleman has at my request taken the trouble to observe several, with a view to ascertain in what direction they rotated, and he says—

* Of which we shall give instances.

"I forgot to communicate to you my observations on the petty whirlwinds which occur in the cold weather, and about which we had talked two years previously, as possibly having a similar origin to the greater ones, which constitute storms and hurricanes. I watched for them with care, but it was a long time during the season, 1839-40, before any of the phænomena in question occurred. There were many whirlwinds, but always in a strong breeze of which I considered them mere eddies. They turned indifferently N. E. S. W. and N. W. S. E., but generally the former. At length one occurred which was to the purpose; the day was cloudless and the air stagnant or nearly so. It turned the right way (for the theory) *i. e.* N. W. S. E., and moved forward at the rate of five miles or so per hour, from N. E. to S. W. I saw no more that season, but the following cold weather, 1840-41, I saw a great number of the kind to be observed upon, viz., those occurring suddenly while the surrounding atmosphere is stagnant and undisturbed. The first three or four were conformable to the theory, and turned N. W. S. E., the next few turned the reverse way, the remainder turned indiscriminately, some one way and some the other, and I became convinced that they depended upon no rule, or at least no rule producing uniformity of motion."

357. In a letter addressed to me from Deesa, on the borders of Kutch and Scinde, Dr. THOM says—

"I have had most extensive observation of the dust-whirls of the deserts of Scinde and this country, and have very copious notes of the circumstances under which they occur.

"They turn in both directions. I have seen twenty in an hour, half a dozen at the same time, and two near one another revolving in opposite directions. They have been equally frequent in dry, clear, and cloudless skies as in approaching storms; in calms as in the sultry air of high winds; never prevailing, however, in a strong gale. They are not seen in the monsoon or winter months, but are most frequent in the transition from the N. E. to S. W. winds, especially in May and June just before the rains set in. I have ridden after them and got *into* their centre on horseback by backing my horse. But I have never had an electrometer to detect the peculiar kind of electricity, which is developed by them."

358. The following notice occurs in the Journal of the late Dr. Griffiths, who was sent as naturalist with the Army to Cabul.

"Whirlwinds are common about Cabul, commencing as soon as the sun has attained a certain degree of power.

"In all cases they assume the shape of a cone, the point of which is a tangent on the earth's surface; the cone varied in shape, is generally of a good diameter, occasionally much pulled out, some being 2,300 feet in height,[*] the currents are most violent at the apex.

[*] An error of the press; 230 must be meant, as ascertained by observation, or "2 *or* 300 ft." probably?

"They come and go in all directions, even after starting, not always preserving the original direction. They are less common on days in which winds prevail from any given direction, and vary much in intensity, from a mere breeze, lightly laden with dust and with no tortuosity, to a violent cone of wind, capable of throwing down a small tent.

"Northerly winds are prevalent here from 1 or 2 P.M. until 8 or 9 P.M. occasionally they only commence in the evening, when they are obviously due to the rarefaction of the air of the valleys by the great heat of the sun, amounting now to 100° at 3 P.M., and the vacuum being supplied by gusts from the high mountains to the north and north-east."

In a report forwarded to the Medical Board of Calcutta by Dr. Baddeley, H.C.S., he says, "Capt. Simpson, Deputy Assistant Quarter Master General, relates, that in Affghanistan he was once witness to one of these columns of dust which remained almost stationary for, he thinks, nearly three hours! This one was several yards in diameter, and he was enabled to approach sufficiently near to remark with accuracy their peculiar phænomena."

359. In one instance, however, carefully observed by myself, and of which the following (abridged) account was sent to the Calcutta *Englishman*, the rotation was according to the law for the Northern Hemisphere, and the incurving of the wind most distinctly marked by the dust.

"On the 2nd April, 1849, at ½ past 1 P.M., I observed the dust rising in a somewhat columnar form, within the inclosure, and just at the gate of the Sudder Board of Revenue Office; and the whirls discernible in the upper part, like those of the smoke of a fire, though as yet thin, convinced me that this was a genuine *Bhoot*, or dust-whirlwind, just forming (or descending, if Mr. Piddington's theory of them be correct). In a moment, trains of whirling dust arose to the north and south of the gates of the Sudder Board, perhaps for about 25 yards each way, in the Chowringhee road, moving up and down to join the main column which now started forward, on about a W.S.W. or W.b.S. track, coming directly towards my carriage, which I had stopped on the cross road in the open Esplanade to observe the phænomenon. The column of dust now made a magnificent and massive, volcano-like, spiral whirl of the dust on the Chowringhee road, but, on crossing it, it became thin and meagre when it reached the grass of the Esplanade, but the whirling and progression were still distinctly seen. When it reached the cross-road, where I was standing ready to watch it, for it crossed only some ten or fifteen yards behind me, it again shewed its extent and power, by raising a large thick whirling column, whirling from right to left outwards (or against the hands of the watch), and distinctly and most evidently forming also what Mr. Piddington calls the incurving, and Mr. Redfield the vorticular spirals; so that at two separate instants perfectly well marked, it resembled at the base a huge turk's-head knot, of red and gray dust, and the column in the midst might well be likened to the strands before they are cut off. I could however allow myself only a moment to note all this, for I ran up and into the midst of it, but could find no central space, and it was not possible to keep the eyes open, and scarcely to breathe in the thick mass of dust, which *seemed* somewhat hotter than the open air I had left. I then stood still, and allowed it to move away from me, which it did at the rate of about six miles an hour. There were also at the edges a number of smaller whirls forming at times, but which did not last, and were evidently drawn into the larger vortex, these were not more than three or four feet high. There was at the time a pleasant breeze blowing from the S.E., and I should estimate the diameter of the *Bhoot* at the base about 10 or 15 yards, and its height at perhaps 25 feet. When within it, there seemed no particular violent wind."

A letter from a friend in Nov. 1852, mentions, that he had also just seen at Calcutta no less than a group of *seven* dust whirlwinds chasing each other in a circle of about 20 yards in diameter. This exhibition lasted nearly two minutes; all were separately revolving from right to left, and the whole "following the leader," also from right to left, and raising considerable dust.

360. In the Journal of the Asiatic Society Beng., Vol. IV. p. 714, Mr. J. STEPHENSON gives several notes on the sand columns which he frequently observed on the sand banks in the bed of the Ganges. He describes them varying from 20 to 100 feet high, and the large ones about twelve feet in diameter, having a whirling motion and remaining perfect several minutes. The natives affirm that persons are sometimes killed or hurt by them. The same account is given of them in the snowy steppes of the Himalaya mountains in MOORCROFT's travels, as follows :

"I had never seen the phænomena of the whirlwind more common than on this plain : It was, perhaps, like that of the Arabian desert on a smaller scale, raising a column of sand suddenly to a great height at one particular spot, whilst all around the air was perfectly calm. In general, these gusts are not at all dangerous, but strange stories are told of their occasional violence in particular spots, and they are said to be sufficiently strong at Digar to carry horses and men off their feet, being accompanied by reports like those of artillery. I can confirm the truth of these last stories to a less exaggerated extent, having heard on the Digar Pass the wind howling through the crags at a very considerable distance, with a noise occasionally like that of a falling stone. Possibly an exposed portion of rock had been blown down."

In Europe also and even in England, whirlwinds, accompanied by loud noises, and which destroy men (and of course cattle) in exposed situations, sometimes occur. In the *Times* of 27th Dec. 1848, p. 6, it is stated that a quarryman of Buttermere and his two sons were leaving their work on Honister Crags, when having reached the top of the mountain they heard the loud sounds of an approaching whirlwind, and being aware of its consequences in their exposed position, threw themselves upon the rocks. The father being unable to retain his hold was carried by the force of the wind some 30 or 40 yards from the summit and dashed to pieces. The Indian accounts then are not exaggerated.

A curious account was given to me of Indian dust-whirlwinds by Mr. Bechendorf, a highly intelligent German resident, educated as a mining engineer, who had run after and penetrated several of them in Upper India, where they are very common, and sometimes of considerable size. He describes them as forming a thick broad wall of dust, through which it was half suffocation to penetrate, but when in the centre it was nearly calm, with nothing but the wall of dust visible. He further told me, that he had seen large ones commence, and that they did so in segments, which afterwards united.

361. In Rich's "Babylon and Persepolis," they are thus described, p. 228:

"Those kinds of *sand-spouts* or whirlwinds which are called in India 'Devils,' are very common in the plain of Shirauz, and often present a very curious appearance, when ten or twelve of them may be seen at once in different places, rising into the air like huge columns. They are generally seen commencing by rising out of the earth with violence like a burst of smoke from a volcano, and gradually extending themselves upwards. The people here say that they are not formed at night, or in the early morning or evening."

362. Col. Reid gives at length the description of the great moving pillars of sand in the Nubian desert, as seen by Bruce, who at one time counted eleven of them together, and describes them as being 200 feet high, though only 10 feet in diameter; and we know, from history and travellers, the undoubted fact that, in the deserts of Africa, armies and caravans have been overwhelmed and destroyed by them from the days of Cambyses to the present time. Colonel Reid also quotes Captain Lyons, describing those of Mexico, which exactly agree with what we have said of them as they occur in India, only that Captain Lyons estimates his as from 200 to 300 feet in height. Lieutenant Fyers describing those of Western India, esti-

mates them at about 18 feet in diameter, and "some hundred" in altitude.

363. The mischievous kinds of these land whirlwinds seem to be nothing more than those just described, but of force enough to destroy houses and men, uproot trees and even to tear, break and throw down buildings, and they may be traced, in accounts from various parts of the world as well as in India, of all sizes, from a few feet up to some hundred yards in diameter, and as occurring in all kinds of weather, and by night as well as by day. Many of these also in passing brooks or ponds, have been known to assume the appearance of water-spouts for the time, and to raise up the water and even the fish with it. HORSBURGH (Introduction, p. 9,) once observed all these phænomena together in a whirlwind. He says—

"I have observed one pass over Canton River, in which the water ascended like a water-spout at sea, and some of the ships that were moored near its path were suddenly turned round by its influence. After passing over the river, it was observed to strip many trees of their leaves, which, with the light covering of some of the houses and sheds, it carried up a considerable way into the atmosphere."

In the Calcutta "Gleanings in Science" for 1829, Vol. I. p. 340, there is the description of a whirlwind near Dacca, after a heavy N. Wester. The writer describes its appearance to be exactly that of a water-spout at sea when seen from a distance, viz., a descending column from a heavy black cloud tapering to a black part near the earth, which gradually increased to the same size as the rest, and appearing from the earth upwards to near the cloud like a fine white smoke. It raised a cloud of dust with bamboos, mats, &c. when it reached the ground in the centre of a village, but upon proceeding to the spot no water had fallen; twelve huts were destroyed in the space of about 50 yards in diameter, but the houses beyond this were uninjured. The roofs of the houses in the centre were crushed flat, those at the sides were thrown onwards. The writer attributes the effect to wind, producing the same effect as "a gun when fired at the earth;" some persons were struck down, but no lives were lost. The writer is also of opinion that the water-spouts at sea, of which he says he had observed many, do not contain any water as usually supposed.

Humboldt, in his "Aspects of Nature," in describing the Llanos of South America, describes the dust-whirlwinds as being occasioned by opposing currents of air, and goes on to speak of "the electricity-charged centre of the whirling current." I have not found upon what observations this epithet is grounded.

In the Philosophical Magazine for August, 1850, No. 248, is a highly interesting paper by Dr. P. BADDELEY, A. C. S., dated from Lahore, shewing conclusively, by electric experiments extending over the hot seasons of 1847-48-49 and 50, that the dust storms, so prevalent in all the North-Western Provinces of India in the dry months, are purely electrical. I extract the following without abridgment from Dr. BADDELEY's paper. The passages marked in italics are so by myself, as exactly coinciding with the views I have submitted since the first edition of this work.

"My observations on this subject have extended as far back as the hot weather of 1847, when I first came to Lahore, and the result is as follows. Dust storms are caused *by spiral columns of the electric fluid passing from the atmosphere to the earth:* they have an onward motion—a revolving motion like revolving storms at sea, *and a peculiar spiral motion from above downwards like a corkscrew.* It seems probable that in an extensive dust storm there are many of these columns moving on together in the same direction; and *during the continuance of the storm many sudden gusts take place at intervals,** during which the electric tension is at its maximum. These storms hereabouts mostly commence from the N. W. or West, and in the course of an hour, more or less, they have nearly completed the circle and have passed onwards.

"Precisely the same phænomena, in kind, are observable in all cases of dust-storms; from the one of a few inches (feet?) in diameter to those that extend for fifty miles and upwards, the phænomena are identical.

"It is a curious fact that some of the smaller dust storms occasionally seen in extensive and arid plains, both in this country and in Affghanistan above the Bolan Pass, called in familiar language "Devils," are stationary for a long time, that is, upwards of an hour, or nearly so, and during the whole of this time the dust and minute bodies on the ground are kept whirling about into the air. In other cases these small dust storms are seen slowly advancing, and when numerous usually proceed in the same direction. Birds, kites and vultures are often seen soaring high up, just above these spots, apparently following the direction of the column as if enjoying it.† My idea is that the phænomena connected with dust storms are identical with those present in waterspouts and white squalls at sea and revolving storms and tornados of all kinds, *and that they originate from the same cause,* viz. moving columns of electricity.

---

* These are exactly the terrific gusts which old sailors so well remember in typhoons and hurricanes, and which are in fact beyond the power of words to describe, nor can any one who has not experienced them imagine their violence.—H. P.

† Looking for prey also no doubt.—H. P.

"In 1847, at Lahore, being desirous of ascertaining the nature of dust storms, I projected into the air an insulated copper wire on a bamboo on the top of my house, and brought the wire into my room, and connected it with a gold-leaf electrometer and a detached wire communicating with the earth. A day or two after, during the passage of a small dust storm, I had the pleasure of observing the electric fluid passing in vivid sparks from one wire to the other, and of course strongly affecting the electrometer. The thing was now explained; and since this I have by the same means observed at least sixty dust storms of various sizes all presenting the same phænomena in kind.

"I have commonly observed that towards the close of a storm of this kind a fall of rain suddenly takes place, and instantly the stream of electricity ceases or is much diminished, and when it continues it seems only on occasions when the storm is severe and continues for some time after. The barometer steadily rises throughout."

The author then goes on to describe his observations on the Dew-point, and his manner of observing it. He calculates that Lahore is about 1150 feet above the level of the sea. And he farther describes the kinds of sparks or brushes, and the manner in which electricity acts upon dust and light bodies, and which I do not extract.

He continues describing the dust storm as follows :—

"Some of them come on with great rapidity, as if at the rate of 40 to 60 miles an hour. They occur at all hours, oftentimes near sunset.

"The sky is clear and not a breath moving; presently a low bank of clouds is seen in the horizon which you are surprised you did not observe before; a few seconds have passed and the cloud has half filled the hemisphere and there is no time to lose —it is a dust storm, and helter-skelter every one rushes to get into the house to escape being caught in it.

"The electric fluid continues to stream down the conducting wire unremittingly during the continuance of the storm, the sparks oftentimes upwards of an inch in length, and emitting a crackling sound; its intensity varying upon (with ?) the force of the storm, and, as before said, more intense during the gusts.

"One that occurred last year in the month of August seemed to have come from the direction of Lica on the Indus, to the West and by South of Lahore, and to have had a north-easterly direction. An officer travelling and at the distance of twenty miles or so from Lica was suddenly caught in it; his tent was blown away and he himself knocked down and nearly suffocated by the sand. He stated to me that he was informed by one resident at Lica that so great was its force at the latter place as to crack the walls of a substantial brick dwelling in which the above officer had lately resided, and to uproot some trees about.

"I have sometimes attempted to test the kind of electricity, and find that it is not invariably in the same state, sometimes appearing +and at other times —; changing during the storm."

We have not yet however on record in India, or in any part of the world, an instance in which North-Westers, as they are called, have been found to blow at the same time, with equal force, in the opposite direction in any other place, and the North-Westers of the plains are the dust storms of Upper India, the rain being more abundant in the North-Westers and African Tornado. The wind, it is true, always veers, and sometimes with the force of a pleasant breeze to the opposite quarter, but this is usually after the squall is over, so that we can only as yet call our North-Westers straight-lined winds. Dr. BADDELEY's observations however do much towards the advancement of the theory of an electric origin for Cyclones of all sizes and classes, and I trust they will be continued.

364. We come next to the singular class of fine weather whirlwinds, or those which appear in comparatively fine weather at sea, which suddenly and seriously damage and dismast ships, committing, like those on shore, all sorts of mischief, from carrying away studding-sail, booms, and royal masts, to downright dismasting; these seem, from the accounts which I have met with in print, to occur mostly (or always) in the day-time, and indeed their whirlwind character is not so well distinguished at night. Most seamen have met with the lighter sorts, but the more serious ones are fortunately not common, and indeed we are not certain, from many of the published accounts, that they *are* whirlwinds forming miniature Cyclones, as they are usually confounded with the white squalls, so well known to seamen, and which rather appear to be straight-lined winds of excessive violence, but which are seen approaching, and in fact named from their whitening the horizon with foam, like a mass of breakers, or from appearing at first like a white cloud.

365. Colonel REID quotes the instance of the ship *Sir Edward Paget*, which lost her fore and main mast in a sudden *squall* shortly after leaving Madras Roads; but the newspaper accounts, to which I have referred, are very incomplete, stating that it was a sudden squall only, and that reports were current that both masts were rotten, &c. so that it is impossible to say what were the circumstances.

366. I am indebted to a friend for a very remarkable instance of an American ship, commanded by a Captain FAIRFIELD, the name of

which is not recollected, a trader between America and Europe, being in the middle of the Atlantic, with so little wind that the Captain was in the act of looking up at the flapping sails, and observing to the mate that they must carry on all they could to make a passage, when on a sudden, and without the least warning, the vessel was dismasted and sunk by a whirlwind, a few of the crew and passengers with the Captain escaping in one boat. The Captain was much blamed by some parties for having quitted the sinking vessel too soon, but the general opinion amongst nautical men was that no blame could be attached to him.* This would appear to have been very distinctly the kind of violent fine weather whirlwind, of which we have all heard accounts.

The *Nautical Standard* of 9th Oct. 1850, gives an account of the Maltese Brig *Lady Flora* being "struck by a water-spout," when eighty miles to the westward of Gozo, and immediately foundered. One man alone was saved by the Brig *Maltese*, which was near. I have also been informed by a very credible witness that some years ago a China junk lying amongst other shipping in Singapore roads, was suddenly struck by a whirlwind in fine weather, and sunk at her anchors.

The following is from a newspaper of 1852, describing a White Squall off New York:

"On the 3d inst., as the revenue cutter *Taney*, Lieut. Martin, was proceeding down the bay, she was suddenly struck by a white squall and capsized, and immediately filled. Lieut. Martin and officers, and some of the crew, who were twelve in number, were taken off by the steamer *Thomas Hunt*, and brought to the city. Five men were drowned. The steam-boat *Hunt*, which was coming close on at the time, and pilot boats, the *Yankee*, and No. 8, then in the vicinity, saved the commander, two lieutenants, and the pilot, with thirteen of the crew. Two arm chests were picked up by the *Yankee* full 150 yards from the wreck. When struck by the squall the *Taney* was about a mile below Governor's Island. The squall seemed to fall aboard almost vertically, causing her to capsize and fill in an instant. So limited was the extent and duration of the squall that the pilot-boats and other vessels, within 150 yards of the spot, were becalmed at the time, and immediately afterwards scarcely more than a breath of air could be perceived. Captain Martin also states that it was so sudden that not a ripple was observed to indicate its approach.—*Express*."

---

* I mention these details, as they may serve to trace out the accounts of this extraordinary meteor, which I am told were published in America, in full detail, in the newspapers of the day.

The American ship *Lightfoot*, on her voyage to Calcutta, in June, 1856, in 16.40 N. 86.08 East, was struck by a whirlwind from aloft, which carried away her three topmasts, jib-boom, cross-jack yard, and mizen-mast head at once. *The wind below remained steady at S.W. as before the accident*, which did not last three minutes, the barometer remaining perfectly steady, being noted only ten minutes before the disaster.

367. The bull's-eye (*olho de bove?*) or bull's-eye squalls, of which we hear and read on the coast of Africa, and which the Portuguese describe as first appearing like a bright white spot at or near the zenith, in a perfectly clear sky and fine weather, and which rapidly descending, brings with it a furious white squall or tornado, may be a strongly marked kind of these squalls. The white cloud is alluded to in M. GOLDSBERRY's account of the tornadoes, see p. 223.

In a very careful log forwarded to me by Captain ROBERT WOOLWARD, of the Royal (W. India) Mail Steamer *Thames*, there occurs, on Sept. 30th, 1856, when the ship had reached fine weather after passing to the Southward of a severe Cyclone travelling to the Eastward in the Northern Atlantic, the following remarkable entry of an appearance which seems to have been an instance of the *Bull's-eye*, which fortunately did not descend on the ship.

"30*th September*. Moderate breeze and fine from North, decreasing towards Noon. At 11¼, A.M. observed a small white cloud subtending a diameter of about 2°, and nearly circular, pass the zenith from West to East at the rate of 10 or 12 miles per hour, whirling round contrary to the hands of a watch, at a height of 3 miles; Bar. at the time 29.74, Ther. 69° in the shade. I pointed it out to all the officers of the ship, who were assembled to take the sun's Mer. Alt. The vessel's position at Noon this day was Lat. 40° 18' N., Long. 19.06. West."

In a large collection of logs and notes sent me by the Hon. East India Company, occurs the following in a letter from Captain JOBLING of Newcastle, who, at the time alluded to, commanded the ship *Kandiana*, which seems as it were to connect the whirlwinds with the white squalls and tornadoes at sea on the one side, and it may be with the larger Cyclones on the other; though we have not, unfortunately, any veering of the wind noted. It is remarkable also, as occurring within the dangerous space which I have named the hurricane tract.

"On Sunday, 4th October, 1840, Lat. 13° South, Long. 86° 9' East, at 9 P.M., a hurricane commenced from North (without any warning; it was a dead calm at 8.15, P.M.), which carried away our three topmasts, jib-boom, and sprung the mizen-mast, laying the ship right on her beam ends. It lasted with the utmost fury for four hours, and then moderated as suddenly as it commenced. My barometer rather rose than fell previous to it, and was never below 29.65. This was a most extraordinary occurrence, and so sudden that the men were up shaking the first reefs out, and I had only just time to call them in off the yards before the masts went over the side. I only gave the orders for them to come in, from observing the extraordinary appearance of the water to windward. *It was like a solid mass of breakers coming down like lightning upon us.*"

In the *Bengal Hurkaru* of November 10th, 1831, is the following account of a whirlwind, which seems nearly similar to that of *Le Paquebot des Mers du Sud*, quoted at p. 38.

DIAMOND HARBOUR, Ship *St. George*, Captain WILLS, November 8th, 1831.

"I must now tell you of our misfortunes. On the 15th September, being in Lat. 35° 30' South, and Long. 44° East, we met with a whirlwind, which in a minute carried away our fore-mast by the board, the main-top mast and mizen top-gallant mast, and we afterwards found our jib-boom and mizen-top mast were sprung. Finding the vessel to steer well under jury-masts, the Captain pushed on for Calcutta instead of the Mauritius, and arrived in four months from pilot to pilot."

I have met with no further details of this whirlwind.

Of that description of whirlwind or tornado-whirlwind on shore, which we might suppose as connecting these with the small Cyclones, there is a capital instance given in Vol. XII. of THOMSON's Annals of Philosophy for 1818, p. 49, in the "Account of a Storm in Sussex" in 1729, taken from a scarce pamphlet by RICHARD BUDGEN, in which, after a description of the weather for some days previous to the 20th May, on which day it occurred, it is said to have commenced its ravages at Bexhill, and to have travelled a little East of North to Newingden in Kent, and a few miles farther; that it was for the first two miles about 30 rods in width, but afterwards more than double. It travelled in all about 12 miles in 20 minutes, and it distinctly *whirled from the right hand to the left*, destroying literally every thing in its progress, but its limits were so exactly defined, that though it tore the largest trees to pieces, those in the neighbourhood suffered no injury. The heaviest bodies were moved and carried to considerable distances by it. It is further noted that the spectators had the surprising horror of seeing, at about 20 miles distant, "such unremitting coruscations, together with such dreadful darting and breaking forth of liquid fire at every flash of lightning (in the way of the hurricane from the sea side into Kent), as perhaps has not been seen in this climate for many ages."

This is a remarkable instance of a small tornado, with all the characters of a true Cyclone, as to its turning like those of the Northern Hemisphere and its progress of 36 miles per hour, and which was, moreover, and this is most important to our views, unquestionably an electric phænomenon in a highly concentrated form.

368. HORSBURGH (Introduction, p. 8) says that he has passed through the vortex of water-spouts that were forming, and that there is a whirling motion. Mr. WALKER, of H. M. Dockyard, Plymouth, quoted by Colonel REID, p. 492, says that he sailed through a water-spout in the Bay of Naples, and that its rotation was with the hands of a watch. Both of these cases relate of course to small water-spouts, the following to larger ones.

369. In Franklin's letters there is a paper of extracts from DAMPIER's voyages, which was read at the Royal Society, and one of the water-spouts described in it, which appears to be cited by DAMPIER as an instance of the danger of their breaking, as it is called, over a ship, is well worth quoting here, as shewing that the sea water-spouts are really whirlwinds, and sometimes dangerous ones.

"And now we are on this subject, I think it not amiss to give you an account of an accident that happened to a ship once on the Coast of Guinea, some time in or about the year 1674. One Captain RECORDS of London, bound for the coast of Guinea, in a ship of three hundred tons, and sixteen guns, called the *Blessing*. When he came into latitude seven or eight degrees north, he saw several spouts, one of which came directly towards the ship, and he having no wind to get out of the way of the spout, made ready to receive it by furling the sails. It came on very swift, and broke a little before it reached the ship, making a great noise, and raising the sea round it, as if a great house, or some such thing, had been cast into the sea. The fury of the wind still lasted, and took the ship on the starboard bow with such violence, that it snapt off the boltsprit and foremast both at once, and blew the ship all along, ready to overset it; but the ship did presently right again, and the wind whirling round, took the ship a second time with the like fury as before, but on the contrary side, and was again like to overset her the other way, the mizen-mast felt the fury of the second blast, and was snapt short off, as the foremast and boltsprit had been before. The mainmast and maintop-mast received no damage, for the fury of the wind (which was presently over) did not reach them. Three men were in the foretop when the foremast broke, and one on the boltsprit, and fell with them into the sea, but all of them were saved. I had this relation from Mr. John Canby, who was then quarter-master and steward of her; one Abraham Wise was chief-mate, and Leonard Jefferies, second-mate."

370. Next to this we may place, from M. PELTIER's work, p. 274, who has taken it from the Physico-chemical Institutions of Father PIANCINI, who had the accounts from an eye-witness, the following instance of a Mediterranean polacre, which was entirely enveloped in a spout in the Ionian sea, opposite to the gulf of Syra. The relation is as follows—

"The wind was E. N. E. and against us. The sea was much agitated, the sky covered with very low black and thick clouds, which, the day being far advanced, made complete night before the usual time. Also at once the wind changed to N.E. and we tacked to the Northward; all the sails were taken in except the four large ones. The wind changed anew, and the Captain tacked to the East, and the wind changed again. These changes arose from our approaching the place where the spout was forming. In an instant we were surprised by thick clouds which passed between the sails and the masts. This was the beginning and arrival of the spout. All sail was taken in as quickly as possible. But now the spout had come upon us, it united itself to the sea, and turned the poor polacre about like a wooden shoe. Her head was in a moment at the 32 points of the compass. We then felt a trembling or shaking from above downwards. Sometimes the wind forced the vessel against the sea, sometimes it lifted it up; as much so at least as its weight would allow. The wind having turned the vessel round continually, now began to heave her strongly on her side and on the sea. Then we felt that the vessel was raised forward and depressed abaft. . . . We thus remained motionless, and trembling and praying during the whole time that the spout lasted, like a person at the bottom of a well, who is looking upwards. The action at last ceased suddenly, together with the violence of the wind, and the spout left us in quiet after a terrible farewell shock. It left us at a distance afterwards, but all the night we had a heavy sea, with strong variable winds."

Making every allowance for the vivacity of an Italian imagination, and the alarming predicament in which the narrator certainly was, there is still here much which gives us an insight into what the state of a vessel so situated must be, and what precautions should be taken in such cases; as well as into what electric processes are going on within a water-spout, for every person acquainted with electricity will see in the apparently fanciful liftings and throwings from side to side of this vessel, and of the one cited by DAMPIER, exactly the alternate attractions and repulsions so familiar to every electrician in many experiments.

In the note of the Tornado Cyclone of the American ship the *Montreal*, in the Pacific, quoted at p. 69, occur the two following passages which singularly tally with Father PIANCINI's impressions, and they are from the pen of the Commander of the *Montreal*, a thorough-bred American sailor, who certainly was not under the influence of any terror at the time.

The vessel was scudding under a fore-top mast stay sail only, and he remarks at 10 A.M.,

"Blowing exceeding heavy, the ship scudding off at a tremendous rate, and apparently forced down in the water *as though the water had a downward tendency*."

Between 11 A.M. and noon, when she had run into the centre, and the Cyclone burst upon her with redoubled fury, so nearly upsetting her that cutting away the masts seemed the only chance—

"Gust succeeded gust with the greatest rapidity, the atmosphere became filled with a mass of indescribable vapour, rendering every object at a short distance invisible, &c."

"The fury of the blast lasted thus about 30 minutes, when we first discerned a break overhead, *the ship appearing to be entombed in a vast crater*."

Compare here with what is noted at p. 270 of the appearances in the Tornado of the *Paquebot des Mers du Sud*, and in Father PIANCINI's description.

371. We have seen that whirlwinds on shore, certainly so far resemble water-spouts, that they lift water and fish. There is equally no doubt, that when sea water-spouts reach the shore they become whirlwinds, of which the following are very clear and distinct instances.

My friend Captain H. HOWE, Naval Assistant to the Commissioner of Arracan, writing from Kyoo Phyoo on the 12th May, 1843, says—

"In the absence of any thing more remarkable, I may mention to you the interest we all derived from witnessing the formation, and full action of four large water-spouts, all appearing at the same time, so that the eye could not sufficiently dwell on, as it were, and admire the one, without being called off to observe the other. The 9th instant, was ushered in with hot sultry weather, and a heavy dense cloud hanging over the offing, from which very suddenly four cones made their appearance, and rapidly elongating, reached the surface of the sea, which under their influence acquired a circular motion; an immense column of whirling spray joined on to the lower ends of the cones, being driven along with each of them before the wind, right on to the shore, from a distance of about 3½ miles. Only one of them, however, reached the land, and that went tearing and revolving along, taking up clouds of dust and leaves, and going at a rate of at least four knots, and had any thing light been in its way, such as a light shed or an insecure roof, I have no doubt it would have carried it off. The description of a water-spout in HORSBURGH's Directory so exactly tallies with our observations here, that nothing new is left to be said, and to enlarge on the subject would be mere plagiary, but I am induced to notice these phænomena from the unfrequency of four large ones being seen in such perfect action at once, and so near to the observer; I have seen hundreds of them at sea, but they generally dispersed before being fully formed, and this made the present ones appear the more remarkable; the sides of the several cones also presented a curious spectacle, dilating and collapsing as though emptying themselves of their surcharged weight of water; they were in full force for about 20 minutes, and after they had disappeared, we had a refreshing thunder-storm with rain."

In a subsequent letter Captain HOWE states that the gyration of the water-spouts was with the sun, or ◯ which is the direction for the Southern hemisphere.

372. It would seem from the foregoing, and from several instances given by M. PELTIER in his work, as also by FRANKLIN, in a letter from W. MERCER, describing the water-spout at Antigua, which was a spout (one of three) which came up the harbour, and on "landing" became a mischievous whirlwind, moving houses, and killing three or four persons by the falls of timber, that the fine weather water-spouts are certainly whirlwinds when they reach the shore; but there is another class of them which I have not seen adverted to yet, but which the sailor should be made fully aware of, and which form altogether a new phænomenon. These are the dangerous water-spouts or whirlwinds which appear in the midst of severe Cyclones, of which the following are instances:

373. In my IIId. Memoir (Jour. As. Soc. Beng. Vol. IX.) in the log of the barque *Tenasserim*, Captain TAPLEY, lying to off Cape Negrais, towards the close of a severe Cyclone,* is the following:

"*29th April*, 1840.—At 1 P.M. up foresail, a very threatening appearance to the Southward. At 2-30 *wore ship to the S. W., at the same time to clear a whirlwind.*

* She was in, and standing to the Northward out of the edge of the Cyclone in which the *Nusserath Shaw, Marion*, and other transports of the China expedition were dismasted and suffered great damage.

By this manœuvre, allowed the whirlwind to pass about 200 yards on the lee quarter. Furled everything to a storm main try-sail, and hove to again."

This phænomenon I thought so extraordinary, that I addressed some queries to Captain TAPLEY on the subject. His reply is as follows:

"I have much pleasure in giving you answers to your inquiries as nearly as I can. At 1 P.M., 30th April, by Nautical time (but by Civil time the 29th), a very threatening appearance to the Southward; ship's head East, a terrific squall from the S.S.E. rising very rapidly, and having a very blowing appearance. When the squall was within two miles of the ship, perceived a heavy whirlwind flying to the N. N. W.; immediately wore ship to the S. W , or first to the Westward, to give the ship way through the water; by doing so, allowed the whirlwind to pass the ship; when passed brought the ship to the wind, clued everything up, and furled all. Soon after, about ten minutes, the squall took the ship from the S. S. E., ship's head about S. W., blowing a complete hurricane, could not see half the length of the vessel on the water, owing to the tops of the sea being blown over us by the force of the wind, and a deluge of rain at the same time. I cannot remember how it was turning, as we were anxious to turn out of it; it was going round at a furious rate, and disappeared in the rain to the N. N. W. I do not recollect any lightning at the time. We could not discern it until it had approached pretty close, and then the most we saw was, the foaming of the water travelling up in a rapid progress. The day had been fine, and a little clear for a few hours, but blowing hard. At the time this squall appeared, the sky all round assumed a threatening appearance, and squalls gathered and rose rapidly. After this severe squall, the weather kept bad during the remainder of the twenty-four hours."

In my Tenth Memoir (Jour. As. Soc. Beng. Vol. XIII.) in the log of the *Coringa Packet*, which vessel experienced a small but severe Tornado-Cyclone off Trincomalee, while a large one was crossing the Bay a few degrees to the North of her, is the following account of it:

" 19*th May*, 1843. Daylight, blowing a tremendous gale from E. b. S., the sea running in pyramids, and the ship labouring very heavily. 8 A.M. Barometer 29-30. 10-30 A.M., *a very large water-spout formed within about two cables' length from the ship, passed across her stern, and hove the ship round head to wind; the fall of water on board was tremendous. Observed the barometer to rise immediately to* 29.45."

In the Sunderbund Cyclone of 1852 (XXIV. Memoir Jour. As. Soc. Vol. XXIV. p. 397), the ship *Amazon*, Captain COOTE, hove to on the Southern quadrant of the advancing Cyclone, which was then leaving him with terrific squalls and much lightning. As the gale diminished at S. W. Captain COOTE says,

" I counted thirteen whirlwinds all in sight at one time, and as high as my mainmast. There was one passed me close to, and it left a very troubled wake. They appeared to blow with terrific violence."

Compare this with the account of the dust-whirlwinds, pp.297—304.

374. But the most tremendous of these whirlwinds or water-spouts occurring in the midst of Cyclones, is the following from the log of the barque *Duncan*, from Cadiz to Calcutta, which vessel experienced, from the 28th to the 30th March, 1846, a severe Cyclone in Lat. 14½° South, and Long. 79° to 78° East. (Track *ii* on Chart No. II.) In the remarks of the 29th, Captain FAWCETT says,

" About 6 A. M. a most singular phænomenon occurred. About two miles astern of the ship, the water was rushing and foaming up to an astonishing height, gyrating round the centre, and passing the track* of the ship with astonishing velocity. *The diameter or breadth of the vortex of this whirlwind could not be less than two miles from the appearance of its spread, and how far the circle of its attraction extended I was unable to guess.*"

The ship was at this time lying to, with the wind at N. E. b. E. and her head to the N. W., so that this whirlwind which bore to the S. E. of her could not be the actual centre of the Cyclone, which in that hemisphere was to the Westward of the ship. The simpiesometer, which was well observed during the Cyclone, seems to have given no particular indication at the time.

A whirlwind or water-spout (or a compound of both, which this seems to have been) of two miles in breadth must bring with it, however, an awful risk to masts, and to ships of which the hatchways are badly secured.

The American brig *Eagle*, Captain LOVETT, was near foundering, by being struck by a whirlwind at the centre of a Cyclone tornado, in the Northern Pacific Ocean, Lat. 40° 10' N., Long. 162° 35' West. She was at the time lying to under a close reefed main top-sail only, with everything well secured for bad weather. The *gale* came on at 1 A. M. from E. S. E., and at 1.36 was blowing severely, then moderated a little, and struck the vessel again in heavy gusts with intervening lulls three times, till it was a flat calm with heavy rain. The clouds then began to roll up in the Westward, having a clear blue sky, but moving with great rapidity, *when a mass of white water, piled up*

* " *Wake*," as she was lying to.

several feet above the ocean, was seen on the port bow! It struck the ship from the N. E., opposite to where the clouds had cleared up, with the violence of a tornado, taking her aback, and sending her stern so far below water that all hands took refuge in the rigging. The sail blew away, and the wind shifting to the Westward, or aft, saved the vessel, the sky by this time becoming clear overhead. All the furled sails were blown from the yards, and the ironwork about the rigging *wrenched and twisted off* in a most singular manner.

I omitted in the first edition of this work to insert, though intending to do so, the following extract from Mr. REDFIELD's valuable Memoir on the Cuba Hurricane of 1846, p. 94:

"*Local Tornado in the Cuba Hurricane.* The accounts from Matauzas mention a destructive phænomenon of this kind as having taken place at Yabu, in the central part of Cuba (Lat. 22°, Long. 79° 34', on the right of the axis line), during the hurricane. It is described as 'a tremendous water-spout which passed through the place, doing much damage,' and confined to a narrow path. 'The effects were the same as if a violent river had run through the town, leaving a kind of channel.' This case has since been mentioned, erroneously, as having occurred in Mexico.

"The appearance of violent tornado-vortices within the body of a great storm is not new nor very unfrequent. A remarkably destructive case occurred at Charleston, S. C., on the 10th of September, 1811, during a great storm which visited our coast. It caused the loss of a great amount of property and about twenty lives. Its track was about one hundred yards wide, and it followed the course of the local storm-wind, from South-east to North-west, transversely to the progression of the great storm. Two very violent tornadoes appeared in New Jersey in a general storm on the 19th of June, 1835, moving in different but nearly parallel paths, at an interval of several hours. These pursued the course of the higher general current which then overlaid the great storm.* Several other tornadoes, together with numerous gusts and severe thunder squalls, appeared on the same day in different places, within the compass of the same general storm. Another tornado occurred on the 13th of August, 1840, at Woodbridge, near New Haven, Ct., during a general storm, and followed the local direction of the Storm-wind from S. S. E. to N. N. W.—transversely to the course pursued by the larger storm.

"These, with other cases which might be adduced, may serve to shew that the small tornadoes which sometimes occur in great storms, have no essential or inherent connection with the vortex of the larger storm, even in those cases in which the courses of progression may chance to coincide."†

---

* One of these was the New Brunswick tornado, described in the American Journal, Vol. XII. pp. 69, 79. See also foot note, Vol. XIII. p. 276.

† In like manner common thunder-storms are often known to appear in or above

See also, p. 78, the account of the tornado in a Cyclone at the Samoan Islands.

But we have even an instance of the transformation of a water-spout into a *storm*, though the word used being *tempête*, we cannot affirm that it was a Cyclone. The following is translated at length from M. PELTIER's work (p. 258), in the chapter giving the detailed relations of water-spouts.

ATLANTIC OCEAN, NOT FAR FROM THE GAMBIA RIVER, 2nd September, 1804.

"No. 41. Day and night water-spout *(trombe)* which lasted fourteen hours ; a luminous column in its whole diameter ; complete calm before its formation ; sky excessively overcast, frightful storm *(tempête)* during its existence ; pleasant breeze from the West afterwards.

"I owe to Doctor LEYMERIE the account of a spout *(trombe)* which he saw on the 2nd September, 1804, being on board of the cutter *Le Vautour*. This vessel sailed with letters of marque (a privateer trader?), and was coming from Cayenne towards the coast of Africa; they were not far from the Gambia when this meteor occurred. Before the spout was formed it was a dead calm: the preceding days had been very hot, and since the morning the sky had been covered with numerous thick clouds. The cutter was chasing an English slaver, when all at once they saw a column of water of about a hundred metres,* which rose up from the sea and went upwards to join a column of vapour which descended from a cloud. At this moment the calm ceased, and the tempest *(tempête)* began to blow with violence. We have preserved Doctor LEYMERIE's expression of a column of water, though we are persuaded that it was not liquid water of which it was formed, but water in the state of a dense vapour, as has repeatedly been shewn. This column was luminous in its whole diameter; it had a phosphorescent appearance, and was slightly yellow or fawn-coloured. The sea itself was on that day resplendent with light, and the vessel left after it a long track of fire. This spout as well as the storm *(tempête)* which accompanied it, lasted fourteen hours, and caused numerous shipwrecks on those coasts. It did not terminate till four in the morning of the following day; so that it had commenced about two in the afternoon, and had lasted a great part of the night.

"*Observations* (by M. PELTIER). I think that the luminous state of the cone of the spout would be often observed if this meteor occurred frequently at night from the local portions of great storms. An examination of this class of storms will shew that the narrow tornadoes and thunder-storms often extend to a greater height than the great gales or hurricanes.

* About 328 feet ; and in diameter no doubt is meant by what follows, though not so expressed.

the preceding facts,* the column was formed of vapour, and not of liquid water, the phosphorescence of the waters of the sea exists no longer in the column; the light by which it was pervaded can then only be derived from the infinite (continued) discharges which constitute an electric current and not from a true phosphorescence."

375. We may remark on this, as to the light, though it is strikingly analogous to what we know has taken place in some Cyclones (p. 214, Note) that, allowing the water at the base to have been mere vapour, if it rose from the sea it might undoubtedly, even in that state, be still phosphorescent; but apart from this, the storm following or accompanying the spout is always of very great interest. Unfortunately being a privateer, and apparently a colonial one, there is no hope now of obtaining a copy of the *Vautour's* log by which we might clear up the question of what kind of a storm *(tempête)* it was; and if, which also is doubtful, the *spout* lasted all the time; for we can hardly suppose, in spite of what is said, that the vessel remained in sight of it.

The expression used, and which I have exactly translated, may be intended to convey that the spout was seen in various places by the vessels shipwrecked?

376. We must not pass by another peculiarity in wind and waterspouts, which is, that great numbers of them are described in the different relations collected by authors, as being accompanied by *noises* of all kinds and degrees; "from the hissing† of a serpent to the noise of heavy carts on a rocky road," says M. PELTIER, or of "a cascade falling in a deep valley," say other accounts; and these noises are usually more violent on land than at sea.

377. In reference to the peculiar noises heard at the passage of the centres of Cyclones, § 246 to 249, p. 207, this is remarkable, and deserving of future attention; as is also the fact frequently referred to by the accounts given in M. PELTIER's work, that the sea often shews, at the commencement of water-spouts and even of tornadoes, a remarkable agitation like a boiling *(bouillonement)*. This on a larger scale may be a pyramidal sea, if we allow electricity to have any part as the cause, or even as the effect, of a tyfoon.

* Adduced by M. PELTIER, in his work, from which he concludes that the lower half of the column of a water-spout is not water, but dense vapour; in which he agrees with most other writers.

† HORSBURGH (Introduction, p. VIII.) mentions the hissing.

378. It may be useful here to give some account of M. PELTIER's able work, *Sur les Trombes*,* which not having been translated into English (that I am aware of) is probably unknown to the majority of my sailor readers at least; yet its contents are often so essential to the right understanding of much that we have to set forth in this Part, that I should not fairly accomplish my purpose of placing before the seaman, in a plain and intelligible form, all that is known of our new Science in all its relations, and of indicating what has yet to be inquired into, did I not make him acquainted as briefly as I can with the contents of the work.

379. M. PELTIER's view then of land and water-spouts (or of whirlwinds and water-spouts, which the French class together in their word *Trombe*, but of which he perfectly preserves the distinction in his descriptions and experiments) is, that they are "but transformations of another meteor." And this meteor he affirms to be an electric action; not simple, but one compounded of the electric state of the cloud, and of that of each separate globule or particle of air or vapour, composing it, and generally speaking analogous to thunderstorms in many points, but acting with greater intensity. He affirms and shews by many proofs, that the visible clouds above are not the only ones in the atmosphere, but that there are also invisible (because transparent and uncondensed) ones; which like the others may be electrically charged, and produce the same phænomena, though perhaps in a weaker degree.

380. M. PELTIER next points out what are, in his opinion, strictly the essential phænomena to constitute a *trombe*, (whirlwind or water-spout,) and what are those which do *not* appear to be so, or which occur only incidentally; and amongst these last he places the whirlwind motion, and currents of air forming a whirlwind, as not being constantly observed in them.

381. He then proceeds by various ingeniously contrived and instructive electric experiments, by conductors with balls and points, or a single point, to reproduce upon leaves of copper-foil, dust and smoke of resins over water; and upon water alone gyratory move-

* *Observations et Recherches expérimentales sur les causes qui concourent à la formation des Trombes. Par M. Ath. Peltier.* Paris, H. Cousin, 1840.

Part V. § 382.]   " *Peltier sur les Trombes.*"   319

ments; repulsions *from* the centre; *horizontal* and vertical whirling clouds in the smoke: bosses or umbels of water;* deep depressions in it;† the raising of light bodies; the production of vapours; and even vapours of pure water from an acidulated mixture; accounting thus, as he fairly observes, for all the essential and usual phænomena of spouts. He adduces then many relations of land and water-spouts, which have shewn all the electric phænomena, and concludes the first part with tabular statements of spouts (*trombes*) and of all other analogous meteors, so as to give a complete and highly convincing body of evidence in a small compass. The second part of the work consists of the detailed accounts of the various instances cited in the tables. The whole is indeed well worth the perusal of meteorologists and of seamen, for the numerous points of relation which it shews with Cyclones, and of inquiry which it suggests to every intelligent man.

382. M. Peltier has collected in his work, in three tables, 137 accounts of water-spouts, beginning with Thevenot's in 1674. The results of these, of which we should note that 66 are properly *waterspouts* seen at sea, or on lakes, rivers, &c., while the rest are whirlwinds or *wind*-spouts, or electric columns acting on land, he states to be as follows—

" In the 137 instances cited—

37 Spouts‡ have had a gyratory movement, continuous, or with some intermittence or but for a short time with others; frequently only a portion of the vapours had this gyratory movement.

25 are noted as having no gyration, or this may be directly inferred from the details.

The accounts of the other spouts (*trombes*) not specifying if there was or was not any gyration, the presumption is in favour of the negative, for an account indicates *what exists*, and not what does not exist.§

* An experiment known since the days of the Abbé Nollet (a contemporary of Franklin) remarks M. Peltier.

† Often seen and noted in accounts of water-spouts at sea, and from M. Peltier's experiments, it would seem that alternate risings and sinkings can be produced, forming miniature waves.

‡ As just remarked, the French use the word *Trombe* to express both a waterspout, and what we should call a whirlwind. Thus, they would have no separate word to express the sea and shore-spouts described at pp. 305—312, assuming, as is there shewn, that the meteor is but one and the same.

§ M. Peltier forgets here, that, in relations of all kinds, everything depends on

33 are stated to have occurred during a more or less perfect calm : we might add to these those which have taken place during a light wind or a regular breeze blowing from one side only.

56 have been accompanied by thunder, lightning, or electric signs of this kind.

10 have occurred beneath a cloudless sky.

19 have exhibited vapours ascending within the column, 7 have shewn descending vapours.

15 accounts are relative to several water-spouts together.

7 spouts have been double, triple, &c. *(multiple)*; some have divided and reunited.

15 have carried up liquid water, and some accounts shew that the sea was hollowed out below in a conical shape.

3 water-spouts at sea, in which ascending vapours were distinguished, having been crossed by ships, have discharged *fresh* water.

6 spouts took place at night.

3 were formed between the clouds."

383. There appears then no reason to doubt, that most, if not all water-spouts are in point of fact whirlwinds, *i. e.* are whirling round at the visible part or column, and they are probably surrounded as Professor ŒRSTED thinks, by an invisible column of air which is also a whirlwind, and this view is moreover confirmed by several examples, adduced by Colonel REID. I believe also, that most sailors would be able to corroborate it. The last well recorded instance we have of its being clearly and indubitably shewn, is that of the water-spout which passed through Sir Robert STOPFORD's fleet near Vourla, of which the rotation was clearly seen; it also "shook the ships' sails violently, and small articles were whirled about as in a land whirlwind." REID.

As to the direction in which they turn, this seems to be quite uncertain, as far as authorities go.

384. The great water-spout which fell on the Island of Teneriffe in Nov. 1826, appears by Mr. ALISON's relation in the Philosophical Magazine, to have been at once a water-spout on the island, discharging enormous volumes of water and tearing up the soil, and even the tufa rock, into deep ravines, while numerous globes of fire were seen

the accuracy, and memory, and habits of mind of the observer and relator; and that in a number of narrators, some from negligence, or from supposing the turning round to be admitted of course, or as of no moment, might omit to state it.

at sea and on shore and a true Cyclone was blowing out at sea. How far all these were connected it is impossible to say, from the imperfect nature of the accounts, but the circumstance of their occurring together and on and about a volcanic island, is of great interest.

That water-spouts at sea are also whirlwinds, then, may be considered as having been indubitably shewn. From many circumstances and well known facts, we have long inferred them to be electric phænomena, and from the following extracts from a letter addressed to me by Captain J. J. CHURCH, of the barque *Rory O'More*, from Launceston to Calcutta, which was accompanied by a capital sketch of the water-spout, it would appear that electric flashes have been seen about them even in daylight. Captain CHURCH says,

"Being in Lat. 2° 4′ N. and Long. 90° 20′ E. Thermometer 84° and Barometer 29.98, wind very light from West, and the weather clear, with the exception of some heavy clouds between W. and S.W., at 9 A.M. a water-spout was observed, bearing W. b. S., distant 1½ or 2 miles. Its altitude where it joined the clouds 10° 30′."

He then goes on to describe it as one of those of uniform thickness, and curved as they are sometimes seen with a double tube as it were at its base, of a transparent appearance, extending till it disappeared about one third of the height of the main tube, the spout being tubular and not conical. The usual boiling of water, &c., was seen at the base, and

"From the upper and outer edges of this foam or mist, sparks appearing in the bright sunshine like stars of the fourth magnitude, were occasionally emitted. Though at some distance, these observations were made with an excellent glass and were distinctly seen. The spout remained in sight and nearly stationary for nearly forty minutes."

In the "Account of the building of the Horsburgh Lighthouse," in the Straits of Singapore, by Mr. J. T. Thomson, F.R.G.S., Government Surveyor at Singapore, from pp. 459 to 464, will be found some very good descriptions of water-spouts and North Westers, and in a note, at p. 460, the author remarks of one spout, as follows:

"In this one I observed what was something new to me, viz., that the particles of vapour contained in the outer and dependent tube (from the cloud), besides being driven in the helical curve round the inner or ascending column, revolved also round the threads of the helix.

385. COMMENCEMENT AND BREAKING UP OF CYCLONES. The most natural questions which arise in the mind, in considering these meteors, are the constantly recurring ones of " *Where and how do they begin ?*" " *Where and how do they end ?*" and we are yet involved in utter mystery as to these two essentials to the history and right understanding of Cyclones. It is clear that there must be *some* beginning, and an ending *somewhere;* but none of our researches have as yet disclosed to us any thing on this subject beyond remote conjectures, and vague probabilities.

386. Mr. THOM thinks, p. 163 to 164, that in the formation of a Cyclone "it may embrace several small vortices formed by the gradual effects of the wide exterior circle in withdrawing air from the internal space."

The following instances, both of which have occurred in the Bay of Bengal, and at or about the commencement of a Cyclone of great severity, seem to me worth recording. The first is from a letter from Captain BUCKTON, of the brig *Algerine*, printed in my Seventh Memoir, Jour. As. Soc. Beng. Vol. XI. This vessel appears, not improbably, to have had a Cyclone overhead, which crossing the Bay from the Southern Andaman to Point Palmiras, afterwards became the Calcutta Cyclone of the 2nd and 3rd June, 1842. There must be an error in the reading or the zero point of the barometer, I presume, and that it was 29 and not 28 inches.

" On the 28th May, in Lat. 10° North, Long. 92° 26' East, the sky became a perfect dense mass of black clouds, with the scud flying rapidly past from N.E., S.E., and W.S.W., the wind light, and the sea rising in bubbles, as if the wind was blowing from every point of the compass, hissing and rising up in bubbles like a boiling cauldron. Here the barometer fell to 28.60.; this being excessively low, for so low a latitude, induced us to make every preparation for severe weather. From this time until the 1st June, Lat. 15° 25' North, Long. 87° 58' East, experienced an increasing gale steady from S.S.W. to S.W. by W. with much lightning, and a very heavy appearance all round. The barometer rising and falling according to the strength of the squalls, or the preponderance of rain from 28.70 to 28.56. On the 2nd (civil time,) the gale increased so as to oblige us to lay to, the barometer having fallen to 28.45 (Lat. 27° 20' North, Long. 87° 6' East.)

" At 9 A.M., experienced a cross sea, setting in from S.W., N.W., and N.E., the former preponderating ; the rain pouring down in torrents, the gale increasing, and the squalls blowing with fearful violence from W.S.W., and shifting suddenly from that to N.W., N.N.W., and as far as North ; the barometer gradually falling until it came down to 28.18; at midnight more moderate, the barometer up to 28.36. Steady gale from S. W. b. W. decreasing towards noon, when Lat. 19° 10', Long. 86° 42', so as to enable us to make sail at 3 P.M., until 11h. 30m. P.M. on Friday, when the barometer again fell to 28.20, during a most severe squall from N.W.; False Point was then bearing about N. W. 12 miles.

387. The next example is that of the ship *Vernon*, Captain Voss, from England bound to Madras, in a note dictated to me and subsequently corrected by him. It commences on the 26th November, 1843, and at this time, or rather between the 27th and 28th, severe Cyclones were raging both in the Northern and Southern hemisphere, and between the same meridians, though distant about 13° of latitude from each other, with a heavy Westerly monsoon blowing along the Equator. These Cyclones form the subject of my eleventh Memoir, (Journal As. Soc. Beng. Vol. XIV.) and on the 28th, the *Vernon* was distant at noon about 400 miles from the centre of the nearest Cyclone then raging, though in two days more another (for there were two in the Northern hemisphere) was developed on this very spot. The *Vernon* is a remarkably fine frigate built East Indiaman of 1,000 tons, belonging to Messrs. GREEN and Co.

"Ship *Vernon, November* 26*th*, 1843. It began to get gloomy and the clouds were whirling about above in a remarkable manner, wind variable from the Eastward below, and in puffs. Barometer not much under 30.00 (about 29.95.)

" *On the* 27*th.* Barometer had fallen to 29.85, dark and gloomy weather, still variable from N. E. to East with squalls, confused swell all round, clouds very low and lowering, with appearance of bad weather, Lat. 9° 6' North, Long. 85° 0' East, Bar. 30.0, Ther. 83°, clouds still moving in all directions; kept snug at night; very squally with rain from East to N. E., sea getting up.

" *On the* 28*th.* At daylight, Bar. at 29.70, every appearance of bad weather, wind increasing, variable and threatening from E. S. E. to N. E. double reefed, &c. and sent down royal yards towards noon. Lat. by acct. 10° 46' North, Long. 84° 7' E., Bar. 29.80, Ther. 78°. We appeared to have got between three clouds, wind then came in hard squalls (ship with top-gallant sails furled and courses up, topsails on the cap and reef-tackles close out.) Forked lightning but not much thunder, squalls from N. E. then North and N.W. and right round, and thus the ship went round *six turns in about* 30 *minutes* following the wind, with after yards square and head yards braced up. The rain falling literally in heavy sheets, so that it was hardly possible to stand; the men obliged to hold on, decks half full of water. The wind not moderating with the rain, but blowing in severe gusts. After this the wind steadier, but still from about N.E. to E.S.E. with sharp squalls obliging us to lower the double reefed topsails, very dark and gloomy.

"*On the* 29*th;* more moderate, still blowing hard with gloomy weather till sunset, when it became finer.

388. With the foregoing examples, and remembering what has been said with respect to whirlwinds and water-spouts, we may venture to consider a little how Cyclones *may* commence. It is much too soon to affirm, or almost to suppose, that they *do* commence in any one particular way, and we must remember also that even to describe clearly the mode or modes of action (for there may be more than one) by which they commence, could we do so, does not at all affirm any thing of what *causes* these actions. In a word, that we are here essaying to describe, or suppose, effects and not causes; and if I give any space to the discussion of what we are yet so imperfectly informed of, it is because I think it very essential that the thoughtful and careful mariner, as well as all who desire to aid us, should, to use a sporting metaphor, be put upon the tracks and scents by which they may furnish us with useful notes.

389. A Cyclone, then, must commence at sea, beneath, upon, or at some distance above the surface of the ocean. That it commences beneath the ocean we have no sort of evidence or of analogy to lead us to suppose; and we are therefore reduced to the last two cases; *i. e.* that it commences at the level of the ocean or in the atmosphere.

390. Now the proposition that Cyclones begin at the surface of the ocean is certainly a highly tenable and probable one; so far as that we may assume that a vortex of any size, or form, or force, may be produced and sustained for any length of time, by counter streams of wind, of sufficient force and dimensions, meeting side by side, and thus generating, as in water, a circular motion. If they met direct in front of each other we should look rather for a calm, or at most a succession of small whirls.

But beyond this we are stopped; for we are much at a loss to imagine how two mere monsoon gales *can* generate forces so far above their own velocity, and then this theory affords us no sort of clue to the progressive motion, which is as extraordinary, if not more so, than even the rotation.

Part V. § 393.] *Commencement of Cyclones. Theory.* 325

391. One would suppose also, that we should have had, far more frequently than we have, some traces of commencement in segments; as in Mr. Rechendorf's dust whirlwinds, (see p. 301) or in vortices as in Mr. Thom's supposed case. We have also another difficulty of no small magnitude in this hypothesis (supposition) that the Cyclones are whirls produced by opposite or lateral streams of air at the meeting of the trades and monsoons, which is, *that it seems to prove too much;* for, if we admit this, why not also allow that, at least for the six months in which a monsoon crosses, or is opposed to, the trades' there should be a constant succession of the Cyclones? while it is notorious that they only happen during certain months in most parts, and sometimes none occur in the season, and they are moreover, to all appearance, and as far as surface winds are concerned, generated both *within* the trades and without their limits and those of the monsoons.

392. The remaining possibility then is this—*Are they originally formed above the earth, and do they descend horizontally (as horizontal disks) or inclined at an angle, and of a considerable size at first?* For we know that they both dilate and contract, as whirlwinds and waterspouts do when in action, at the surface of the earth.

393. Every theory which pretends to account for any natural process, such as a meteor like our Cyclones, should account fully for or admit *all* the principal phenomena of it. If it does not do that, it cannot answer the questions put to it. Let us see how far the theory, that Cyclones may be formed in the atmosphere and are in fact disk-shaped vortices, descending from above, will do so.

1. The Cyclones *may* be so formed, either electrically or by the mere dynamical (mechanical) forces of currents of air causing whirls or whirlpools.

2. They may be formed, both at the sides and at the upper and lower surfaces of different strata of clouds. In the last case the action is more probably electric than the former.

3. Being once formed and in action, they may either rise higher, or descend or plunge downwards, from various causes.

4. They may also be admitted to extend themselves greatly, or to diminish in diameter; and finally to exhaust themselves by their own

violence; or to break up and disappear on reaching the land from seaward.

5. They may be supposed, as in the analogous cases of waterspouts, to have a calm centre, however this is formed and sustained.

6. They might have great developments of clouds and rain and sometimes of electricity, both at the centre and circumferences.

7. They would be *more* apt to be generated in parts where the upper currents of the atmosphere, of various temperatures, degrees of moisture, electric states and other differences, may be supposed to meet in large bodies, than in those parts where perhaps these currents are more regular, and less liable to be in states widely differing from each other, or to be interfered with by their contact with, or approach to the earth, and its constant electricity.

8. They might also, from many causes, not be disks parallel (truly tangential) to the surface of the globe, but be inclined forwards.

9. As they *are* certainly disks, if we suppose them at first truly horizontal above and below, they must when pressed down to contact with the earth, over a space of 300 miles, be much thinner at the centre than at the edges. A flat disk of 300 miles and ten miles high at its edges, would be but 7·2 miles high in the centre; supposing that it did not curve, but remained horizontal above and only curving below; becoming thus what is called in optics a plano-concave lens; and one of three miles in height at the edges would be only two-tenths of a mile thick at the centre.

10. If generated above the surface of the earth, and by electric causes, there might be, according to the nature of the electric action, as we have seen by M. PELTIER's experiments, at § 381, a disk or umbel of water raised at the centre, which would move onwards with the electrified clouds above; or in some cases there might be a depression and circular currents.*

See also as to this at p. 172, Mr. MARTIN's account of the electric action of a thunder-storm on the waters of the sea, as registered at Ramsgate Harbour.

11. If so generated there might be more than one Cyclone formed at the same time, or they might divide or unite, and a Cyclone might

---

* M. PELTIER's were *outward* circular currents, it should be noticed.

at first, like a water-spout and *its* cloud, be stationary, and then move on at a greater or less rate.

12. A Cyclone in full action, being as we have assumed a disk only, might be lifted up by an obstacle, as mountains, hills, &c., and cross them as well as intervening valleys, so that the clouds might be seen by the inhabitants "whirling about in an extraordinary manner,"* and then descend again to a low flat country or on the ocean. There is no doubt that this has very frequently occurred in storms crossing the Peninsula of India, or travelling inland from the head of the Bay of Bengal.

13. If a Cyclone was formed, as whirlwinds and water-spouts certainly are, above, and from the same causes, it might, like them, produce in a stronger degree, and especially towards the centre, peculiar noises, which in the water-spout are described as rumblings and hissings, see p. 317, and in the Cyclone as "roarings," "thunderings," "yellings and screamings." See p. 208.

14. The unequal duration of the different sides of a Cyclone may also in a great measure be accounted for, by supposing them to be descending disks. The fact that the front or advancing side of a Cyclone is almost always of much longer duration than the latter part, is too well known to require any instances in proof of it.

15. Now if we suppose a disk to descend horizontally, no doubt its force and duration may be equal on all sides; but if we suppose it at all inclined, in a plane of which the lowest part is in front, we must equally suppose a part of it lifted up from behind. We know almost nothing of the actual facts, and can therefore only adduce suppositions.

394. We thus see that if we suppose a Cyclone to be formed in the atmosphere between strata of clouds, in the same or opposite states of electricity, and then from any cause to descend or settle down upon the ocean, it would fulfil all the conditions which we have set down; and these are not only the principal, but most of the secondary ones. It is very true that these might also be fulfilled by Cyclones formed at the surface of the earth or ocean, by contrary or

* This expression occurs in a report printed in one of my Memoirs, and the equivalent to it in others, and in those of Mr. REDFIELD.

side-long winds, but then we have the objections that I have before cited, § 391, *i.e.* that they should be constant during the opposing or crossing monsoons, and should occur much more frequently than we know them to do. I desire to be understood here as having been simply desirous to set forth the probabilities on both sides, and only as offering this one amongst others which may be proposed; waiting for evidence to confirm or refute it. In other words this theory is yet "lying at single-anchor."

395. The relation of Mr. REDFIELD's explanation of the fall of the barometer, and of the actual barometric curves, as shewn by our Plate at p. 246, to this view of the possible mode of the origin and continuance of Cyclones is closer than would be at first imagined;[*] for we may easily suppose a storm-disk, formed above, and not yet descending to the surface of the earth or ocean, but still affecting the Barometer. At its first formation, it is clear that it may be, if formed between two horizontal strata of clouds or gases, very thin; or if formed as a vertical column between the *sides* of masses of cloud or gas, be yet very short, till it extends and thickens (or lengthens) before it reaches the earth; and all these modes of origin, increase, and action might go on with fine, or at most in cloudy weather below, and be only indicated by the Barometer; and when the Cyclone descended, its effects would be felt at once in various places at great distances from each other, and this seems to be what really occurs; for we have nothing which will allow us to say that we have traced a Cyclone from a mere whirlwind to a full storm. We may even go a step farther, and suppose that as (apparently) the greatest condensation is at the outer part of the circle, the disk may reach the earth in a curved form, so as to be felt at the outer zones. All this, it is true, is conjecture, and highly uncertain, but observation may do much for us in clearing up many of the questions I have adverted to, and a single one may throw a flood of light upon a dozen others.

396. In reference to this, I translate from the Italian of RAMUSIO, not having the original Spanish work to refer to, the following

---

[*] I am aware, that I am, in part, supporting here one theoretical view by another, and that such is but weak scientific logic; but the barometric curves are indubitable facts, and the calm centres are equally so.

PART V. § 397.]   *St. Domingo Cyclone of* 1508.   829

remarkable passage from the *Historia General de las Indias* of FER-
NANDO DE OVIEDO, Book VI. Chap. III., describing the great Hurricane of St. Domingo, 1508, abridging it a little in matters not
essential to our purpose, and distinguishing a few passages by italics.

"Hurricane *(Huracane)* in the language of this island, means properly an excessively tempestuous storm, for indeed it is nothing else but a violent gale, together
with rain.

"Now it happened that on Tuesday, the 3rd of August, 1508, (Father NICOLAS
OVIEDO being then governor of the island,) about noon an exceeding great wind
with rain came on at once, *which was felt at the same time in many sites in the
island*, and there arose from it suddenly great damage, and many estates were ruined.
In this city of St. Domingo, all the straw houses were prostrated, and some even
of those of stone were much shaken and damaged. At Buena Ventura, all the
houses were destroyed, so that for the many who were ruined there, it might more
properly be called Mala Ventura.*

"And what was worse and more grievous, was that in the harbour of the city
more than twenty ships, caravels, and other vessels were lost. The *Northerly* wind
was so strong that as soon as it began to blow hard, the seamen did everything in
their power, by laying out more anchors and fastening by more ropes, to secure their
vessels ; but the wind was so violent that no precautions could withstand it, every
thing was carried away, and the force of the wind drove all the vessels, large and
small, out of the port, down the river, and they perished in various ways. *But the
wind changing suddenly to the opposite quarter, and with not less impetus and fury,
blew from the South as violently as before from the North*, when some vessels were
driven furiously back into the port, and as the North wind had driven them out to
sea, so this opposite one drove them back to the port and up to the river. They were
afterwards seen drifting down again with only the tops in sight above water. Many
persons perished in this calamity, and the most violent part of the tempest lasted
twenty-four hours, until the next day at noon, *but it did not cease all at once as
suddenly as it came on.*"

The author then goes on to describe the frightful appearances and
damage occasioned by the storm, and adds, that the Indians (and the
Caribs were then a numerous people) said that they had frequently
experienced hurricanes, but that neither they nor their fathers had
ever before experienced the like for its extreme violence.

397. There can be no doubt, first, that this was a true Cyclone,
and one of extreme violence, but the remarkable part of this relation

* *Buena ventura* " good luck." *Mala ventura* " bad luck."

is the stress which is laid upon the very sudden commencement of it in many places at once; and the Spaniards even at that early period (16 years after the discovery of America) had many settlements along the coasts, and probably far inland, for the gold washings and minings.*

If we suppose the Cyclone to descend as a disk, this sudden onset of its fury in many places at once is exactly what should occur, and its gradual cessation so clearly and pointedly contrasted with the sudden beginning, is also such a description as might have been written in our own days of a Cyclone of the usual diameter of those of the West Indies, or about 100 or 200 miles, descending with its centre on the Mona Passage, and travelling along on an East and West track to the Westward, at a rate of 10 or 12 miles an hour.

In reference to what has been said on the settling down of Cyclones as disks from above, and on the lifting up of the rear or following part of a Cyclone, I have in my 18th Memoir, Jour. As. Soc. Beng. Vol. XVIII. been able to establish with tolerable certainty from the Logs of 22 vessels, that the advancing semi-diameters of the Cyclone of 12th and 13th October, 1848, in the Bay of Bengal, were respectively, as compared with the following or rear ones, as follows, viz.

	Advancing semi-diameter in front of the Cyclone track.	Following semi-diameter, behind the track.
	Miles.	Miles.
12th Oct.	140	90
13th Oct.	115	65

and that, as Mr. REDFIELD suggests, the Monsoon or surface wind evidently forced its way beneath the following or uplifted half of the Cyclone. I have given on Chart No. III. which is that of the Bay of Bengal, and part of the Arabian Sea, an imaginary section of the latter half of a Cyclone lifting up on its rear quadrant, with a double scale, so that the reader may consider it as one of 150 or of 300

* In 1495, only three years after the discovery, the Spaniards were working gold mines at 60 leagues distance from St. Domingo. See RAMUSIO, p. 9, vol. III. P. MARTYR's History.

Part V. § 398.] *Descent of the October Cyclone of* 1848. 331

miles in diameter and of 5 or 10 miles in height. If we take it for example as a disk of 150 miles in diameter, we can easily see that while the 75 miles of the van* or front semi-diameter, if it was moving at the rate of 6 miles an hour, would take 12½ hours to pass over a ship hove to, or an island in its track, a slight inclination of the disk might elevate the rear half sufficiently for it to leave 20 or 30 miles of that part free from its surface action, so that in such a case the centre of the Cyclone would have 75 miles of the storm disk, with its nearly regular circular winds, before it, representing 12½ hours duration, and only 45 miles or 7½ hours of duration behind it; and so on in any proportion to which we may extend the size of our Cyclone disk. The occurrence of lightning in that part only of the disk, where we may suppose the Cyclone to have been lifting up, as noticed at p. 216, § 260, will forcibly remind the scientific sailor of the sparks from the condensing disk of the electrician.

398. The descent or settling down of Cyclones was also distinctly proved in this instance, for there were ships on the 10th and 11th of October, so situated between the Eastern and Western shores of the Bay of Bengal, and to the Southward, that had any Cyclone existed on those days, we must have had full evidence of it; and yet on the 12th of October, in the very middle of the Bay (Lat. 17° 48′ N., Long. 89° 18′ East), we had three ships with light variable winds within a circle of about 50 miles in diameter, while for a space of about 300 miles in diameter round the same centre, a true Cyclone-hurricane had commenced, which is fully traced to Point Palmiras, and in which seven vessels disappeared, and fourteen were more or less dismasted. There is no doubt also that this Cyclone before settling down in the Bay had crossed some other land, the Andamans or Point Negrais, for it brought with it in its calm centre, and when it first settled down, vast numbers of land birds, insects, &c. It is possible, as to time, that this Cyclone, came originally from the China Sea, where a severe one, in which H. M. S. *Childers* was driven on the Pratas Shoal, was raging on the 9th; but if it did so, it was lifted by the high land of Cochin

---

* The Merchant sailor will recollect that this term is strictly nautical, as we have the van, rear, and contre of a fleet.

China. In this last case it would not be exactly that of an originally-formed Cyclone settling down, it is true; but if it did not come from the China sea then it is so.

399. So far as to the origin and continuation of these tempests. With respect to their termination, we may well suppose it to arise from the causes above being exhausted* or from the storm's "rising up," or from its friction on the surface of the ocean or land, which must be a very considerable element in the exhaustion of the motive force.

As regards the possibility that one of the modes of termination of Cyclones may be by "rising up," as well as that they may be formed above and descend to the surface of the globe, I give here, without abridgment, a remarkable extract from the Log of the Ship *Swithamley*, Captain JENNINGS, printed by Mr. THOM in his Memoir on the *Cleopatra's* (Malabar Coast) Cyclone of April, 1847,† premising that the position of the *Swithamley* on this day, the 20th, is exactly upon the produced line of the track of the Cyclone, which however I have not been able to trace satisfactorily as a Cyclone, at the surface of the Ocean, beyond the 18th, when it was of terrific violence, dismasting the ship *Buckinghamshire*, after probably destroying the Steamer *Cleopatra*, and committing other extensive ravages. The position of the centre on the 18th, was about 3° to the S.S.E. of the *Swithamley's* on the 20th.

"Experienced on the 20th (Lat. 17° 31' N., Long. 72° 00' East) the height of the gale: it blew at times very hard. Obliged to lower the double reefed topsails on the cap, and stood to the S. S. W. (he was bound to Bombay and his direct course was N. N. E. or thereabouts) when I invariably found I ran out of the gale. These violent squalls always came on from 8 to 9 P.M., and continued two or three hours from N.E. to East, accompanied by heavy rain, lurid lightning, and a long rumbling noise like stifled or distant thunder, *but (which) appeared quite above head.*‡ These N.E. squalls forced themselves up against a strong double-reefed breeze from

---

* As a thunder-storm ceases, when the electricity of the clouds is discharged, which, admitting Cyclones to be electric phænomena, offers a strong analogy : and we may note by the way that it also furnishes an argument against the dynamical theory ; for as soon as the Cyclone is over, the monsoon again returns as before in opposition to the trade-wind.

† Transactions of the Bombay Geographical Society.   ‡ Italics are mine.

the N.W. and West, leaving actually no calm space between the two winds, but took us flat aback. On the 20th it blew quite a gale of wind from the West to N.W., washed a man from the jib-boom, and lost him. With very hard carrying could bear double reefed top-sails. I consider I was in one of the spokes that led to the centre of the hurricane, and had I continued standing to the E. S. Eastward in the N.E. squalls would in all probability have been dismasted. It always kept clear to the Westward, and very black to the Eastward, thus I chose to stand to the first."

Capt. JENNINGS, who appears to have understood the law of storms according to Mr. ESPY's theory "of spokes leading to the centre," evidently speaks here of the squalls of the 19th and 20th, and perhaps part of the 21st, during which days he was, as I should judge, running close to, or right beneath the lifted vortex, of which the sudden N. E. squalls and noise overhead were strong evidence? We have unfortunately no Barometer with this Log; the indications of that instrument and of a Simpiesometer in these squalls would have been of high interest.

The singular facts, of which we have abundant instance both in the Eastern and Western hemispheres of Cyclones, passing over large extents of land or over mountain chains, on their route from one sea to another, while others break up soon after they reach land, would seem to indicate that the electricity of those which are not affected by land is of one kind, while that of the Cyclones which are so, is of the same kind as that of the earth or mountain ridges, and this would account in part, if not wholly, for their "lifting up" as I have termed it, this being in fact the repulsion between the two similar electricities. The actual effects of the passage and discharge of electric storms over high mountains have been frequently, and of late years very carefully observed at Geneva and in Switzerland, and I abridge here, from the *Bibliotheque de Geneve*, for October, 1852, page 101, an interesting notice of these phænomena as observed in Switzerland. Canon RION, the author, is speaking of the Valley of Sion.[*]

"As regards atmospheric electricity, storms are frequent, but the inhabitants know that no matter how loud the thunder is, they have nothing to fear from the lightning. In no instance has it been known to do mischief in the lower part of the valley. In one case alone in the last century, a powder magazine on the Tourbillion (a neigh-

---

[*] Address on the opening of the 37th Session of the Helvetic Society of Natural Sciences, by the Canon RION, President at Sion in the Valais.

bearing eminence) was blown up. At elevations of 1100 metres, (1177 yards), steeples are sometimes struck; higher up the larches are frequently so; but the lightning falls chiefly on the ridges of the mountains. Our terrestrial electricity exhibits the same phænomena. No signs of electric tension, sparks, or luminous brushes are seen in the plain. At a certain elevation on the slopes of the mountains, as for example, at the Mayens near Sion, pretty strong shocks produced by the return stroke are not unfrequently felt after an electric discharge. It is upon the ridges that the electric action is most evident, as witness the case of the surveying Engineer, who having reached Mittelhorn just at the approach of a storm cloud, dared not touch his instruments, which were giving out large sparks, and wisely left so dangerous a post. The ridges of the mountains with their thousands of sharp peaks are our conductors, which discharge the atmospheric electricity, while that of the earth is constantly tending towards them."

When to these facts we add, that all the rocks of which the peaks of mountain ranges are composed, are of different electric powers, the scientific sailor will at once see how, admitting the Cyclone to be an electric disk, land, ranges of hills, and peaks of mountains may variously affect it.

400. We should not conclude these views, however, without remarking that while almost all the phænomena of Cyclones can be accounted for by either mode of formation, *i. e.* at the surface, or far above it, and afterwards descending, there is another great law which is wholly unaccounted for by either of them, and this is the LAW OF THEIR ROTATIONS; so invariable in each hemisphere that we find nothing approaching to a doubt of it in all the numerous investigations which have been made; but we *do* find that the mere water-spouts and the dust whirlwinds, which we may suppose to be generated near the earth's surface, turn indifferently either one way or the other.

401. Now this constancy of revolution for the great Cyclones, and violent, though smaller, tornado-Cyclones, would incline us to believe that their motions are dependent upon some invariable system of atmospheric influences beyond the reach of mere terrestrial causes; and the dynamical theory (of the Cyclones being occasioned by opposite or crossing *forces* of winds) seems wholly to fail, even where its simplicity at first sight recommends it; for if we suppose a strong N.W. stream of air, and allow it to rush into the S.E. trade, so that it passes on the *Northern* side of a strong S. Easterly stream, the two may doubtless produce at their sides a whirl, which might increase to a

PART V. § 403.] *Law of Rotations not accounted for.*   335

Cyclone, revolving ◯ or with the hands of a watch, as in the plate given, p. 159, by Mr. THOM ; and if the forces continued it might also continue for days together *in the same place.* But nothing here would account for its moving forward across the trade, and its doing this is much against the supposition that the forces sustaining it are the mere surface winds ; for supposing a Cyclone to be so set in motion in 10° South, and 90° East, the N. W. stream of air must force its way across, and through some 10 or 15 degrees of latitude, to be still acting in the same way when the Cyclone reached the Mauritius ; and must moreover propagate a new kind of motion for itself, *sideways*, to follow the Cyclone on its track, and to continue to supply it with forces. If we say, that the Cyclone is set in motion by one cause, and then that it generates a force of some *other* kind, as electricity, to keep it in motion, we depart wholly from all good rules of scientific logic, and we had better at once go back to the *other* supposed cause —whatever that may be.

402. Again : If we take the N. W. stream of air to be blowing from the N. N. W., and the S. E. trade from the S. S. E., and to meet so that the N. N. W. stream of air is to the *South* of the S. S. E. one, it is clear that they may produce a revolution *against* the hands of a watch, or ◯ ; and we should then have a contradiction in the known law of revolution for the Southern hemisphere. We all know that trades and monsoons vary infinitely between the points from which they take their names.

403. The Bay of Bengal and China Seas will give apt illustrations of this. In them the S. W. monsoon meets the N. East trade (the trade being called a monsoon while it lasts), and each prevail during six months of the year. It is about the period of the changes that the Cyclones are felt in greatest number and violence, as will be seen by reference to the table at p. 296.

Now if the N. E. wind meets the S. W. one "dead on end," either a calm, or a whirl, or vortex, or a Cyclone, turning either way may be produced. If the S. W. monsoon be a W. S. W. one, and the N. E. monsoon an E. N. E. one to the north of it, we shall have a Cyclone turning the right way for the Northern hemisphere, *i. e. against* the hands of the watch, but if the N. E. monsoon be a N. N. E. one and blowing strongest on the *Eastern* side of the Bay, and the S. W. monsoon a S. S. W. one, blowing strongest on the *Western* side of the Bay, it will turn the contrary way, or *with* the hands of the watch, like those of the Southern hemisphere. Now this never occurs, and hence it seems a forced conclusion to insist on the forces of the counter currents of wind at the surface of the ocean as the first cause of Cyclones.

404. From all these difficulties, as well as from that great one which arises from considering that the effects of opposing winds *may* tend, after all, to produce as much calm as storm, we are freed by supposing the Cyclone to form above, and to descend as a whirling disk to the surface of the ocean; when, as Mr. THOM remarks, "their diameter is (may be) from 400 to 500 or even 600 miles." With respect to the rotation, also, we may then easily allow that *if* they are formed far above us, and are mainly electric phænomena, then they may be subject to influences and laws—the influences and laws of the electric state of the atmosphere of great heights—of which we are as yet as profoundly ignorant as we were of the electricity of steam ten years ago.

405. When this passage was written for the first Edition, I had not seen SIR JOHN HERSCHEL's *Astronomical Observations at the Cape of Good Hope*, the following passage in which, Chap. VII. *Observations on the Solar Spots*, will occur to the scientific reader, and is well worth the attention of the sailor:

"The spots, in this view of the subject, would come to be assimilated to those regions in the earth's surface, in which for the moment hurricanes and tornadoes prevail—the upper stratum being temporarily carried downwards, displacing by its impetus the two strata of luminous matter beneath, (which may be conceived as forming an habitually tranquil limit between the opposite upper and under currents,) the upper of course to a greater extent than the lower, and these wholly or partially denuding the opaque surface of the sun below. Such processes cannot be unaccompanied with vortical motions which, left to themselves, die away by degrees and dissipate ; with this peculiarity, that these lower portions come to rest more speedily than the upper, by reason of the greater resistance below, as well as the remoteness from the point of action which is in a higher region ; or that their centre (as seen in our water-spouts, which are nothing but small tornadoes) appears to retreat upwards.

"Now this agrees perfectly with what is observed during the obliteration of the solar spots, which appear as if filled in by the collapse of their sides, the penumbra closing in upon the spot and disappearing after it."

When we recollect that the solar spots appear only in two zones of about 35° North and South of the sun's equator, and are separated by an equatorial belt on which spots are very seldom found, and that the existence of an atmosphere about the sun is now almost universally allowed, we are struck with the close assimilation so well noticed here by this illustrious writer, and we can easily conceive how our Cyclones seen from another planet, or from the moon, might also have the appearance of spots more or less circular, or oval, or elongated, according to their places on the earth's disk and the inclination of the vortex towards the earth.

406. For the present, then, we must content ourselves with carefully registered observations, and these aided by the resources of science, may one day reveal to us the true cause or causes of Cyclones, as the carefully registered experiments of FRANKLIN did to him the magnificent mystery of the lightning. We cannot in this research *make* the experiments, but we can vary them a little by trials, and our business is to observe them patiently and carefully in all their phases.

407. A remarkable, and at first sight a very strong objection has been raised to the whole rotatory and progressive theory by asking, *how is it accounted for, if we suppose a body of air to be at the same time whirling round and moving forward, that the side which is moving in its rotation in the same line with the track, has not the wind of double, or treble the force of that in the opposite side? Since it must have at the same time a double set of forces, i. e. that of the rotation and that of the progression acting upon it.*

408. This is simply enough answered if we suppose the whole Cyclone to be an electric meteor, composed of one, or many, close and nearly horizontal, but yet slightly spiral streams of electric fluid descending thus from the higher regions; and in its (or their) descent giving rise to currents in all the air it successively passes through, but not carrying this *same* air forward with it. It is evident that such currents might be spirally inwards or outwards, and we know from M. PELTIER's experiments, quoted p. 318, that such rotatory currents can be created, both in the smoke of resins and in water, by electricity.*

* I do not overlook the fact, that in the smoke the whirls were both horizontal and vertical, and that the currents in the water were, though circular, *from* the centre out-

409. It is evident that the objection quoted above tells with great force against the dynamical theory, or that which would suppose Cyclones generated at the surface by the forces of crossing currents of wind, because this affirms, that the whole body of the whirling Cyclone is supplied with its air (wind) by these currents, which then begin to move onwards, the streams of air doing the same, and that side of the circle which has its motion with the path of the Cyclone, to be therefore moving at a highly increased rate. We have also the difficulty of getting rid of the huge volume of air which by this theory must be at every moment poured in to supply the forces of the Cyclone, and for this Mr. ESPY and Mr. THOM assume their up-moving currents.

410. It appears to me that a simple, flattened spiral stream of electric fluid generated above in a broad disk, and descending to the surface of the earth, may amply, and simply, account for the commencement of a Cyclone; and that its gradual propagation onwards, in such direction as the laws of the forces generating it in the upper regions may give to it, will *as* simply account for its continuance and progression, and the exhaustion of the forces for its termination. And nothing of all this is gratuitous supposition; for we have all seen, if we have not carefully remarked, the actions of opposing thunder clouds and storms, which *must* be generating some action somewhere; their passage over great tracts of country and appearance over hundreds of places at once, or successively, which is certainly like a descent and a progression, and which can be foretold, and vaguely accounted for, and finally after a longer or shorter duration their disappearance.* If we are asked with all this, *why* the effect of a stream of electric fluid should produce storms of air (wind) so as to form Cyclones? we must reply that as yet we only suppose that it is so—and in this as in every other theory wait for farther facts to guide

wards; but M. PELTIER remarks in more than one place on our imperfect means, and on the extremely insignificant scale on which we can imitate the processes of nature, and that when we come to those of the atmosphere we are forthwith confined, for the most part, to experiment in and upon the air at the surface only.

* See Kaemtz, p. 383 and 384, on M. TESSIER's Hail-storms in two parallel lines, &c.

us to a new one; or to confirm it. It may be remarked also in reference to what I have said, at p. 327, on the unequal duration of the sides of Cyclones, that this also may be perhaps accounted for, by assuming, as I have done, the Cyclone to be formed by a spiral stream of electric discharges, for however flattened the descending spiral may be, one part of it *must* be lower than the other, and this may be on the advancing side. It must also part with more of its electricity to the ocean on that side, and thus have less for the forces required in the following side of the whirl.

This assumption of a *spiral* stream of electric fluid has also nothing extraordinary or excessive in it. All zig-zag lightning is supposed to be really spiral, and to assume that appearance to us, because we see it sideways. The plain seaman has only to turn a corkscrew perpendicularly, on a level with his eye, to understand that a flash descending corkscrew-wise (spirally) would appear to be a zig-zag one.

411. The general impression which evidently exists from China to the West Indies, and which we find from the days of COLUMBUS[*] to our own, in all the " Hurricane Countries," that however threatening the weather may appear, there will be no Cyclone, *if it thunders* as well as lightens *at the commencement*, is favourable to the view of the Cyclone being a purely electrical phænomenon. For if we conceive a disk to be formed or forming, and to discharge its electricity *before* it descended, then we might expect both thunder and lightning. If, on the contrary, the discharge only took place when it *had* descended, it would then be the silent transfer of vast quantities of the electric fluid from the clouds to the earth (or vice versa) creating the circular currents of air which M. PELTIER has shewn, can be so generated, and which then become the circular winds of the Cyclone.

In Mr. Crank's notes already quoted, p. 209, he particularly states that there was no thunder or lightning, and again that when endeavouring to proceed about 100 yards from his house to save his boat in the height of a Cyclone,

"The strife of the elements was awful : ever and anon such a pressure of the wind *from above*, that I was almost crushed to the earth ; then it would strike me on one side, then on another, and I came to the conclusion that I was in the centre of a

[*] In the old Spanish relations.

whirlwind. I was, for a time, stupefied. To turn my face to the Eastward for more than a few seconds was impossible. The rain was driven with the force of arrows into my face, and the oppression was similar to what one feels on riding a fast horse at a racing pace."

412. Mr. REDFIELD's description of the clouds during a Cyclone, p. 268, adds also some probability to this view of their being formed above, and descending as disks to the surface of the ocean: for the true disk we may suppose to be the great sheet of stratus cloud which he describes; not in actual contact with the earth, but at that distance from it which would induce the discharge, and that the Cyclone is nothing more than the effects of that discharge; of which the wind and the whirling and fast flying clouds (cumuli and broken strata) and the rain are the effects. We evidently *see* this in hail-storms, though these *may* be phænomena of a different class.

413. We may advert here also to some other effects which have been undoubtedly felt close to, or at the centres of Cyclones, and which it is very difficult to account for, except by the agency of electricity. Amongst these are the accounts given by Mr. SEYMOUR, master of the brig *Judith and Esther*, in his letters, pp. 74 and 76, of Col. REID's work, where he states that when thrown for the third time on their beam ends at the centre,

"For nearly an hour we could not observe each other, or anything, but merely the light; and, most astonishing, every one of our finger nails turned quite black, and remained so nearly five weeks afterwards."

And again writing to Col. REID in explanation, he says—

"Thirdly, as to the cause of not being able to see each other?

"The cause of this I cannot well tell; but while running before the wind the vessel was hove the third time on her beam-ends, and while on her beam-ends, the atmosphere had quite a different appearance, darker, but not so dark that (I should imagine) would hinder one from seeing the other, or from seeing a greater distance, were it not that our eyes were affected. It was about this time our finger-nails had turned black; and whether it was from the firm grasp we had on the rigging or rails, I cannot tell, but my opinion is, that the whole was caused by an electric body in the element. Every one of the crew were affected in the same way.

WILLIAM SEYMOUR."

414. The Cyclone of October, 1848, in the Bay of Bengal, to which I have already alluded, appeared to offer a favourable instance for

obtaining some information on this subject. I took much pains to inquire of most of the Captains of ships, who sent me their logs, but except the distant lightnings before the onset of the Cyclone, four ships only on its South and S. Eastern quadrants (or on that part on which I suppose it to have been lifted up), of twenty-two whose logs were examined, experienced heavy electrical discharges,* the reply from the others was that there was no lightning or "little to speak of."

415. Analogous to this is the following from my Tenth Memoir, Jour. As. Soc. Beng. Vol. XIII., on the Madras and Masulipatam Cyclone of 21st to 23rd May, 1843. Track X. on Chart No. III., in which Captain CORNEY, of the ship *Lord Lyndoch*,when close to the centre of the Cyclone, and but shortly before the shift of wind, says—

"The strongest gusts were about 1 P. M., when there were intermitting severe gusts, *accompanied by great and terrible heat—and there were alternate gusts of heat and cold after the hurricane veered to S. W.*"

In the October Cyclone of 1848, Capt. ARROW of the *Wellesley*, while hove to on its South Western quadrant, to avoid running into it, distinctly says in his notes, and he was about 3° to the E. N. E. of the *Lord Lyndoch's* position,

"Hot and cold blasts were distinctly felt. *I can compare the hot blasts only to the Scirocco of the Mediterranean.*" Other ships distinctly describe *sleet, i. e.* snow or hail with rain.

Having no thermometric data, we cannot say what the extent of these variations of temperature was, but they are very remarkable as occurring at this time, and while the fury of the Cyclone was from seaward, the N. E. Nevertheless as these vessels were not very far from the land some uncertainty exists as to the origin of this heated air, but if it came originally from the land, it must have been carried up and then forced down again without alteration or mixing! which we can barely allow.

416. Captain MILLER, whose able management of the *Lady Clifford*, I have alluded to at p. 166, says in a letter to Captain BIDEN, with his log of the Madras storm of 1836—in the ship *William Wilson*—

* The analogy of this to the spark from the condensing disk of the electrician is obvious, but we want more evidence before we reason upon it.

"In the storm under notice there was an extraordinary change took place suddenly in the temperature of the air, but I regret that I cannot state if a corresponding change took place in the Thermometer, as I had not looked at that instrument until directed to it by my own feelings, and the complaint of cold from all my crew."

417. Captain RUNDLE also, whose log I have quoted so often, says incidentally:—"The rain water exceedingly cold, the sea water very warm, much more so than usually;" and in another place "fresh gale with furious squalls and rain as cold as ice." This vessel was in 5° or 6° South latitude. The *William Wilson* was in from 11° to 12° North. Captain Ross, in his Notes from the Keeling and Cocos Islands, says, that while the Barque *Harriet*, on her voyage from the Straits of Sunda to the Cocos, was upon the verge of a Cyclone which she hove to to avoid, and in which are noted "heavy dashes of cold rain," the wind at the Cocos had been strong from the S. E., "with cold drizzling rain, so cold that it withered all the blossoms on the orange trees." These cold and hot blasts and rains thus become matter of curious speculation as to their origin, and are probably connected with the descending theory of Cyclones and their relation to hailstorms, as already described.

Mr. CRANK, whom I have already quoted (pp. 209, 339), says that at the time of the lull of his Cyclone, it became so exceedingly cold that every person in his family remarked it, and on looking at the Thermometer he found it standing at 65° or 66°, being at daylight 90° or a little less, so that the fall was not far from 20°.

PART VI.

1. DIRECTIONS FOR STUDYING THE SCIENCE. 2. POINTS OF INQUIRY AND DIRECTIONS FOR OBSERVERS. 3. MISCELLANEA AND ADDITIONS. 4. CONCLUSION.

418. DIRECTIONS FOR STUDYING THE SCIENCE. I trust there is no part of this book which the plainest seaman will not be able, with a little attention, clearly to understand, but it may nevertheless

be useful to some to inform them how to use it as a sort of Grammar of the Science; for a general understanding of the Law of Storms does not suffice to make it so ready to the mind that an error may not be committed in the first application of the rules to a difficult case.

419. To study the Science then, lesson by lesson as it were, the sailor should, after his first perusal of the whole book read over again carefully the definitions of the words at pp. 7 to 16 in Part I. from § 16 to § 26. He should next read Part III. from the beginning of § 110, p. 101, to § 114, p. 104. And then turn to the details of Mr. REDFIELD's tumbler experiment at p. 233, § 286, without paying much attention to the Barometer part, but carefully considering and varying over and over again, and backwards and forwards (with and against the sun) for both hemispheres, the passage of the miniature Cyclone for one, two, or more ships, and upon all sorts of tracks; noting carefully the winds which his ship, or island, or port, may experience as he makes the Cyclone travel down upon and over it. And he should remark how truly the track may be judged of by the average of the shift. He should then substitute the Horn Card for the glass, and he will instantly see that, if he supposes the arrows upon it in motion, it exactly supplies its place.

420. The next lesson must be how to estimate the bearing of the centre in an actual case of being at sea with a Cyclone evidently commencing, and for this he must carefully study the sections on the bearing of the centre, and on the Wind Points and Compass Points at p. 104, from § 115 to § 119, with the Storm Card before him. And then the method of estimating the bearing of the centre and the probable track by the winds and ship's run, copying afresh the projection which I have given at pp. 116—122, § 130 to 135, and then selecting some case from an old Log, or supposing one, make the projection of it for himself.

421. Having thus arrived at knowing where the dreaded centre lies, and upon what track it is moving, and why the motion of the Cyclone and his run make the wind veer in a particular way, he has to consider (always in actual cases if possible) what should be, or has been, the proper line of management, and here the figures at pp. 124

and 129, should be carefully considered for both hemispheres, and the questions of scudding or heaving to, with all their various conditions, as shewn at p. 126, § 138 to 143, should be carefully pondered.

422. If the seaman, old or young, will but follow these directions and patiently examine a few actual cases for both hemispheres, I think he will find that nothing is, in reality, so simple and so ready of application when the mind has been a little exercised upon it. And when he will but take the trouble also to consider (for this cannot be repeated too often when life and property are at stake) what his ship and officers and crew can bear and do, when the hour of need and danger may come, he will find I think, also, that in most cases he will—and I do but quote here the words of one of the many correspondents who have acknowledged to me the advantage and relief from anxiety which the Science has afforded them—have exchanged dread for confidence.

423. POINTS OF INQUIRY, AND DIRECTIONS FOR OBSERVERS. I have so often alluded in the course of this work to the many matters we have yet to inquire into, that at first it may seem a repetition to go over this again; but it is always advantageous to have such matters brought closer together under distinct heads, and there may be some which I have at heart, but have not been able to suggest. Let us take the inquiries in their natural order :—Before; during; and after a Cyclone.

424. BEFORE A CYCLONE. *a.* The signs of all kinds, celestial, terrestrial (if in port), and those of the ocean, as in our table at pp. 278-282, being sun, moon, stars, sky, clouds, light, air, wind, lightning, thunder, noises, effects on animals, &c. Appearances of the stars should be carefully noticed, distinguishing between the fixed stars and planets, and between those near the zenith and at the horizon.

b. The signs considered by Natives or European inhabitants of different places as announcing an approaching Cyclone, and especially those connected with volcanoes and earthquakes, and their subsequent relations to the Cyclone when in action, if any. Pilots and fishermen are often very well informed on these matters.

c. State of the barometer and simpiesometer, both on shore and on board ships if in port; for the one might, for instance, oscillate and the other not.

d. States of all other meteorological instruments, such as hygrometers and the like; of any sorts attainable.

e. Exact observations as to the state of the clouds, the scud, &c. above; or that of the weather in elevated spots, so as to elucidate as much as possible the question of whether the Cyclone is formed at the surface of the earth or in the higher regions of the atmosphere.

f. The bearings and angular altitudes of all banks of clouds, and of lightnings with the distances of them, as nearly as can be measured by the directions given by KAEMTZ (§ 330), the time of the observations being accurately noted. The bearings of the extremes and centres should be taken. If bands of light or colours are well defined, the angles subtended by these should be measured by a quadrant or sextant.

g. In the neighbourhood of volcanoes their peculiar actions should be noted, and the opinions of the residents inquired into, as also where volcanic lakes, and the like, are known to become agitated or phosphorescent. When volcanoes are near the sea, it is by no means unlikely that they may shew some forewarning indications.

h. The state of the surfs on coasts, peculiar swells arising from the sea, its phosphorescence and smell, the appearance or flights of peculiar birds, or even the peculiarities known to fishermen of the habits of fish should not be neglected. At first sight many of these things seem mere fancies or local superstitions; but if we found on opposite sides of the globe, for instance, that on the approach of a Cyclone, certain fish, abundant at other times, are not to be caught for three or four days before its outbreak; or that certain others are seen to be much agitated and to come constantly to the surface; we might fairly say, that this was a fact derived from observation, and if verified by competent observers, it might be a clue to further knowledge.*

---

* So far the first edition. I have since found in a valuable note on the *London's* Cyclone of October, 1832, in the Bay of Bengal already alluded to, pp. 154, 181, the following curious passage. After describing five days of entirely oppressive calm, in which "the sea became a perfect mirror, the glare intolerable and the zenith appeared as if it had lowered within a few feet of an horizon, of a greenish and yellow sickening unnatural tinge, and the air felt as if it were surcharged with brimstone

i. Observers on shore might render great service by carefully pursuing any *one* single branch of research or observation only, and of these, some do not even require instruments, or very simple ones. As an instance, I may cite a series of careful cloud, or (to use the Greek word, which Mr. ESPY has usefully adopted) NEPHELOGICAL observations, which might be of high importance, both as shewing for a long period, what the various currents above and below are, and what are the peculiar appearances which forewarn of the approach of Cyclones at that place. This scud, as every sailor knows, is often flying in a direction which differs from one to four points from that of the surface wind which the ship has, and the difference may even be greater on shore.

j. The various names given by the Natives or European residents to various kinds of gales; as an instance see note p. 70, and in the next section, the remarks on the names Hurricane and Tyfoon.

k. Great attention, both on sea and on shore, should be paid to the choice of the *words* used to describe the different phænomena of Cyclones. For example, the words "veered" and "shifted" are often carelessly used for each other, yet they express, the one, a gradual, and the other a sudden change of wind.

l. The intervals of *time* between the various changes in the storm or the states of the instruments should also be marked. As thus: "The wind veered between 4 A. M. and half-past 6 A. M. from North to W. b. S.," or "The barometer sunk from 8h. 30′ to 10h. 45′ A. M. from 29.05 to 28.70 as per register."

m. Sudden heavy squalls and whirlwinds, and especially such as give little or no warning of their approach; as white squalls, the bull's eye squalls on the coast of Africa, &c. (see § 367), should all be noted and carefully described as soon after their occurrence as

and charcoal almost enough to suffocate one," Captain McLEOD adds, "Turtle innumerable were floating about us, and as far as our vision extended were seen equally numerous at the distance. I need not say that we caught as many as we required. Whether it was fancy on my part or not, I thought they were in a state of stupor, as they never made the least attempt to escape when the boat approached them, *nor when they were taken hold of.*" Sleeping turtle we have all caught, but more have been missed perhaps by their struggles, and this passage settles the question of the stupor.—H. P.

possible. We have yet much to learn of these dangers, and especially in their relations, if any, to Cyclones.

n. Whirlwinds and waterspouts, and peculiar veerings of winds with them and alternations of calms, and puffs, and flaws. The appearances of whirlwinds and waterspouts, and their manner of turning, whether ◯ or ◯ should be noted with all details, particularly if occurring at night.

o. Rising and falling of wind, and if any moaning, or distant roaring noises are heard before or after the Cyclone, and if these noises are certainly in the atmosphere.

425. DURING A CYCLONE. *a.* Atmospheric states and changes as often as can be noted, and particularly at the changes of wind, and at the calm centre, if this should unfortunately reach the ship.

b. State of the barometers, simpiesometers, and thermometers, as often as possible, their oscillations, and if (which should be observed in the dark) any flashes of light can be *certainly* distinguished in the vacuum of the tube, and if these are constant for a time, or by fits, and if connected in any way with the oscillations or squalls.

c. The same at the shifts or calm.

d. Lightning and thunder; and particularly remarkable flashes at the shifts.

e. Phosphoric light of the sea.

f. Temperature of the rain; if colder or warmer than the sea; form and size of hailstones, if any; relation of the fall of hail to the lightning and state of the barometer.

g. Whirlwinds or waterspouts occurring in the Cyclone to be carefully noted, as to their appearances, whirlings, tracks, size, &c.

h. Circles of light, or clear sky overhead, in the midst of the Cyclone, to be estimated or measured as to how many degrees in diameter: stars, moon, or sun, if seen at any time in the Cyclone, to be also noticed if of peculiar brilliancy or colours.

i. State of the clouds; appearance, velocity, and direction of the scud, &c.

j. State of the swell and sea, as to regularity, rising, breaking, &c. particularly at the onset and centre.

k. Veerings or oscillations of the wind, and, if possible, the exact intervals of time in which they occur, should be measured. Thus if we knew that the wind veered backwards and forwards six points about every 15 minutes, we might, with the drift, calculate what the amount of incurving was.

l. Moderating of the wind for an hour or two or more after the gale or Cyclone has appeared to commence, and the state of the barometer, simpiesometer, and sky at this time.

m. An exact account should be, if possible, kept of the vessel's coming up and falling off, and the log should be hove if possible, to ascertain with the utmost care* the direction and rate of the ship's drift if lying to ; so as to be enabled after the Cyclone to calculate with precision what may be due to the storm-wave and storm-currents.

n. Blasts of hot and cold air. Extraordinary light or darkness.

426. AT THE CLOSE OF A CYCLONE. *a.* Appearances of the clouds when going off. If forming banks, to be noted and measured, as at the approach of a Cyclone.

b. Gradual rising of the clouds at the horizon or zenith to be noticed if it occurs.

c. Barometer and simpiesometer, rate of rising to be as carefully noted as during the Cyclone; and how long the oscillation continues after the passage of the centre.

d. Any effects of the Cyclone on the ship's compasses, or on the respiration, or on animals on board, &c. to be carefully noted.

e. On shore any accounts of inundations, and if by sudden sea waves or gradual rising. Old and foreign accounts of those also to be collected if obtainable. Directions for preserving records of the state of the weather in the British Colonies have been issued some years ago ; but it is not known what these have produced. The importance of obtaining the observations of foreign settlements is therein also alluded to.

427. DIRECTIONS FOR OBSERVERS. We have four classes of observers who may be useful to us. 1, Those on shore, landsmen or sailors. 2, Sailors in harbour, where they have much more leisure and advantage. 3, Sailors at sea; and 4, Surgeons and Passengers. As it may, however, be useful for all to know what the others might

* By a leeway circle on the taffrail, if this can be managed. I mention every thing, though I well know that in merchantmen all hands are fully employed ; but in Men-of-War much may be accomplished by a little zeal.

observe, if they pleased (for they will thus know how to question them, or what to ask for in the way of copies of papers, &c.) I set down the whole together, and each class will easily distinguish what belongs to them.

428. I commence with pointing out what may be done with the observations when made or obtained from others; for I know that many are deterred from, or rendered indolent-minded about such matters by the feeling of "*What is the use of all this trouble in making notes? I do not know where and how to send them to any one who will make use of them.*" I trust that none will have read this book without being convinced that, speaking here for my brother Storm-meteorologists,[*] as well as myself, we shall always be glad to turn every scrap of information to account, and to forward to each other such as may assist our common research in those quarters of the globe where we may be situated: and those who have read all the works on the subject can, I hope, also testify how usefully and faithfully every line yet obtained *has* been brought to account for the great family of mankind.

429. For myself, I shall be glad to receive notes, memoranda, or extracts from books, from any part of the world, on storms or matters relative to them, in any language; as detailed or brief, and as plain or as scientific as may be.

430. It will be seen from the foregoing pages that, independent of the particular points of inquiry, which I have specified in the first part of this section, there are wide blanks to fill up in the chapter on the tracks of the Cyclones alone; and that we ought gradually to be able to produce charts for every separate tract of sea, like the four which we have now given; and that this can only be done by the workmen being furnished with materials; and a single experiment with a storm card will shew that if a hurricane moved, (for instance across the Atlantic or Bay of Bengal,) from West to East instead of the contrary ways, which we now *know* they do, all the changes of

---

[*] W. C. REDFIELD, Esq., New York. Lt.-Col. REID, R.E., V.P.R.S. Dr. A. THOM, H.M. 86th Regiment, Poonah, Bombay. M. H. BOUSQUET, Port Louis, Mauritius.

winds and rules for management would be different. Hence the great utility, tedious and sometimes almost repulsive as the labour is, of collecting the data and investigating every separate storm till, as in all other branches of the physical sciences, we are able to say with comparative or absolute certainty, what is the LAW by which their tracks are governed, since we know that of their rotations.

431. As I have before said, for these investigations we must have data, and these data are log-books, journals, memoranda, newspaper notices, and the like; and the more clearly to explain what we require, I set down here in separate paragraphs, what occurs to me; premising always, that *the more details, the better.*

1. We require all the accounts, registers, and notices, logs, journals, memoranda, and even *references* to books, (in any language,*) which can be obtained. The full log should always be sent if possible, as well as notes and extracts.

2. These may be new or old, for the phænomena of the Cyclone of yesterday, may be corroborated by those of a hurricane a hundred years ago.

3. They may be plain common-sense notes or narratives, or as scientific as they can be made; but the plain common-sense accounts are often *quite as valuable* as the scientific descriptions; and it is a great mistake, and one I fear which has deprived us of much good material, to suppose that we *must* have scientific data, because the research requires the aid of science to develop the laws to which they lead.

4. A mere note of the times of commencement, violence of the wind, from what quarter, how changing, how ending, and the position of the observer on shore, or the latitudes and longitudes at sea, with the run, &c. are all that are strictly required. If the barometer, thermometer, simpiesometer, and all other points noticed are added, so much the better.

5. The place of the observer at noon each day, before, during, and after the storm, should be carefully given at sea; even from mere

* Many valuable hints are to be found in the old navigators, English as well as foreign, but few or none of their works are obtainable in India.

estimation, if nothing better can be obtained; because no one can, from a log, estimate a vessel's drift in heavy weather so well as those on board of her. And the exact positions before and after the storm are always important.

6. When extracts from Private Logs or Note books are sent (and these are always valuable, if containing notes and observations), the ship's complete Log if possible should always be sent with them for the corresponding days, for it is often necessary to lay down her run or to calculate positions at other hours than noon, before arriving at satisfactory results.

7. On shore, the latitude and longitude of the place, or its distance and bearing from the nearest well known station or city should be given; and some notions of its position as to mountains, hills, rivers, valleys, &c. will be always useful.

8. Mariners may very usefully employ their own leisure hours, *or the leisure hours of any boy on board who can write*, during a passage, by copying from their *old* log-books, no matter *how* old, the logs of any old storms, or the logs which shew a ship to have been *near* the place of any known storm; and especially to copy the log of any storm they may have had on the actual voyage, so that it may be ready to send off on their arrival in port. I have reason to believe, that many Commanders are willing, and even desirous of aiding us, but when they come into port, they are hurried and anxious, do not like to send their log-book out of the ship, or it is left a long time with the Notary for the protest, and finally leave without carrying their good intentions into effect. Much valuable information is thus lost. As to the old logs, *I have* a great deal of scattered information thus collected, which might become far more valuable, with the addition of a few more corroborations.

9. Another very mistaken notion is, that some are apt to fancy that their particular log, or note, or memorandum on shore is "*of no great consequence*." This is a very mischievous notion, and I entreat those who may entertain it to consider first, that it is impossible to say beforehand, in any research of this kind, *what* is, and what is not, important; and next that twenty proofs are always better than ten.

10. Some Commanders, too, are apt to suppose that unless a Cyclone amounts with them to a furious and damaging Hurricane or Tyfoon, it is useless to trouble us with the details. This is also a great mistake. Whenever there is anything *Cyclonic* in a breeze, or even in the appearances of the weather, I shall be glad to have details of it. And especially when other vessels have also felt it. A moderate Cyclone may be as instructive as the most violent one for the great object of tracing out the tracks in that locality or at that season of the year; or there may have been a Cyclone overhead not yet settled down.

11. Some again, I fear, who may not have looked into the details, excuse themselves by saying they do not believe in the truth of the Law of Storms. There is no harm in this opinion—if it does not cost the dismasting or loss of a ship—and I am old enough to have heard Barometers, Lunars and Chronometers sneered at, as "newfangled notions;" but I think I do not exaggerate when I say, that in the writings already published on the subject, there is to be found, for every impartial mind, abundant and almost mathematical proof of its correctness. And if these researches are not useful, it certainly can do no harm to publish all the facts relative to storms for scientific men to make some *other* use of.

12. I fear also, that some Shipmasters, who may not have enjoyed a good education in early life, may feel a little embarrassed and unwilling to submit their log-books or extracts to the eye of a stranger. I can only assure such, that I have had many log-books in which great deficiencies were observable, but have never, and should on no account, think of ridiculing, or even of criticising them beyond what is strictly necessary to establish the truth of a question.

13. Copies of public documents, such as reports, registers, and the like should be inquired for.

14. Details relating to inundations and the manner and time at which they occur should be collected, and the *gradual* risings of rivers carefully noted, apart from the *sudden* ones, or sudden waves rolling in from the ocean or from great lakes.

15. Cases of the compasses being affected during or after a Cyclone.

16. All notes regarding the thunder and lightning of a Cyclone, *i. e.* before, during, and after it, are of the highest interest.

432. *Persons residing on shore should,—*

1. Set up a vane or weathercock if there are none in sight, and if they are not old seamen enough accurately to estimate the point from which the wind blows.

2. They should also, from various parts of their dwelling, take North and South and East and West marks *in fine weather*, and not at too great a distance, such as trees, chimneys, &c. In a storm they will find these of essential service in estimating the course of the wind, and that of the clouds.

3. If furnished with instruments the registry of all of them before, during, and after a storm, will of course be most acceptable; and if compared with standards so much the better, but a plain common-sense account of all the phænomena of a storm; or of the weather when one is passing near, the driving of the clouds, the *times* of the changes or veerings of the wind, calm interval, &c. &c. are all that are required in most cases, and will always be eminently useful.

4. The collection of reports from other parts is also a most useful assistance to us; even if mere native ones, they are better than none.

5. Accounts of tornados or wind-spouts, dust whirlwinds, &c. their tracks, the direction in which they turn, whether *with, or against* the *hands of a watch*, and notes of their formation, progress, ravages and disappearance, are all of great interest and utility.

6. In observing the oscillations of the barometer so frequently alluded to, we should be careful as to the observations made in the height of a Cyclone, that the room or cabin be not too close; for it might happen that the heavy gusts might force a volume of air into a closed apartment which would momentarily force up the mercury by its pressure; care should be taken on this account to have the instruments placed so as to obviate this objection, as by opening a window or door to leeward, or the like. Of course this does not apply to the oscillations at the approach of the Cyclones.

7. Every commander of a ship on his departure from and arrival in port should endeavour to procure a comparison with a known standard Barometer for one or more days. This is easily done by leaving the

hour at which the standard is observed, say at 9.50 A.M. and 4 P.M. and noting the ship's Barometer exactly and as often as possible at those hours, paying due attention to note at the same time the temperature, so as to render the comparison as accurate as possible.

433. MISCELLANEA AND ADDITIONS. THE LOG-BOOK. On the label of a ship's log-book sent to me, and which was, it appeared, printed and sold by CLARK and Co., 72, Gracechurch Street, London, I observed below the usual title and form for entering the ship's name, voyage, &c., the following very proper caution—

"N.B.—On the accuracy of a log-book depends the recovery of the owner's and shipper's property, in the event of loss. No erasures or obliterations ought to be allowed, but every incident recorded in a clear and explicit manner, and, on no account, should the log-book of a former voyage be permitted to remain in the ship, but should be deposited for reference with the owner or broker."*

434. I have elsewhere said (XI. Memoir, Jour. As. Soc. Beng. Vol. XIV.) that—

"It has often struck me to remark on the absurd practice of keeping a ship's log-book without entering the longitude. It is quite possible that a case might arise in which, at least ignorance of his true position, if not wilful destruction of his vessel might be alleged, if not proved, in a court of law against the master of a ship through this omission; and his insurance thereby become vitiated or his character destroyed, though really a good navigator, in case of accident. The private log, or 'lunar and chronometer book' of a Captain, would barely be held as a legitimate document when the book which *should* contain the vessel's place at noon is a blank."

435. But passing over all this, none will question that a well kept log is as high a testimony of a commander's ability, as a well kept set of books is of the commercial talent of the merchant for whom he sails. But when we see (as I have seen scores, if not hundreds of) log-books, even of voyages to India, in which, if the log is hove it is not registered every hour, or it is registered by mere guess; which by the way is almost a fraud; or the barbarous and careless form of the

---

* Sailors would demur greatly to this: many capital runs and safe passages through difficult straits, good and essential recognizances of headlands, &c., are made with the help of old logs. No ship should be allowed to go the same voyage twice, without having a copy of her old log-book, which may often serve to correct an erroneous, or supply a deficient chart.

coaster's logs (marking it every two hours only) is adopted; leeway, variation, barometer and simpiesometer apparently not thought of,—though the opening of a cask of beef or peas is carefully set down,—while the stock of water on board, even with troops, often seems a mystery! and in fine everything, to all appearance, is left depending on the latitude and *the* (one) chronometer; and this even on the approach to land—when we see all this, I say, we must regret to avow that in this respect the French are far before us; for not only are their merchant ships' logs properly filled up with everything that a good log-book should contain, but moreover, in most of them,* the log of each watch has the signature or initials of the officers by whom it is written; so that it thus becomes what Messrs. CLARK and Co. properly recommend that it should be.†

The French Logs contain a column for "*route corrigée*," which is the actual "course made good," after all corrections for leeway, variation, &c. according to the estimation of the officer in charge at the time. This is very useful, particularly as regards leeway—and as regards the due attention of the officer of the watch to the ship. In LORIMER's Letters to a Young Master Mariner, p. 45, of Edition of 1849, he says—

"Generally speaking the Log-Books of British Merchantmen are a disgrace to the writers, and to so great a commercial nation. From being allowed to remain on board

---

* I speak both of original logs and copies. In these last the signatures alluded to are usually omitted.

† Nautical Booksellers would much aid in the advancement of our science, and furnish their customers with good hints, if they were to adopt Messrs. CLARK's useful plan, and their N. B., might run something as follows:—

"N.B.—On the accuracy of a log-book depends the recovery of the owner's and shipper's property in the event of loss, and the good names of the Captain and his Officers. The ship's latitude and longitude should be carefully noted at every noon by observation and account, chronometers and lunars, and leeway, variation, deviation, barometer and simpiesometer duly entered. All changes of wind and weather, and peculiar appearances on the approach of bad weather should be registered. Erasures and obliterations should be avoided, and alterations certified by a note in the margin. A duplicate log should be kept and daily signed, and on the ship's return to port, all bad weather or extraordinary phænomena should be reported to any person known to take an interest in such matters."

voyage after voyage, they are frequently torn, defaced, and eked out with different papers ; presenting a slovenly appearance, and too frequently in case of disputes giving room to suspicion, uncertainty, altercations, &c."

He then goes on to describe in a note the custom of the French and Danes, who receive stamped log-books from their Government, and are bound to produce them for inspection and deposit at the end of the voyage. I hope this will fully justify my remarks on this most important subject, which are made for the furtherance of our science.

EFFECTS OF CYCLONES ON THE COMPASSES. The following is an extract from my Seventeenth Memoir, Jour. As. Soc. Beng. Vol. XVIII. being part of a note by Captain SHIBE, of the Bark *Easurain*, who experienced a very severe Tyfoon-Cyclone in the China Sea, as noted at p. 58. Captain SHIBE says :—

"I will mention another circumstance that may be perhaps interesting as it was certainly new and startling to me ; and that was, that we could get none of our compasses to remain steady, but at every succeeding burst of the heavy squalls, they spun round and round, eight points at a time, and we had no other means of steering, but by the roll of the sea, and the feel of the wind on the back part of the head, and this continued for some time after the gale had passed and rendered the approach to the land upon any safe course very precarious."

In an account of a succession of Cyclones experienced by the ship *Lord George Bentinck*, Captain EDGELE, from London to Calcutta, by Dr. GORDON, of H. M. 10th Regt., it is stated that upon two occasions, one on the approach of a Cyclone, and the other in the midst of it, "the ship's compasses were remarkably unsteady, oscillating and moving about several points," and Dr. GORDON at once attributes this phænomenon to electro-magnetism.

We have many detached notices of this phænomenon of the whirling motions of the Compass Cards in squalls and even in Cyclones, which till now was inexplicable, but which is perfectly explained by Dr. BADDELEY's experiments.

In the Naut. Magazine for 1843, p. 99, Captain T. B. SIMPSON, of the barque *Giraffe*, from Sydney to Manilla by the Eastern passage, says, speaking of his passage between Cape Deliverance and Pleasant Island, that he had violent squalls from the N. West to S. West, with much rain and vivid lightning.

" During one of these heavy squalls on the 27th January, in about Lat. 6° 9'

PART VI. § 435.] *Effects of Cyclones on Compasses.* 357

South, Long. 164° 15' East, at 11h. 45 A.M. observed the compass card to revolve several times without any apparent cause; the phænomenon might probably be occasioned by the effect of electricity on the magnet, the squall being charged at the time with a large quantity of electric fluid. It is worthy of remark that during these heavy squalls there was no perceptible alteration in the Barometer, it shewing 29.75."

The Editor of the Naut. Magazine remarks in a note, that Captain GRAAH in his Voyage to Greenland, p. 18, notices a similar phænomenon, and at p. 310, we have full evidence from an excellent observer, Captain SHIRE, that it occurs in Cyclones! The inference indubitably is that Cyclones are great spiral currents of electricity. If we admit this, our next inquiry is, " Whence can these be derived?" And it is only observation and long and careful registry of these observations which can afford us any information. The following note from Humboldt's Cosmos is curious, as shewing that the idea of a vast electric atmosphere surrounding the aerial atmosphere of our globe has already been entertained to account for another phænomenon as mysterious in its origin as the Cyclone. After describing the peculiar black metallic crust which is found on stony Aerolites, and forms their most distinguishing feature, and remarking on the intense heat necessary to fuse it, far beyond that of our most intense porcelain furnaces, he adds in a note that the celebrated astronomer POISSON had endeavoured to account for this fusion, which must take place where the density of our atmosphere is almost null, in the following manner: " May we not suppose that the electric fluid in a neutral condition forms a kind of atmosphere extending far beyond the mass of our atmosphere, yet subject to terrestrial attraction, though physically imponderable, and consequently following our globe in its motion? According to this hypothesis the Aerolites would on entering this imponderable atmosphere decompose the neutral fluid by their unequal action on the two electricities, and they would thus be heated and in a state of incandescence, by becoming electrified." (POISSON, *Recherches sur la Probabilité des Jugements*, etc. 1837, p. 6). Compare here with what has been said of the Solar spots, at p. 336, § 405, and there will seem much analogy to support our views of the electric origin of Cyclones.

In the Naut. Magazine for Jan. 1852, p. 330, is an account of the deviation on board the iron steamboat *Keera*, Captain A. M. SAINT-HILL, in which is remarked the increasing amount of and alteration in the deviation, when arrived in 27° South in the Atlantic, so that in 40° South and 83° East the vessel steered N. E. to make an East course. In Lat. 42° South, Long. 100° East, in December, it is added " hove to in a heavy gale ; the compass frequently *changed ends*, sometimes continuing reversed two or four hours and then vibrating a long time, in fact useless. They continued so for three days, when they were partly corrected by reversing the poles of the antagonist magnets."

The H. C. Steamer *Fire Queen*, Captain BOON, proceeding from Moulmein to Rangoon, on the evening of the 14th May, 1852, wind S. Westerly, weather from noon gloomy and thick with squalls, "observed that the steering compass was revolving to such a degree as to render it useless. Examined the ridge-pole compass, elevated eight feet, and the bridge compass, elevated 11 feet from the deck, and found they were all affected in the same manner. Barometer 29.87. Heavy squalls from the S. West with a heavy sea running."

In 1840, the H. C. Schooner *Mahi*, from the Persian Gulf to Bombay, was surrounded by beautiful groups of whirlwinds and waterspouts, which were seen raging about her in all directions, when suddenly the needle lost its polarity and for some time continued useless. In April, 1846, the H. C. Steamer *Queen*, when about 300 miles from Bombay coming from Aden, was surrounded with clouds of a strange lurid appearance indicating bad weather. Bye and bye strange turbulent looking vapours, commonly attendant on a whirlwind or waterspout, were seen overhead and a slight whirlwind followed. At this period the magnetic virtue of the compass seemed to vanish, the needle losing its polarity and traversing equally all round.

In the *Cleopatra's* hurricane off the Malabar Coast (see pp. 199, 212) in April, 1847, the magnets in the Bombay Observatory were violently disturbed. *Naut. Magazine*, July, 1854.

In the same work for August, p. 433, is an account of an electric storm passing over the ship *Aries*, in Lat. 35° South, and 27° East,

PART VI. § 435.]   *Effects of Cyclones on Compasses.*   359

in which the lightning was seen to dart upwards from the sea to the clouds. The thunder and lightning which followed were very severe, but the wind during the Cyclone-like storm which followed did not exceed 8 or 9 of the scale. It was found however the next day that the ship had been set 98 miles to the westward in 16½ hours, or nearly six miles an hour, the current on the following day being Easterly off Cape St. Francis. This the Editor and Sir Wm. SNOW HARRIS both consider to be an instance of an electric current, such as I have long ago described the Storm Wave to be, see page 168.

In a letter from Captain SILVER, quoted p. 144, occurs the following passage :—

"I found running down my easting from passing Crozet's Islands, that the compasses were very much affected, and vibrated considerably, so much so that not one compass on board at times could be used. I tried all means with them, but of no avail, until I put two compass cards together with their needles exactly over each other, and rivetted the cards together with lead, which answered admirably. I had no trouble afterwards. I mentioned it to many masters, and some that had often been out here, and they all complain of the same."

In a paper in the *Edinburgh Philosophical Journal* for 1821, is an instance quoted by Lieut. MUDGE, when the compasses of a Hudson's Bay Company's vessel became suddenly affected, in Lat. 62° N. and 93° W., but it is not spoken of there, as in the letter I quote from, as a local disturbance which appears to be permanent.

It is not clear if the following extract, from the *Naut. Magazine* for June 1852, refers to a local or an electric affection of the compass, for the state of the weather is not mentioned, but it is probably a local one as one compass only was affected. Nevertheless, and because sailors cannot be too much on their guard against these sudden deviations, and because it is most important to draw attention to them, I give it somewhat abridged, as follows :—

"It will be remembered that the unfortunate *Birkenhead* was lost on the same night that the strange circumstance took place in reference to the compasses on board the *Propontis*. Some days before making the Cape Land* on board the *Propontis*, in Feby. 1852, we found nearly six points difference between our standard and binnacle compasses, the standard having nearly three points West and the

---

* The *Propontis* is an (iron ?) screw steamer belonging to the Screw Steam Nav. Company, like the *Birkenhead*, and was on a voyage from England to the Cape.

binnacle compass nearly three points East variation. On approaching the land the night of Feby. 25th, we found the binnacle compass so unsteady and oscillating so much, *at times taking even a round turn,* that we could not steer by it; but conned the ship by the standard, which remained steady."

436. MODERATING OF THE WIND AT THE COMMENCEMENT OF CYCLONES. The following extract from the Horn Book of Storms for the Indian and China Seas, describes this remarkable peculiarity; of which I have again and again met with notices in numerous logs, and in very many instances I have remarked that commanders have been deceived by it. The rule is simple, *" Make no sail till your Barometer rises, except what may be absolutely necessary to steady the ship."*

" In the hurricanes of the Bay of Bengal and China Seas, it seems not an uncommon circumstance, that in a few hours after their commencement, there is a lull for an hour or two, or more; after which it comes on to blow harder than before from the same quarter. I observe that this is also noted in the hurricanes of the Isle of France and in the West Indies. This treacherous peculiarity might, without attention to the barometer, deceive those unaccustomed to our tempests. I have not met with an instance of these lulls occurring more than once at the commencement of a storm. I do not allude here to the lull or calm which precedes the shift of wind, which occurs when the centre of a storm is passing over a ship or place, but to a sort *of promise of fine weather* which occurs at the beginning of a hurricane."

437. There is also a kind of treacherous fair weather interval when no *blowing* weather has occurred, and the only indications have been dark, lowering clouds, and a fall in the barometer. It is perfectly well described in the following note appended by Captain BIDEN to the log of the ship *Princess Charlotte of Wales* in her Cyclone of March, 1828, in the Southern Indian Ocean. Track *k.* on Chart II.

"Previous to the gale, the weather had been unusually close, Ther. 82°, and for 48 hours, the clouds were dark and fixed, with an appearance indicative of heavy rain. Barometer steady. The approaching change of the moon in its Perigee caused an anxious attention to the Barometer, which gave the first warning at 11 A. M. on the 14th, falling gradually until 8 P. M., when it sunk with more rapidity than I ever witnessed, the sure presage of a storm! Without this excellent and almost infallible instrument, the most experienced seaman might have been lulled by the sudden change at 2 A. M., stars bright, clouds fixed, and every appearance of settled weather. The Barometer fell from 29.27 at midnight to 29.08, the certain precursor of a hurricane. The heavy rain had so tired and exhausted our men, that at midnight I relieved the watch in hopes of a few hours daylight to get the top-gallant masts upon deck ; we had just handed the main top-sail and rounded to on the larboard tack with scarce a breath of wind, when the gale came on at W. S. W., and at 4 it blew a perfect hurricane."

438. In the *Singapore Free Press*, I have found an instance in which the brig *Guess*, with a falling Barometer at 29.50., and wind N. N. W. in the China Sea, *in the month of August*, 1847, (the height of the S. W. monsoon) was seduced by an interval of moderate weather into making sail so far as to let out reefs and set top-gallant sails at 3 p. m.; having double reefed at 2 p. m. By 6 it was a strong gale, and by midnight a Tyfoon Cyclone had commenced. See 17th Memoir, Jour. As. Soc. Beng. Vol. XVIII.

439. ETYMOLOGY (DERIVATION) OF THE WORDS HURRICANE AND TYFOON. This word Hurricane seems originally to have been a Carib or Indian one, for in the *Relacion summaria de la Historia Natural de las Indias, &c. &c.*, addressed to the Emperor Charles V. by Captain FERNANDO DE OVIEDO, speaking of the superstitions of the Indians (Caribs) of Tierra Firme, probably about Yucatan, the author says—

"So also when the Devil wishes to terrify them (the Indians) he promises them the *Huracan*, which means Tempest. This he raises so powerfully that it overturns houses and tears up many and very large trees ; and I have seen in thick forests, and those of very large trees, for the space of half a league, and continuing for a quarter of a league in length, the forest quite overthrown, and all the trees large and small, torn up by the roots ; the roots of many being uppermost, and the whole so fearful to see, that it doubtless appeared to be the Devil's work, and could not be looked on without terror."*

The author here exactly describes the passage of a tornado in a thick forest, and by the explanation he gives of the word *Huracan*,

* I give here the original old Spanish. " Asimismo, quando el Demonio los quiere espantar, promcteles el *Huracan*, que quiere decir Tempestad : la qual hace tan grande, que deriba Casas ; i arranca muchos, i mui grandes Arboles ; i yo he visto en montes mui espesos, i de grandisimos Arboles, en espacio de media legua, i de vn quarto de legua continuado, estar todo el monte trastornedo, i derribados todos los Arboles, chicos i grandes, i las raices de muchos de ellos para arriba. i tan espantosa cosa de ver, que sin duda parecia cosa del Diablo, i no de poderse mirar sin mucho espanto."

we see that it was an Indian one. The Dictionary of the Spanish Academy does not give the derivation of words. In the extract from RAMUSIO, given at p. 329, we find also that the same author gives the same word as the Hatian name for Cyclones.

440. TYFOON. This word is undoubtedly Chinese, and by no means derived from the Greek "Typhon," as has been supposed. Dr. MORRISON, in his Notices concerning China and the Port of Canton, says that—

> "At Hainam and the Peninsula opposite (to the North of it), they have temples dedicated to the Tyfoon, the god (goddess ?) of which they call *keu noo*, 'the tyfoon mother,' in allusion to its producing a gale from every point of the compass, and this mother-gale with her numerous offspring or a union of gales from the four quarters of heaven makes conjointly a *taefung* or tyfoon."
>
> In a work called "*Kw'an Tung Sin yu*" the tyfoon is called either *kow-fung* or *fung-kow*. A severe one is called *teé hwuy* or *Teo keu*, an iron whirlwind.*
>
> They have also separate names for whirlwinds, which I am informed are called "*Yung-kok-foong*," and "*Suing-foong*."

441. These names are not questions of mere curiosity, for apart from the use to the sailor of knowing the native terms for everything relative to his profession in every country, the mere fact of such special names existing for various kinds of gales, shews at once that there are certain well known and recognised phænomena, or effects, distinguishing the Cyclones. Thus in one of the South Sea Islands, they are called "the wind that breaks the banana-trees," and I have mentioned at p. 70, the names *bagyo* and *sigua* as those given by the natives of the Philippines to Cyclones and monsoon gales. And in Col. REID's recent work, p. 2, he says that he was much struck when he heard the inhabitants of Bermuda "call all the gales in which the wind veered and the Barometer fell, *roundabouts*." Ships that visit the Marianas, Carolinas and other groups, islands and coasts of the Pacific, from the Kuriles to New Zealand and Chiloe, may collect much information by attending to the details of natives and residents relative to their different classes of storms: as, for instance, we might learn from the pearl fishers and coasters of California, and from the

---

* Captain DOUTTY, of the *Runimede*, in his log, exactly uses the same metaphor, likening the wind ripping up the front of the poop to "a metallic body!"

coast Indians, South of Valparaiso, what the usual changes of winds are in their storms, and thence deduce the tracks.

442. SURGEONS AND PASSENGERS. Having said so much to the sailors (more indeed, perhaps, than will be acceptable to some of them ?) I must address a word to another class of persons often found on board of ships, and who may materially assist us in many things, and that is to THE DOCTORS AND THE PASSENGERS.

These gentlemen may often usefully get over some of the *ennui*, of which they all, for the most part, complain, by careful registry of the meteorological phænomena, and especially at the approach of gloomy or bad weather. If they honour this book with a perusal, they will see that there are very many points which *can* only be attended to in bad weather on board of Men-of-War, or of first-rate East India-men, and yet which might throw a great light upon our researches; and I hope that every educated man will feel with me how serious a duty it is, morally and politically considered, which their country and mankind can claim from them in these cases. If they merely note any one point, such as the course and appearance of the clouds, and the exact direction of the wind, or the electrical appearances, they may greatly aid future research, and they will find in what has been said (from p. 348) an ample choice of observations to make.

443. TORNADO-CYCLONES. These appear to exist also between the Azores and the Chops of the Channel, and within the Channel, in considerable force. In April, 1845, the *Monarch*, Captain WALKER, homeward bound from India,* was at 10 A. M., on the 22nd April, with barometer at 29.70 under double reefs. At 2 P. M., breeze freshened from the S. W., and every appearance of bad weather. Barometer at 29.50, and all preparations were made, ship steering to the E. N. E. At 7. P. M. Bar. 29.30, blowing very hard, high sea, and atmosphere very threatening. At 8 P. M. Bar. 28.95, furled everything but storm-mizen try-sail. At 8.30, wind suddenly lulled to a dead calm, which lasted a quarter of an hour, ship not steering, and

* The *Monarch* is a first rate Indian passenger ship of 1400 tons, and was then on her first voyage. I regret that I have been unable to give the vessel's position, having mislaid the memorandum, but it was about midway between the Azores and the Lizard.

the sea striking the counter in an awful way, shaking her fore and aft, the appearance of the weather stormy in the extreme, with rain *and lightning*. At 9 P. M. instantaneously from a dead calm it blew a most terrific gale from the North with rain and hail. Fortunately every sail was firmly secured, or Capt. WALKER was convinced the ship would have been dismasted; this continued for one hour, and settled into a strange gale which lasted until sunrise, the wind having gradually veered to N. W., and the Barometer steadily rising.

The following account of one of these, which was experienced by the ship *Windsor*, is reprinted from Mr. REDFIELD's Memoir on the Cuba Hurricane of October, 1844, p. 68; it is a recent instance of a tornado-like Cyclone in the Channel! It originally appeared in the Calcutta papers.

"Oct. 7th, 1844, A. M. wind N. N. W. moderate; noon, Lat. 49° 47', Long. 4° 23' W. Bar. 30 in.; P.M. wind S.S.W., increasing. Oct. 8th, noon, Lat. 49° 20', Long. 7°. Bar. 29° 50', the whole 24 hours very unsettled weather and barometer falling, wind S.W.; unsteady and increasing. Oct. 9th, wind had veered to N.N.W., coming in hard gusts with sudden intermissions, and barometer had fallen to 28.60; sent down small spars, double-reefed, &c. at 8. A.M. Bar. 28.50, and weather clearing up to the eastward, *thought of making sail, notwithstanding the low state of the barometer*.* Saw a schooner close to us with a good spread of canvass. But the wind suddenly flew out from N. E. and back to E. S. E., then E., and we observed the water blown up like clouds of dust, and the sea in frightful commotion. This took place at 1-30 P.M.; the Barometer having been 28.44 at noon, 28.35 at 0-30 P.M., and lowest at 1 P.M. 28.12. It was now (1-30 P.M.) at 28.14, and before the canvass could be got in the hurricane had struck the ship from the northward with extreme violence, driving her forecastle under; at 2 P.M. Bar. 28.40, at 3 P.M. with great difficulty brought the ship to, on the starboard tack; 3 P.M. Bar. 28.60; at 5 P.M. Hurricane less violent, and settled into a heavy gale, veering to N.W. and W.N.W.; at 8 P.M. severe gale, with sleet, hail, rain, and vivid lightning; at midnight Bar. 29.99, gale blowing with unabated violence. The schooner which was close to us disappeared suddenly, and there is little doubt that she foundered. Oct. 10th, heavy gale from N. N. W. with a high sea; noon, Lat. 48° 51', Long. 8° 4' W., Bar. 29.40, midnight, no alteration."

444. BAROMETER SIGNALS TO SHIPS. The Captain is on shore and the Barometer on board. The Mate is embarrassed with cargo unstowed in the hold, loose cargo upon deck and between decks, and

* Another instance of the temporary clearing up, just described at p. 360.

cargo-boats alongside; or he is a young officer, who is afraid of being thought over-cautious or over weather-wise; or is an old one, but not upon the best terms with his Captain; or the Captain is what the French expressively call a *vieux loup-de-mer*, and might growl at the time being lost, or at boats being returned to allow preparations to be made for expected bad weather, without his orders;—or, in short, a hundred circumstances, well known to all merchant sailors, and many which will readily occur to the minds of those of H. M. service, may combine to place a very deserving officer in the unfortunate situation of being blamed for not having his ship ready for bad weather; either to slip and go to sea, or to ride it out; and serious loss and damage to valuable property, if not frightful loss of life have been, and still will be the results, for want of a right understanding, or from an over-confidence in the security of the port. We cannot provide against all these contingencies in detail, but it is a duty to point out to Governments, Chambers of Commerce, and the Mercantile Community, the advantage to all parties derivable from a properly organised system of signals, indicating the approach of bad weather, and the state of the Barometer, as declared by a standard instrument, and by the opinion of the authorities of the port, who are in all cases supposed to be competent judges, and acting from disinterested motives. Every port must have its own particular signals, adapted to its own localities, but the following, as relating to this subject in the Western Hemisphere, may be worth perusal and imitation, in ports subject to Cyclones.*

445. BAROMETER SIGNALS AT BARBADOES, BY H. E. COLONEL REID, GOVERNOR OF THAT ISLAND.

The following Memoranda are published by order of His Excellency the Governor.

JAMES WALKER,
Colonial Secretary.

MEMORANDA RELATIVE TO THE HURRICANE SEASON AT BARBADOES.—A barometer will be kept and registered at the Principal Police Station at Bridge-town;

---

* At Madras they have a very good code of signals for warning ships to be ready to put to sea, but I do not know if they have extended this to shewing the state of the Barometer also, which, as just explained, is so useful as a sort of authority to young officers.

and notice will be given to the Captain of the Port when it falls. On the Captain of the Port will rest the responsibility of causing Signals to be hoisted, that the barometer indicates bad weather.

One ball at the mast head of the signal posts is to signify that the barometer is falling, and should be carefully watched.

If the barometer continues to fall, and the weather appears threatening, a second ball will be hoisted at the mast head.

As the indications of the weather become alarming, these two balls will be gradually lowered, until they are only half-mast high.

As soon as the barometer begins to rise again, the two balls will begin to be slowly re-hoisted, so as to be again at the mast head, when the barometer shall have risen one-tenth of an inch.

When the barometer shall have risen two-tenths of an inch, then one ball will be taken off, and the other be left until the storm shall have passed over.

Hurricanes being whirlwinds, the wind in the circuit of its revolution, blows from every point of the compass within the circuit of the whirlwind. The veering of the wind is owing to the whirlwind's progress. Hence, the reason why the trade wind is often reversed during these tempests.

Ships riding at anchor in Carlisle Bay, unless their commanders prefer to remain there, cannot put to sea too early after the first indication of a hurricane.

When the wind veers from N. E. towards East with a falling barometer, it may be expected to become S. E. and S. S. East; and in this case the centre of the storm would be passing to the southward of the island.

When the trade wind veers from North to N. by West with a falling barometer, the wind may be expected to become N. W. and West, and perhaps even S. W. and in this case the centre of the storm would be passing to the Northward of the island.

When the wind changes to North, blows steadily from that quarter, it may be expected to change suddenly to the South; in this case the centre of the storm would pass over the island.

In either case, ships remaining too long at anchor, would be in danger of becoming embayed on a lee shore.

The hurricanes which have passed over Barbadoes, and of which we have any precise records, have all come from the Eastward. When the centre is expected to pass to the Northward of the island, ships quitting Carlisle Bay should endeavour to run to the Southward, and South Eastward, by scudding in the first instance. But when the centre is expected to pass to the Southward of Barbadoes, a ship should go to the Northward, and come to the wind on the Starboard tack. By keeping to the Eastward, whilst the storm is moving Westward, the ship will sooner be out of the hurricane.

The earliest indication of a coming storm is sometimes a heavy swell of the sea, caused by the storm at a distance.

NORTES OF THE GULF OF MEXICO, (p. 189). As we have now

PART VI. § 445.]   *Revolving Winds true Cyclones.*   367

one undoubted proof, in the October Cyclone of 1848, of the Bay of Bengal (See p. 330, and Jour. As. Soc. Beng. 18th Memoir, Vol. XVIII.), that Cyclones, wherever and however they may originate, certainly *descend* and settle down on the Ocean after passing over high land, it may be well worth while to consider whether these Nortes, which often cannot be traced up from the Eastward and South Eastward, may not be at times Cyclones from the Pacific Ocean, passing over the high land of Guatimala and Mexico, and settling down on the coast of Honduras and Mexico when the central part is clear of the land, which would then give a *Norte* along the coast?

REVOLVING WINDS TRUE CYCLONES, BUT NOT OF HURRICANE FORCE. The note, p. 133, announces this, and in Colonel REID'S new work, p. 245, he has fully shewn by an instructive series of registers, which "*are divided according to the barometric oscillations*," and not as usual periodically, that at the Bermudas in Lat. 32° N., Long. 64° West,

"The revolving winds which pass over them are of various degrees of force, from breezes to storms. In the summer season the winds then are light and usually steady for a considerable time, blowing in straight lines and without veering, with a little fluctuation in the Barometer. But after the commencement of November veering winds of various degrees of force set in and gradually become frequent; yet they seldom follow in such rapid succession as that one gale becomes confounded with another. Light winds and very fine weather usually intervene between the passages of revolving winds, while at other times hard blowing straight-line winds with a high barometer intervene. The arrival of each succeeding progressive circuit of the wind is indicated by the Barometer falling as well as by the increase of the wind's force."

The registers given most instructively prove and illustrate this, as well as the propriety of the quaint old English name of *roundabouts*. When we obtain more of the statistics of winds and of the connections of the Barometer and Hygrometer with them in various parts of the world, as within and on the limits of monsoon winds, as well as of trades, and *on both sides* of these, and again in high latitudes, we shall perhaps be able to deduce the Law of their succession. And hence again the practical rules arising out of it.

In September, 1856, we had a complete instance of these *Roundabouts* at Calcutta, of which the following, which I abridge of its preliminary notice, which merely explains the term to general readers, and regrets the shameful persecution to which all scientific research in India exposes those who venture on it, is a copy:

"On the 31st of August, the weather was clear with flying clouds, but the Barometer was falling. On the 1st September, wind from the Northward with cloudy weather and slight squalls with drizzling rain, and Barometer always falling. On the 2nd of September at 9 A.M. the Barometer (uncorrected) had fallen to 29.41, with squally weather and rain, the scud flying with tolerable rapidity from the North, the wind from North to N. N. E. By 4 P.M. the barometer had fallen to 29.31, with a gloomy scud from the North East, wind N.N. East to North with drizzling rain, showers and squalls at times, all the appearances being those of a Cyclone at some distance to the Eastward, but travelling very slowly towards us. During the night much the same weather, and on the 3rd September in the morning at 6 A.M. the barometer had fallen to 29.21, the thermometer 84½; the wind N. East to N. N. E. blowing a moderate breeze with squally rain and frequent intervals of calm, the scud travelling moderately fast from the North East as yesterday. Between this and half past 10 A.M., it fell calm with gleams of sunshine, rain and gloomy weather, but the Barometer rose to 29.27, Thermometer at 84⅜, and the wind had shifted to light breeze from S. S. E., and the whole of the scud was now travelling up from the S. East as fast as it before did from the Northward. The Barometer continuing to rise, and the wind to haul to the Southward and Westward of South with heavy rain in the afternoon, evening, and at night, clearing up on the morning of the 4th, with the Barometer rising to its ordinary height, exactly as at the close of a Cyclone, which this curious breeze resembled also in another respect, namely, the entire absence of thunder and lightning.

It may be quite possible that this may have been a tyfoon in the China Sea modified to us to the force of a strong breeze (for it was at no time a gale here) by its passage over the southern part of China and northward of Burmah. If this should prove to be the case, it will be a remarkable instance of a Cyclone, decreasing in force by its passage over an extensive tract of land, but yet preserving its peculiar powers and character so perfectly, as to reverse the monsoon here for at least two days!"

NELSON AND VILLENEUVE'S GALE IN JANUARY, 1805, (p. 230). It may be surmised indeed that this "gale" was a Cyclone, at least for the French Fleet, as I shall now shew from Sir N. H. NICOLAS, "Despatches and Letters of Lord Nelson," Vol. VI.*

First we find (p. 339) that the French "came out of Toulon on the 17th January, 1805, with gentle breezes at N. N. W. and lay (laid) between Giens and the Hieres Islands, till the gale set in on the 18th in the afternoon," they then "sailed with a strong gale at N. W. and N. N. W. steering South or S. W. on the 19th." At this time (p. 324,

* I shall quote the pages in parentheses, as there are extracts from several letters.

PART VI. § 445.]   *Nelson's Gale a Cyclone.*   369

Nelson's Diary) 19th, the British Fleet were lying in Maddalena Harbour with hard gales also at N. W. They weighed and put to sea, passing through the Biche (Biscie) passage by seven in the evening, and bore away, wind W.N.W. (p. 327), along the eastern side of Sardinia. " During the night it was squally and unsettled weather," and by Noon 20th, " Mount Santo bore N.W. 6 leagues. All night very hard gales from S. S. W. to S. W. which continued through the next day; during great part of the time we were under storm stay-sails."

Now if we suppose a Cyclone travelling down from Genoa, between Minorca and the coast of Sardinia, or between Minorca and the Spanish coast, say upon a track to the W. S. W. or S. W. (true) it may give a strong N. N. W. breeze off Toulon and N. W. gales at the Maddalena Islands on the 18th; making of course due allowance for the obstruction of the very high land of Corsica. Then as the central portion* passed to the N. W. of Maddalena it might be squally and unsettled, varying to W. N. W. and to the S. S. W. as it passed on; while it might be a violent gale at N.W and W.N.W. veering also to the Southward with the French Fleet between Sardinia and Minorca, so as to enable them to fetch Toulon again when crippled, as they did: and it would seem certain that it *was* on the day and night of the 18th, 19th, that they suffered, since (p. 332) it was on the evening of the 19th, that a disabled French 80 gun-ship had put into Ajaccio.†

There are however two difficulties in the consideration of this question. The first is the unusual track (though in truth we know nothing of the tracks in this sea) and the second that the S. S. Westerly gale

* For we are not now investigating a tropical hurricane with a distinct calm central space, which is the meteor in its most concentrated form, but a sudden gale in a narrow sea and a temperate climate, which might or might not have been Cyclonal.

† As the French fleet was in the latitude of Ajaccio steering to the Southward with a N. W. gale under a heavy press of sail at 10 P.M. on the night of the 18th (p. 327), when a British look-out frigate was close to them going 13 knots, it is possible that, as the central portion neared them, they may have been taken aback by a shift to S. S. W. and so have been dismasted; and this with Admiral VILLENEUVE'S avowal of their bad management and raw crews, accounts for the number of ships which suffered, as in the case of Admiral GRAVES' fleet of merchantmen and ill-manned prizes.

B B

lasted with Nelson's fleet to the East of Sardinia (p. 330) till the morning of the 25th: but it is always possible that a Cyclone may precede a monsoon, or winter gale, or other right-lined wind; and vague or forced as this deduction may now appear, it is always worth while to direct attention to such questions; for whatever be the answers which a better knowledge may bring to them we shall always profit when that knowledge is obtained, be it affirmative or negative of our present impressions.

TRACKS; NEW CALEDONIA. (See above, p. 76.) The Barque *Nimrod*, in February, 1849, experienced a terrific Cyclone commencing in Lat. 17° 38′ S., Long. 161° 26′ East; or about 80 miles to the W. $\frac{1}{4}$ S. of Huon Island at the North point of New Caledonia, in which her Barometer fell from 29.70 to 28.20, and she was obliged to cut away her main-mast. This Cyclone was either very extensive or one of a very slow progression, for it lasted with the *Nimrod* from the 11th to the 14th of February, though hove to the whole time. Its track, so far as the imperfect newspaper report and the want of the vessel's log will allow us to judge, appears to have been a very anomalous one, for it seems to have travelled down upon the vessel from the N. N. W. or N. W. and then to have curved back to the Westward, the ship being from S. S. W. and S. W. to N. West. This ship's log is given in the Nautical Magazine for 1850, p. 667, but unfortunately also without any drift or the amount of falling off or coming up, or her run when she "kept away N. E., to run out of the gale, for two hours and a half, (wind at W.S.W.) when the shift took place to N. W.," so that it is impossible to do more than to estimate roughly the track. The Barque *Scamander* was also wrecked on the Southern Reefs of New Caledonia to the S. S. E. of the Isle of Pines, on the 15th February, in "a hurricane," which beginning at East veered to South (as if it came down from the N. W.), but we can scarcely infer this Cyclone to be identical with that of the *Nimrod*. I give these imperfect details indeed rather to put the seaman on his guard; and I hope also thereby to remind him of how much may be done to aid us, and to forward his own interests, by a little trouble on his part.

CYCLONES OF ICELAND. At p. 68, I have mentioned that the

PART VI. § 445.]   *Cyclones of Iceland.*   371

Tchukutskoi of Behring's Straits described to KOTZEBUE in the *Rurick* storms, which as to violence might be considered Cyclones. I find in Vol. XIV. p. 297, of JAMIESON'S Edinburgh Phil. Jour. in a paper on the Glaciers and Climate of Iceland by W. SARTORIUS VON WALTERSHAUSEN, the following description of the Icelandic storms, which certainly appear to be at times a sort of stationary tornadoes and at times Cyclonal, from their extent, duration, and veering. When we consider that Iceland is in fact one huge mass of active or extinct Volcanic formations, and that it is moreover (see p. 23) in the terminal line of the West Indian and N. American Coast Cyclones, when their tracks tend much to the Northward, these facts from the pen of an independent and scientific observer become of much value.

After speaking of the changeableness of the Icelandic climate, in which tranquil calm weather forms the exception, and storm the usual rule, he says—

"Storms of the most terrific character and fearfully devastating force, and which carry every thing before them, are very common. These often place the traveller in very critical and dangerous positions, or at least in circumstances accompanied with many hardships and difficulties.

We experienced one of the most terrible of these storms on the 8th of June, on the Hvalfiorder at Thyrill, in the region which is notorious for them, and which has already been pointed out by OLAFSEN* as dangerous. The one which we experienced would seem exaggerated, or even almost incredible, were it not that our description of it agrees in substance with the description of such storms given by that eminent Icelandic traveller. In the morning, when we left Reynivellir, a violent wind was already blowing, and this increased more and more until we reached, a little before noon, a height which divides the Svinadal from the Hvalfiord. Here

* *Reise durch Island*, Vol. i. § 4. Thyrill is a round, very high, steep, and prominent hill-top, at the inner extremity of the inlet which has just been mentioned. It is so named because the air frequently whirls around it, and thus causes dreadful whirlwinds from the north and north-east, against which travellers would require to be on their guard. § 186. At the inner extremity of the Hvalfiord, especially around the hill Thyrill, violent whirlwinds blow. These storms last always for several days; and they are such as to carry up the sea-water like snow into the air, whilst, at the same time, in the southern country, beyond the rocks in the Bogarfiord, there is but little wind, or none at all. From this reason, the district at the Hvalfiord is called by the neighbouring inhabitants Wedra-Kista, that is, Box or chest of Winds, which implies that this inlet is, as it were, the abode of violent storms.

the storm began to blow in such a fearful and indescribable manner, that we could scarcely advance, and that we sometimes thought we should lose our breath. Our circumstances became hazardous in the extreme, as we proceeded down the steep declivity on our way to Botusdalr, the eastern extremity of the Hvalfiord. The storm blew from the south-east with such violence that it threw one of our attendants from his horse, and threatened to hurl us over steep precipices into the abysses below. While the storm raged over the water of the fiord, the surface was converted into a cloud of spray, which reached even to us, having passed over hills 2000 feet high. In it there floated a rainbow of the most brilliant colours, which appeared like a bridge uniting the two sides of the dark-green fiord. During the afternoon the storm still continued to rage with equal fury, and it was only towards the evening, and during the following night that it began to abate. The storm was not confined, however, in accordance with the representations of OLAFSEN, to a very limited space; but, on the contrary, it was felt along the whole south-west coast of the island; and also, on the same morning's ship bound for Reykjavik ran ashore at Oereback (Eyran Bakki). According to the statements of some Icelandic traders a perfect calm prevailed during this time on the sea, at about 5 miles (27 English miles) from the coast.

Other storms of a similar kind occurred repeatedly during our journey, and were still more destructive than the one which has just been described; as they were accompanied with rain, hail, and impenetrable clouds of fog, or dust. The neighbourhood of the Hecla is peculiarly subject to storms accompanied with dust; the dust being carried up by the wind from the extensive fields of volcanic ashes, which are spread out around that mountain."

Colonel REID remarks in his first Volume, though I have been unable to find the passage, that some of the great Atlantic Storms, which have a very Northerly route, must terminate between Greenland and Iceland, and it seems probable, if they really travel so far, these tornado-like tempests are the large ones breaking up into smaller ones, which I have already shewn* certainly occurs in Tropical hurricanes. If they are not so, then their occurrence with such severity and frequency in a Volcanic Island is a fact of great interest, when we connect it with what has been said of the tracks of the Cyclones from one Volcanic focus to another, the Teneriffe and Isle of France hurricanes and meteors, and the perpetually recurring Willy-waws and Storms of Tierra del Fuego, which, to use the Icelander's pithy term, is the *Wedra Kista* of that part of the globe.

HAIL, IN ITS RELATION TO ELECTRICITY AND WHIRLWINDS. Mr.

* Page 99; and VII. Memoir, Journal As. Soc. Beng. Vol. XI.

J. C. MARTIN of Pulborough, in Sussex, (speaking of a remarkable hailstorm, in which some of the masses of ice were five, six, and even seven inches in circumference;) says (London and Edin. Journal of Science for 1840, p. 86) after admitting that the congelation of large drops of rain, the formation of ordinary hail, and even a considerable accretion of ice to the original globule in its passage downwards, is not very difficult of comprehension and explanation:—

"But there is only one way in which I can suppose such masses of ice as these can be suspended long enough in the atmosphere to grow to such enormous sizes; and that is by the assistance of a nubilar whirlwind or waterspout, (*Trombe aériennes*) with sufficient power to keep them in its whirl and to resist the earth's attraction whilst the concretive action is going on, till their momentum overcomes the suspending power, or till they are thrown beyond the range of its intensity. That such operations are amongst the reciprocal electrical phenomena of the clouds, distinct from, though allied to the waterspout, is, perhaps, well-known; and I was myself once witness to an appearance of this sort, between a higher and a lower cloud, that had a strongly electric aspect before they had resolved themselves into nimbus. It was a bent narrow column of dark vapour, which I could distinctly observe to be in rapid rotatory motion, passing from one cloud to the other, continuing for some minutes, and then gradually disappearing. During this time it emitted no sound, and had no visible connection with the earth whatever.

"The above theory of hailstones will be further corroborated, if we consider the form of the stones in this instance, viz., a sphere flattened at its poles as the result of a rotatory motion; especially if it be a law, as perhaps it is, that all solids in rapid gyration acquire, *per seipsos*, a rotation on their own axes."

In reference to the yet scanty notices we have of hot and cold blasts of air, and of hail and sleet at or near the centres, these views are of some interest; and the sailor will, I hope, see from them how essentially useful and necessary it is that we should be furnished with all the facts attendant on the progress of a Cyclone.

In some hailstorms excessive falls of the Barometer, like those in Cyclones, seem also to occur. In the Transactions of the Royal Society of Edinburgh, Vol. IX. part I. p. 199, is an account of one which passed over the Orkney Islands, devastating principally Stronsa and Sanda, and which is described rather as a shower of masses of ice, with intense and vivid discharges of *ball*-lightning which seemed, it is said, to dash up the sea as high as a ship's mast. It was not more than a mile and a half in breadth, and the fall of the

Barometer, from the register at the Start Point Lighthouse in Sanda, was from 29.72 to 27.76, or very nearly two inches!

Mr. B. Douglas, of the Coast Guard Service, in a letter to me describes the fall of some remarkable pieces (cakes or disks, or as the sailors would say *trucks*) of ice, which fell on board of his vessel in Southampton Water, at the time of a hail-storm on the Isle of Wight and Coast of Hampshire, which with him was only a smart rain squall: he says,

"The force of the wind lessening and the rain subsiding, I was surprised by perceiving some large substances fall into the water and on the deck of my little craft. They proved to be immense hail-stones or pieces of ice, of irregular but principally of flattened spherical forms, several measured four inches in circumference and one inch and upwards in thickness. The great peculiarity was that the hail-stones had evidently been formed by some spiral or circular motion, as the outer rims were quite white and opaque like frozen snow, gradually becoming clearer and more transparent towards the centre, in which, however, appeared '*the calm*' or a small round spot as opaque as the outer edge."

Mr. Douglas then gives a sketch of a stone, and he adds,

"There was another singular circumstance attending this phænomenon; that of the stones falling as far apart as 3 or 4 yards, thus giving one the idea of the hail-stones being smaller ones, condensed as it were by some centripetal action, in the highly electrified state of the atmosphere."

In Mr. Martin's work, p. 69, will be found a copy of the account of the hail-storm Tornado, (called there a Cyclone), which devastated the city of Dublin, on the 18th April, 1850, in which it is said that,

"Some persons who observed the phænomena from a distance, were able to distinguish the two strata of oppositely electrical clouds, and to see the electrical discharges passing between them.

"The hail-stones, which were as large as a pigeon's egg, were formed by a nucleus of snow or sleet surrounded by transparent ice, and this again was succeeded by an opaque white layer, followed by a second coating of ice. In some of them I counted five alternations."

The tornado was accompanied by almost continuous forked lightning, and the roar of the thunder was continuous. This at once distinguishes the phænomenon as a tornado, and as I think puts it out of the class of Cyclones; though there is no doubt of the whistling of the wind.

Nothing is so common as to find in the Logs of ships, which have been involved in Cyclones, and especially if near the centre, the remarks of "rain as cold as ice." "Sea-water warm, rain bitterly cold;" "rain and sleet," and the like. In a published account of the *Duke of York's* Cyclone, which though of small extent, was of terrific violence, "Cold most intense during the hurricane," occurs twice. Yet in some it evidently does not occur, as the Barometer and Thermometer are carefully marked, and though the Barometer may fall an inch or more, the Thermometer does not shew more than from $3°$ to $5°$, as in a common thunder shower or North Wester.

THE ELECTRIC TELEGRAPH WARNING OF THE APPROACH OF CYCLONES. In my Sixth Memoir, Jour. As. Soc. Beng. Vol. XI. p. 703, published in 1842, I have suggested that if China was a country under European dominion, a telegraph on the Eastern Coast might warn the shipping at Hong-Kong to be prepared for a Cyclone. I find in a recent newspaper notice in the *Atlas* of a work entitled "The Emigrant Churchman in Canada," the following passage, which I print without alteration, though the writer evidently is no Cyclonist:

"An extract given from a Canadian paper brings out a practical use of science which our fathers would have stared at, could it have been foretold in their days. It is the use of the electric telegraph in giving notice of the approach of the whirlwind (Cyclone) :—

"The telegraph now gives notice of storms ! For example, the telegraph at Chicago and Toledo now gives notice to shipmasters at Cleveland and Buffalo, and also on Lake Ontario, of the approach of a north-west storm. The result is practically of great importance. A hurricane storm traverses the atmosphere at about the rate of a carrier pigeon, viz. 60 miles an hour. Our north-west winds come apparently from the sources of the Lakes, and sweeping over Lakes Superior, Michigan and Erie, spend themselves in the interior of the country. Our south-west winds come apparently from the Gulf of Mexico, where the force is very great, and pass up the general direction of the Mississippi and Ohio. Commencing at these remote points, it is obvious that if telegraphic offices are established at the extremes of the line, notice of the approach of a violent wind may be given to distant ports from 12 to 20 hours before it will be felt there. The practical effect will be that a vessel in the port of New York, about to sail for New Orleans, may be telegraphed 20 hours in advancing on the coast from the Gulf of Mexico. We are only on the threshold of the real substantial advantages which may be rendered by the electro-telegraph. Already have notices of storms on the lakes been given from Chicago and Toledo to Buffalo."

446. CONCLUSION. Sailors who have not looked into the literature of their profession, will be not a little surprised to hear that books yet exist in print in which the Variation of the Compass is treated as an absurdity! and MERCATOR's projection as only fit for land-lubbers! but this is really the case, as the following extract from Mr. WILLIAM BURROUGHS' "*Discourse of the variation of the Compass or Magnetical Needle,*" appended to "*The Newe Attractive*" (London, 1581) of ROBERT NORMAN, (who announces in it his discovery of the Magnetic Dip,) will shew—

"But as I have alreadie sufficientlie declared, the cumpas sheweth not alwaies the pole of the worlde, but varieth from the same diversly, and in sayling describeth circles accordingly. Which thing, if PETRUS NONIUS, and the rest that have written of Navigation, had jointly considered in the tractation of their rules and instruments, then might they have been more available to the use of Navigation; but they perceivyng the difficulties of the thyng, and that if they had dealt therewith, it would have utterly overwhelmed their former plausible conceits, with PEDRO DE MEDINA,* (who, as it appeareth, havyng some small suspicion of the matter, reasoneth very cleverly, that *it is not necessary that such an absurdity as the variation should be admitted in such an excellent art as Navigation is*) they have thought best to passe it over with silence."

447. And BURROUGHS himself, though a sailor and a hydrographer, and Comptroller of the Navy to QUEEN ELIZABETH, says of MERCATOR'S projection, that—

"By augmenting his degrees of latitude towards the poles, the same is more fitte for such to beholde, as studie in cosmographie by readyng authors upon the lande, than to be used in Navigation at the sea."

Now nothing is more probable than that for a time our new science will be treated by many much after the same style, as "utterly overwhelming all former plausible conceits," and as only "fitte for such to beholde as studie in cosmographie by readyng authors upon the lande," though it is not probable it will be so in print, the facts being too numerous to admit of contradiction; and, in truth, more or less within every seaman's knowledge.

448. Still we must expect to find many "of the old school" who do not like "new-fangled notions;" many who "do not like to be

---

* The earliest Spanish author on Navigation; he had published at Valladolid in 1545, the "*Arte de Navegar.*"

put out of their way;" many who "think the old plan is good enough;" and that "hit or miss, for luck's all," is quite enough with a stout ship and good crew; and to conclude, many who are too proud, too ignorant, or too indolent to take the trouble to learn, and who will for a time keep their seats amongst the scorners. From such we can only entreat a patient hearing; and if by chance they have lost a mast or two in a Cyclone, quote to them the assurance of POOR RICHARD, that "they that will not be counselled cannot be helped," and further that "if you will not hear reason she will assuredly rap your knuckles"—in the next Cyclone. We trust, however, that even these persons, admitting that if the knowledge we profess to hold out be not perchance exactly correct, yet the search for it may be useful in some other way, will not withhold from us the *materials* we desire. The Cyclone in all its various forms, and in all the parts of the world in which it prevails, is not perhaps less a scourge to human life (and we will leave out property) than the scurvy of old. We have, so to say, exterminated the disease; and few who know the science of the Law of Storms will, I think, refuse to say that it will finally reduce to a comparatively trifling amount the frightful list of the victims of the Cyclone; the most fearful and the most frequent enemy* which the mariner's perilous calling obliges him to encounter.

449. As a mere question of humanity then, we claim on this, the highest ground, the support of all, whether sailors or landsmen. As a scientific question we are certain of the aid of men of science in every quarter of the globe; and as one of mere pecuniary calculation we may boldly say that merchants, ship-owners, under-writers, and sailors themselves, must be wilfully blind and deaf to their plainest interests when they do not encourage the researches relating to

* I do not here forget FIRE, but this is so rare that its danger is far below that of the Cyclone, by which, perhaps, in all parts of the world, two or three hundred lives are lost against one by fire. I trust also that we shall ere long do something towards diminishing, if not entirely preventing the chances of this accident from spontaneous combustion. See a pamphlet recently printed for the use of the Lords Commissioners of the Admiralty, "On the means of obtaining early warning of any approach to spontaneous combustion, &c." by the Author, and reprinted in the Nautical Magazine for October, 1847.

Storms; and not less so Governments, until they adopt some organized system for collecting, registering and digesting, the vast masses of scattered information which already exist, and which are, from the countless ships of England, America, and France alone, hourly accumulating and passing into oblivion. No single individual or association, without official authority and support, and adequate means, can do this; but the nation and the ministry that accomplishes the great tasks which I have from section to section sketched out, and truly and fully gives to the Mariner of all nations a Code of LAWS FOR THE STORM in every quarter of the globe, will fairly claim a share in the honours awarded to the noblest benefactors of the human race.

INDEX.

THE NAMES OF SHIPS ARE IN ITALICS.

A

	PAGE
Acre	88
Adelaide	78
ADIE, Mr.	234
Ægean Sea	86
Akbar	66
Alacranes	190
Albion's Cyclone *(H. M. S.)*	48, 92
Albion	92, 154, 288
Algerine, Brig	322
ALISON, Mr.	320
American ship suddenly sunk	305
Amherst	67
Amoy	58
Anamba Sea	62
Andaman Sea, Tracks	54, 56
Rates of Travelling	91
Andros Island	191
Antigua water-spouts	312
Hurricane	178, 216, 227, 288
Apia Harbour	73
Appearance of the Stars	290
Application of the Law of Storms	101
Arabian Sea, Cyclone Tracts	51
Rates of travelling	90
Lessons	151
Arched Squalls	219, 293
ARCHER, Lieut.	200
Ardent, H. M. S.	125
Ariel's Tyfoon . . . . . *Note*	244
ARNOLD, Dr.	227

	PAGE
ARROW, Capt.	341
Ascertaining the Track of a Cyclone	115
Ashley River	98
ASHMORE, Capt.	75
Asia	263
Aurora-Borealis-like lightning	293
Australia, N. W. Point	50
Coast	50
South Coast	64
Average Tracks	25
Rates of travelling of Cyclones	89
Avoiding a hurricane	102
Avon	31
Axes upon deck	252
Axial Oscillation	209

B

Baboo	165
BADDELEY, Dr.	299, 303
Bagyo and *Sigua*	70
Bahamas	191
Balasore	278
Banda, Celebes, and Sooloo Seas	62
Bangalore	52
Banks of Clouds	272
Barbadoes	193
hurricanes of 1780 and 1832	1, 287
BARCKLEY, Mr.	198
Barham	26, 38, 238, 289, 291
BARLOW, Professor	237
BARNETT, Mr.	284
Barometer and Simpiesometer	229
Causes of their motions	231
Names *Note*	232
Experiment to shew the fall of	233
good ones, and comparisons	234
rise before Storms	235
Oscillations	236
Water Barometer	237
KAEMTZ on	237
tides interrupted	240
in following Cyclones	240

INDEX.   381

	PAGE
Barometer may be driven up	240
sudden fall between parallel Storms, *Eliza*	241
flashes of light in the tube	239
rise of, before the strength of Hurricane is over	241
a measure of distance from the centre	242
Conditions of the Problem	243
Table of the rate of fall	248
fall explained by REDFIELD	233
Barometer-signals to ships	364
at Barbadoes	365
Barometrical Chart	246
Described	246
Table explained	247
Rule	250
Cautions with Rule	252
Table of examples	253
Explanation of Table of examples	255
Notes to Table of examples	260
Excessive falls, Table	264
Barrier Reefs	73
Bashee Passage	58, 69
Bass' Straits, Cyclones, Storm-Currents	64, 186
Baticolo	*Note* 214
Bay of Bengal, Cyclone	54, 216, 291
Practical lessons for	162
Average rate of travelling of Cyclones	90
Bay of Islands	78
Bay of Biscay sea	34
Bayonnaise, corvette	71
Bermuda, Cyclone	193
BERRY, Sir John	164
Beagle, H. M. S.	50, 51, 221, 242
Bearing of the Centre	104
Table of	105
Beaver	81
Behring's Straits Storms	68, 371
Beirout	88
Belize	178
BECQUEREL, M.	284
Belen	83
BELL, Capt.	43
Belle Poule's Cyclone	46
Bellerophon, H. M. S.	89

	PAGE
Berceau	47
Bermuda Storm, 1839	30, 193, 204
Bexhill	306
Bhoot, or dust Whirlwind in India	299
BIDEN, Capt. Chr.	33, 59, 111, 208, 285, 360
Birds at the centres of Cyclones	113
Birkenhead, steamer	359
Black Sea	86, 144
Black Nymph	275
BLANE, Dr.	1, 227
BLAY, Capt.	204
Blanche	263
Blessing's water-spout	309
Boiling of the Sea	317
Bombay	151
BOND, Capt.	273
Bonin Islands	66, 185
BOON, Capt.	358
Bores of the Hooghly	172
Boston Storm, 1815	3
BOUSQUET, M. H.	38, 46, 161
BOWATER, Admiral	184
Boyne's Cyclone	90
Braemar	166
BRANDE, Professor	5
BREWSTER, Sir David	5
Brest	188
Bridgetown	193
British Channel	88
Islands	88
BRIDGMAN, Mr. T.	297
Briton and *Runnimede*	56, 112, 160, 191, 240
BRUCE	301
Buccleugh, H. C. S.	71, 121, 244, 286
Bucephalus	26, 38
Buckinghamshire	51, 332
BUCKTON, Captain	322
BUDGEN, Richard	308
Buena Ventura	329
Buffalo, H. M. S.	78
BUIST, Dr.	277
Bull's Eye	307

INDEX.   383

	PAGE
BURROUGHS, William	376
BURN, Capt.	212
Burrampooter	91
Burrisal and Backergunge	91, 198

C

Cabul Whirlwinds	298
Cacique	67
Calcutta Whirlwind	298
Storm of 1842	100
Caledonia	113, 212
California, Coast of	84
CAMBYSES	301
CAMPBELL, Dr.	69
Camaras Bay	82
Camperdown	229
Candahar	50
Canada, H. M. S.	125
Cape Leuwin	50
Horn	34
Recif	40
St. Marie	49
Tres Puntas	81
Finisterre	186
St. Francis	359
Capucin Friar	164
CAPPER, on Storms of the Coromandel Coast	2
Cards, Storm	12, 132
Carribean Sea	114
Carolinas	72
Castlereagh	76
Castries	272
Cashmere Merchant	55
Castle Huntley, H. C. S.	183
Cataraqui	188
Caton, H. M. S.	125
CATTERMOLE, Captain	208
CAVENDISH	272
Cedar Keys	178
Central America	23
Centre or Focus of Cyclone	5, 16, 104
of a Tyfoon	*Note* 156

	PAGE
Centaur, H. M. S.	125
Ceylon Tornado-Cyclones	54
rates of travelling	91
Chagos Archipelago	49
Channel, and Chops of the Channel	88
Channel Currents	185
Chanticleer	222
CHAPMAN, Captain	177
Charles Heddle	47, 93, 108, 179
Charlestown	315
Tornados; Meeting of them	96
Charleston	208
Charles Kerr	287
Chevalier	52
China Sea, Tracks	57
rates of travelling of Cyclones	91
Cyclones occurring in May	296
Lessons in	173
Storm-wave and Current	181
Childers, H. M. S.	67, 331
Chili, Coast of	83
Chops of the Channel	88
CHURCH, Capt. J. J.	321
Chusan	67
Circular, or high curved winds	10
City of London	293
Claims of the Science	377
CLAPPERTON, Captain	79
Clarendon	189
CLARK, Captain Stanley	231
CLARK and Co.	354
CLARKE, Dr. E. D.	86
Rev. W. B., Sydney	*Note* 80
Cleopatra, H. C. Steamer	199, 212, 332, 359
Clouds	267, 272, 294
at centre	210
Coast of Guinea, DAMPIER's water-spout	309
Cochin China	59
Cocos or Keeling Islands	47
COLLINGWOOD	229
COLLINS, Dr.	33
Columbine, H. M. S.	67

INDEX.

	PAGE
COLUMBUS	339
Commencement and breaking up of Cyclones	322
Comoro Islands	49
COMPTON, Captain	214
Conditions of Theory to account for Cyclones	324
Constantinople	87
Contemporaneous Cyclones	96
Contest, H. M. S.	31
COODE, Mr.	173
Cook's Straits	78
Coolies . . . . . *Note*	229
Cooper's River	98
COOTE, Captain	313
Coringa . . . . . *Note* 89, 195,	197
Coringa Packet water-spout	313
CORNEY, Captain	341
Cornwallis	226
Coronation, H. M. S.	164
Cosseguina	23
Counter-Tides	177
COURTNEY, Captain	75
CRANK, F., Esq. . . . 209, 226, 339,	342
Crozet's Islands	359
Cuba Cyclone 23, 177,	261
coast of	190
Cuffnells, H. C. S.	121
Culloden, H. M. S.	240
Currents, Storm	12
Cuttack 195,	197
Cyclonal wave	204
Cyclonal Points; Table of them	107
Cyclone; new name for rotatory storms	10
Cyclones, their causes . . . 17,	21
beginning beneath the ocean	324
at the surface	324
in the atmosphere	325
descending as disks	330
thickness of disks	266
of Iceland	371
commencement and breaking up . 322,	338
termination	332
in the China sea	296
effect on Compasses	357

C C

	PAGE
Cyclones, rate of travelling	89
height above the surface of the ocean	265
their sizes	93
stationary	91
parallel	96

D

Dacca whirlwind	302
DALRYMPLE, A.	196
DAMPIER	272, 286, 309
DANA, R. H., Junior	85
DANDO, Captain	67
DANIELL, Professor	237
Data for Barometer problem	244
DE LA BECHE, Sir Henry	170
Deliverance, Cape	357
DE LUC	239
Department of the Loiret	99
Desjeada	82
Deviation of the compass	185
Devonshire	33
Directions for observers	348
for studying the Science	342
Disks of Cyclones	266, 268
Dividing Cyclone	96
DON JUAN DE ULLOA	2, 11, 82, 84
Dorre Island	50
Double arched Squalls	293
DOUTTY, Captain	362
DOVE, Professor	5, 8, 26, 133, 187, 247
Dover, H. M. S.	164
DRAKE	272
Driver, H. M. Str.	71, 78
Duke of Buccleugh, H. C. S.	121
Manchester	274
Wellington	292
York, H. C. S.	273
Duncan's whirlwind	314
DUNCAN, Captain	292
DUPETIT THOUARS	70
Dust whirlwinds,	
Africa, Mexico	301
Calcutta	299

Dust whirlwinds	
Dacca	302
Deesa	298
India	297
Shirauz	301

E

Eagle, Brig	314
Earl of Hardwicke	148, 275, 285
Earthquake in Cyclones	225
Easurain	58, 296, 356
Effects of Cyclones on the Compasses	356
Electric Telegraph in Cyclones	375
Electrical whirlwind in Sussex	308
Electricity and Magnetism	213
Electricity in Dust Storms	303
Electricity of Storms	217, 338, 339
Electro-Magnetism	219
Eliza	214, 241
ELLIOTT, Mr. R. N.	67
ELLISON, Mr.	189
EMY	12
England	35
Eolian Researches	285
Erin	56
ERSKINE, Captain, R. N.	40, 79
Esperanza	83
ESPY, Professor T. P.	6, 16, 18, 23, 25, 211, 237, 267, 333, 338
Etymology of the words "Hurricane and Tyfoon"	361
Exmouth	207, 284
Experimenting on Storms	210
Explanation of words used	7
Eyafjeld Yokul	23
Eye of the Storm	16, 270

F

FAIRFIELD, Captain	305
Fairlie	217
Falkland Islands	36
False Point, Palmiras	198, 271
FARADAY, Professor	294
FARRAR, Professor, on the Boston Storm of Sept 1815	3, 270
Favourite	74, 191

	PAGE
FAWCETT, Captain	314
Feejee Group	73
FERNANDO DE OVIEDO'S *Historia de las Indias*	329, 361
Fine weather whirlwinds	305
Finisterre	186
Fire on board ship	377
Fire Queen, H. C. Steamer	358
FISHBOURNE, Captain	38, 276
FITZROY, Captain, R. N.	32
Futtay Salam	76
Futtle Oheb	208
Futtle Rozack	48, 200, 260
Flattening-in of Cyclones	114
Formosa	198
Straits	65
Channel	58
Fort Royal	192
FRANKLIN on N. E. storms of American Coast	3
Letters	309, 312
Lightning	337
Freak	72, 153, 238
French log-books	355
Friendly Islands	75
FYERS, Lieutenant	301

G

Gale	9
Gambia River, water-spout and storm	316
Ganges	91, 225
General Kydd	165, 166
Geneva *Bibliotheque Universelle*	228
General Palmer	31
Geographe	51, 239
GIMBLETT, Captain	42
Giraffe Barque	357
GITTENS, Mr.	207
Glorieux, H. M. S.	125
Godavery	195
Golconda	51, 58, 96, 117, 157, 236
GOLDSBERRY, M.	223, 307
GORDON, DR.	356
Goruckpore district	297

	PAGE
Governor Reid	194
GRAAU, Captain	357
Grand Cayman	178
——— Ladrone	182
Grand Dusquesne	176
Grande Terre	273
GRAVES', Admiral, disaster	125
GRAVIERE, M. de la	71, 72, 230
Great Western, Steamer	140
Great Liverpool, Steamer	185
GREEN, and Co., Messrs.	323
Green Light, or Sky	292
GREENFIELD, Admiral Sir R.	93
Great Storm of 1703 *Note*	184
GREY, Captain	50
GRIFFITHS, Dr.	298
Guadaloupe	177, 273
Guess	361
Guinea Coast, DAMPIER's water-spout	309
Gulf of Tonkin	58, 182
——— Lyons	86
——— California	84
——— Martaban	57
——— Mexico	84, 114, 367
——— Stream	176

H

Hail-storms	11, 98
——— in relation to Whirlwinds	372
Hainam	182, 198
Halifax tornado	96
——— harbour	261
HALL, Captain	191, 275
Harbinger and Navarin Rocks	188
Harbour	162
HAMMACK, Captain	77
HARE, Professor	16, 211
HARRIS, Sir Wm. Snow	359
Harvey Islands	75
HASSENFRATZ	284
Havannah	190

	PAGE
Havannah, H. M. S.	26, 42, 79
Havering	26
HAUKSBEE, F.	236
HAY, Commander	67
HAYES, Comr. C. O.	78
Hecla	23
Hector, H. M. S.	125
Height of Cyclones, above the Ocean	265
On Barometrical Chart	269
Horn Card Disk	268
KAEMTZ, ESPY, REDFIELD	266
Hermes, H. M. S.	39
HERSCHEL, Sir John	21, 260, 336
HEYWOOD, Captain Peter	221
Hindostan, Steamer	260
History of the Law of Storms	1—6
HOPKINS, Mr.	19, 220, 224
HORSBURGH	2, 53, 64, 182, 201, 220, 309
Horn Cards, Use of them	132
Examples as disks	268
Hot and cold blasts	341
Houtman's Abrolhos	51
HOWARD, Luke	25, 184, 215, 274
HOWE, Capt.; Kyook Phyoo water-spouts	311
HUMBOLDT	23, 84, 302, 357
HUDSON, Capt.	292
HUNT, Mr.	32
Howqua	63
Huracan	329, 361
Hurricane	9
Tract	48
derivation of	361
making one	234
HURST, Mr.	176
HUTCHINSON, Capt.	172, 288
Havalfiord	371

I

Iberia, Steamer	187
Iceland, Cyclones of	68, 370
Ida	207
Illustrious, H. M. S.	261
Incurving winds at the centre	108, 113

	PAGE
Independence, whaler	75
Intensities of Cyclones	245
Inundation from the Storm-wave	190
Andamans, Barbadoes	193
Bermuda	193
Burrisal and Backergunge	199
Coasts of China	198
Coringa	195
Cuttack and Balasore	197
Ingeram	196
False Point	198
Grand Cayman	178
Savanna La Mar	192
Martinique	193
New York	194
St. Petersburg . . *Note*	199
West Indies, St. Kitts, St. Croix, Antigua	192
Invisible clouds	266, 284
Involved in a hurricane	103
Isabella Anna	77

J

Jamaica	163, 227
James	208
JAMIESON's Edinburgh Journal	371
Jane	194
Janet Wallis	36
Japan	66
Java and Anamba Seas	62
JENKINS, Capt.	293
JENNINGS, Capt.	332
JERVIS, Admiral	230
JOBLING, Captain	307
John McViccar	59, 288
John Fleming	260
John Ritson	207
JONES, Capt.	292
Juan Fernandez	84
Judith and *Esther*	214, 275, 340
Jumna, H. M. S.	43, 209
Juno	296
JURIEN DE LA GRAVIÈRE, M.	230

K

	PAGE
KAEMTZ on whirls of wind	21
Thunder Storms	217, 266, 292
Barometer	237
Kolpeni	199
Kandiana	307
Key West	259, 261
Kilblain	238
Kingsmill Group	75
KIRKSOPP, Captain	296
KOTZEBUE	68, 73, 371
KRUSENSTERN	66, 265

L

La Vallette harbour	87
LA PLACE, M. de	195
Lady Clifford	166, 275, 341
Lady Feversham	265
Lady Flora	306
Lady Jocelyn	290
Laccadives	52, 73, 199
LAIRD and OLDFIELD	224
LANGFORD, Captain	2, 163, 291, 295
Lark, H. M. S.	92, 191
Latent heat	*Note* 18
Law of Rotation of Cyclones	334
Law of Storms, term explained	7
LAY, Captain M. J.	179
Le Vautour's spout	316
Le Naturaliste	239
La Berceau	47
Lefoo	76
LEIGHTON, Captain	207
LEMEYRIE, Dr.	316
Light of the Age	37
Lightfoot	307
Lightning; single flashes	214
Lightning, Aurora-Borealis-like	293
LINSCHOTEN	93
Log Book	354
Log Book Currents	185
London's Cyclone—and Barometer	154, 181, 250, 288

INDEX. 393

	PAGE
London, H. C. S.	183
Loo Choos	185
Loo Choo Sea	65
Loopuyt	207
Lord George Bentinck	356
Lord Lyndoch	79, 341
Lord of the Isles	37
LORIMER's letters	356
Louisiana, Barque	66
LOVETT, Captain	69, 314
Low Archipelago	73
Loyalty Islands	76
Lull	16
LUSHINGTON, Commodore	52
Luxon	71
LYNN, Mr.	286
Star Tables	121
LYONS, Captain	301

M

Mace Rock	187
Macqueen, H. C. S.	48
Madagascar, practical lesson	151
Maddalena Islands	87
Madras Roads, Example	173, 174
Observatory	260
Cyclone of 1807	95
Magellan	202, 295
Magicienne	59, 88
Mahi, H. C. Schooner	358
Making a Hurricane	234
Mala Ventura	329
Malacca Straits	152, 153
MALCOLMSON, Dr.	270, 275
Maldives	53
Malta	87
Maltese, brig	306
Manar	213, 226
Mandane	172, 288
Mangeea	76
Manilla	225
Maria Somes	106, 148

	PAGE
Mariana Islands	72, 185
Marion	161
Marlborough	26
MARTIN, Dr.	225
Mr. J. C. on Hail Storms	373
Mr. K. B. on the tide guage in a thunder storm	172
Mr. F. P. B. on Equinoctial Storms of March and April, 1850	33
Martinique	177, 227
Mary	23
Matauzas	315
Mauritius, practical lessons	150
Mauritius Cyclone, 1840, 1844	113, 161, 285
MAURY, Lieutenant, M. T.	66
McCARTHY, Captain	214, 241
McLEOD, Captain N.	59, 154, 181, 288
McQUEEN, Mr.	208
Mediterranean	87, 172
mariner's invocation	270
Medway, Steamer	236
Meeting of Cyclones; Charleston tornado	97, 98
Men of War and Merchantmen	131
MERCATOR'S projection	376
MERCER, Mr. W.	312
Mercury Bay, N. Zealand	78
Mermaid	274
Merope, Whaler	80
MILLER, Captain	166, 275, 290, 341
MILLS, Rev. W.	74
MILNE, Captain D.	76, 88
MILNE, MR., on African Tornadoes	224
Minerva, Brig	62
Miscellanea and additions	354
MITCHELL, Dr.	4, 276
Moderating of the wind at the beginning of Cyclones	360
Moderato	Note 86
Mona Passage	330
Monarch	363
MONDEL, Captain	272
Montreal	69, 311
MOODY, Captain R. E.	36
MOORCROFT	300

INDEX. 395

	PAGE
MORRISON, Dr.	215, 362
Mosambique Channel, tracks	49
rate of travelling	90
Moulmein, Barque	59
MUDGE, Lieutenant	359

N

Nagore	166
Names, native, given to Storms	362
Narcondam	160
Natal Coast	39
Naturaliste	339
Navigator's Islands	73
Negrais Cape	161, 312
Negril South Point	163
NELSON's gales	86, 230, 368
Nephelogical Observations	346
NEVILLE, Commander, R. N.	46, 202
Newingden in Kent	308
New Caledonia	77, 370
Hebrides	76
Zealand	78
York	194
Nicaragua	84
NICHOL, Professor . . . *Note*	11
NICOLAS, Sir N. H.	231
NICOLAS OVIEDO, Father	329
Nimrod	370
Noise of Cyclones	204
Waterspouts	317
NOLLET, L'Abbé	319
NORMAN, Robert	376
North Westers of Bengal	219
of the Straits of Malacca	219
North American Review for 1844	4
rates of travelling	90
Northern Pacific	65
Northers, *Nortes* of the gulf of Mexico	82, 84, 114, 178, 189, 366
Northumberland	113, 214, 285
N. W. Point of Australia	50
Note on winds for Bermuda vessels, Colonel REID's	138
Nurserath Shaw	152, 161

396   INDEX.

                                                              PAGE
O

Octavia, Whaler 63
ŒRSTED, Professor 101, 320
OLAFSEN 371
OLDFIELD 224
Olho de bove 307
Ongole Storm 226
ONSLOW, Mr. 189
Open Coasts 162
Open Roadsteads . . . 162
Opposition to the science for a time . . 376
Orient 148
Orissa, Coast of . . . 155
Orkney Islands 373
Orwell, H. C. S. 49
Oscillations of Barometer and Simpiesometer . . 236
       of centre of Cyclone . . • . 209
OSGOOD, Captain 292
Otway Cape 188
OVIEDO, Fernando, Historia de las Indias . 329, 361

P

Pacific . . . . . • 287
Pacific Ocean, rates of travelling . . . 90
Palawan Passage, Tracks . . . 59
Pallas, H. M. S. 125
PALMER, Mr. J. 287
PALMER, Captain 49
Palmiras Point 161, 273
Pamperos 10, 219
Papagallos 84
Paquebot des Mers du Sud . . 32, 214, 270, 274
Paradox of Revolving winds . . . 15
Parallel Cyclones 96
PARSONS, Captain 190
    Mr. 196
PASCO, Lieutenant Crawford . . . 59
Passage of the Centre . . . 207
Passengers, observations may be made by . . 363
Patagonia 36, 81
Patchung San 292
Patriot 161

Pedro Point	213, 226
PEDRO DE MEDINA	376
Pelorus, H. M. S.	63
PELTIER, M., *Sur les Trombes*	266, 284, 310, 312, 316, 318, 319, 337, 339
Table of Spouts	319
Peninsular and Oriental Steam N. Company	*Note* 186
PERON, M.	51, 215, 239
Persia and Arabia, Coast of	51, 151
PETER MARTYR's History	330
PETRUS NONIUS	376
PEYSSONNEL, Dr.	177, 214, 216, 273
Philippines, and native names for storms	*Note* 70
Phœnix, H. M. S.	200
PIANCINI, Father, his water-spout	310
PINKERTON	51
Pioneer, U. S. B.	262
Plowden's Island	198
Pluto, H. C. Steamer	57, 106, 157, 249
Point Pedro	213, 226
Points of Enquiry and Directions	344
before a Cyclone	344
during a Cyclone	347
at the close of a Cyclone	348
POISSON, *Sur la Probabilité*, etc.	358
Pondicherry	166, 173, 236
POOR RICHARD	377
Pooree and Cuttack Cyclone	214
Pororoca of the Amazon	172
Port Essington	63, 179
Port Leschenault	78
Portland Breakwater	173
Practical applications	101
Practical Lessons with Storm Cards	135
Arabian Sea	151
Barbadoes to Bermuda	140
Bay of Bengal	152
Bermuda to New York	139
Black Sea	144
China Sea	155
Coasts and Seas of Europe	141
England to Bermuda	140

398 INDEX.

	PAGE
Practical Lessons with Storm Cards,	
Europe to New York	141
Halifax and Bermuda	139
Mauritius	146
New York to Bermuda	139
Northern Atlantic	138
Northern and Southern Pacific	158
Recapitulated	159
Southern Indian Ocean	144
West Indies	135
Prince Albert	160
Princess Charlotte of Wales	208, 285, 360
Profiting by a hurricane	102
Proofs of the Accuracy of the Rules	160
Providence, H. M. S.	121
PROWD, Captain	286
Pulicat Hills	260
Pumps, crushed	157
PURCHAS' Pilgrims	93
PURDY, Atlantic Memoir	21, 31, 185, 221, 223
Pyramidal and cross seas	200

Q

Queen, H. C. Steamer	358
QUEEN ELIZABETH	376
Queen Victoria	176

R

Radack Islands	73
Rajasthan	293
Ramillies, H. M. S.	125
RAMUSIO	328, 362
Raratonga	75
Ras (or *Raz*) de Marée	177, 192
Rates of travelling	89
Andaman Sea	91
Arabian Sea	90
Bay of Bengal	90
China Sea	91
Coast of Ceylon	91
Mosambique Channel	90
Pacific Ocean	91

INDEX. 399

	PAGE
Rates of travelling,	
Southern Indian Ocean	90
West Indies and North America	90
Rawlins	208, 276
Recapitulation; Cases of error and good management	159
RECORDS, Captain	309
RECHENDORF, Mr.	301, 325
Red Sky and Light	283, 289
REDFIELD, Mr. 3, 4, 5, 6, 9, 15, 16, 22, 26, 34, 50, 70, 94, 108, 115, 122, 161, 177, 178, 187, 194, 209, 228, 237, 241, 267, 276, 328, 340, 343	
REDFIELD, REID, and DOVE's Theory of tracks	26
REDFIELD and REID, ESPY and THOM,	
on motion of the winds	15
on the causes of hurricanes	18
References, Chart No. II.	43
No. III.	55
No. IV.	59
Pacific Ocean	59
REID, ESPY	17
REID's Hurricane Cards	12
REID, 5, 7, 16, 26, 30, 46, 75, 84, 88, 92, 138, 161, 167, 176, 193, 204, 225, 233, 236, 240, 264, 270, 272, 274, 276, 301, 305, 320, 365, 372	
Remarks on the Tables of Signs	283
RENNELL, Major	185
Repulse, H. M. S.	187
Resultant curve of winds	109
Revenge	93
Revolving Winds, true Cyclones	367
RICH, Mr.	301
RICHARDSON, Mr.	273
RIDLEY, Captain	50
Rifleman, Barque	77
Right-lined winds	10
Ringdove, H. M. S.	67
Rio de la Plata	219
RION, Canon	333
Rising up of Cyclones	332
Rise of Barometer before Cyclones commence	235
before they terminate	241
RITSON, Captain	207

		PAGE
River Thames, New Zealand	Note	78
Rivers		162, 173
Roadsteads, Harbours, &c.		162
Robin Gray		201
ROBINSON, Captain		238, 271
ROCHON, L'Abbé, Theory of Cyclones	Note	20
RODNEY's, Admiral, prizes		125
RODNEY, Lieutenant		209
Rodney, H. M. S.		88
ROMME on whirlwind storms, and Dictionary		3, 193
Rory O'More		321
Rose, H. C. S.		49
ROSS, Captain		34, 48, 63, 206, 342
Rotatory and progressive theory		264
objections answered		337
Rottnest Island		51
ROYER, Captain		179
Rule for lying to		122
RUNDLE, Captain		200, 204, 291, 342
Runnimede and *Briton*		56, 112, 160
Rurick		68, 73
Russell, Whaler		50
Ryacottah Observatory		260

S

SABINE, Lieutenant-Colonel	22
Saint Lawrence's Bay	68
Samoan Group	73
San Blas	84
Sand Heads	155, 181, 241
Sand-spouts, Persia	301
Santa Barbara	85
Santa Cruz	179, 215
Sappho	31
Sarah	76
Sarangani Island	62
Saratoga	278, 292
Savage Island	79
Savaii Island	74
Savannah La Mar	192
Scaleby Castle	183
Scamander	370
Scinde whirlwinds	298

INDEX. 401

	PAGE
Scirocco	341
Scudding or heaving to	126
Seasons at which Cyclones occur	295
Seaton	275
Secret Barque	207
SEDGWICK	7
Seringapatam	49
Serpent, H. M. S.	46, 151, 202
Settling down of Cyclones	331
Seychelles	49
SEYMOUR, Captain	340
Shakespear	272, 288
Shanghae	67
Shark's Bay	50
SHAW, Captain	238, 280
Sheerness, H. M. S.	55
Ships sent out to experiment on Storms	*Note* 210
Ships prevented from running into the Storm Circle	159
SHIRE, Captain	58, 291, 356
Siberia	68
Signs of Approaching Cyclones	276, 295
SILVER, Captain	93, 144, 359
SIMONS, Mr.	276
Simpiesometers	229
Improved Tropical Tempest	234
SIMPSON, Captain T. B.	72, 356
Singapore	220
Sir Edward Paget	305
Sir Edward Parry	33
Sizes of Centre of Cyclones	211
Sizes of Cyclones	93
Arabian Sea	95
Atlantic Ocean	93
Bay of Bengal	95
China Sea	96
In the West Indies	94
Malabar Coast	95
Southern Indian Ocean	95
Smell of the sea before Cyclones	295
SMITH, Captain	208, 274
SMOULT, Captain	238
Snow Clouds	284

	PAGE
SOLANO, Admiral	190
Sophia Fraser	113, 229
Sophia Reid	237
South Coast of Australia	64
Southern Ocean, Cape of Good Hope, and Van Diemen's Land	40
Southern Indian Ocean, rates of travelling	90
Spey, H. M. S.	227
Spiral streams of electric fluid causing Cyclones	338
Spontaneous Combustion	377
Spots on the sun may be Cyclones	336
Spouts, noises with them	317
SPROULE, Captain	202, 295
St. Croix	192, 215
St. Domingo Hurricane of 1508	329
St. Elmo fires	217
St. Eustatius	177
St. George's whirlwind	308
St. Lucia	227
St. Pierre	192
St. Petersburgh inundation	199
St. Thomas	227
Standard	204
Stars, scintillation of	290
Stationary Cyclones	24, 48, 91
Steam, Mr. HOPKINS' use of the word	19
STEPHENSON, Mr.	300
Stern-way	103
STEWART, Captain	293
STOKES, Captain	81, 188
STOPFORD's, Sir R., fleet	320
Storm, A.D. 1703	9, 184
Cards	12, 133
Current	12, 168
Disk	16
Tract	48, 49
Wave	12, 168
Storm-wave and Currents, Instances	172, 174
Bass' Straits	186
Bay of Bengal	180
China Sea	181
Coasts of Australia	179, 180
Sand Heads	181

	PAGE
Storm-wave and Currents, Instances,	
Southern Indian Ocean	179
West Indies	176
Straits of Malacca	152, 220
Singapore	220
STUART, Captain A.	165
Sulimany	285
SULLIVAN, Captain	36, 76
Sumatras	*Note* 220
Surat Roads	53
Surgeons and Passengers	363
Sussex Whirlwind	308
Swan River	51, 180
SWANEY, Mr.	76
Swell felt at distance	203
Swift, H. M. S.	71, 121
Swithamley	332
Sydney Herald	80

T

Table for Bearing of Centre	105
of Cyclone Points	107
of fall of Barometer and rates per hour, &c.	248
of Results of Barometric rule	258
of Celestial and Atmospheric Signs	278
of Terrestrial and Oceanic Signs	281
of Excessive Falls of Barometer	264
of average number of Cyclones in various months	296
of Water-spouts from M. PELTIER	319
Tack to lie to on; Colonel REID's Rule	122
Tahiti	76
TAPLEY, Captain	312
Tartar, H. M. S.	10, 222
Taunton Castle, H. C. S.	121
Tchukutskoi	68, 371
Tehuantepec	84
Tenasserim's Whirlwind	312
Teneriffe, Waterspout at	320
Tercera, Cyclone at	93
Termination of Cyclones	332
Ternate	52
TESSIER, M.	338

	PAGE
Thalia	111, 112
Thames, Mail Steamer	307
Theories of the motion of winds in hurricanes . .	5
Theory and Law . . . . *Note*	7
Theory of the commencement of Cyclones . .	325
Thetis of London, and of Calcutta . . .	97, 157
THEVENOT	319
THIOT, M., Physical Geography . . .	177
THOM, Dr. A. . 4, 6, 17, 46, 48, 76, 86, 113, 161, 193, 201,	
217, 237, 243, 263, 284, 325, 332	
Theory of Storms, S. Indian Ocean . .	20
Thomas Grenville	112, 215
THOMSON's Annals of Philosophy . . .	308
THOMSON, Mr. J. T.	321
THORNHILL, Captain	215
Thunder and Lightning	212, 339
Thunder storm, effect on tide guage . .	171, 172
Thunder, H. M. S.	92
Thyrill	371
Tigris	238, 271
Timor Sea, Cyclones of	63
Tobago Cyclone	28, 216, 227
Tonkin, Gulf of	58, 182
Tornado in the Cuba Cyclone of 1846 . .	315
Tornadoes	219
Tornado-Cyclones, Azores . . .	93
and in the Channel . .	363
Tornado-like Cyclone	54, 308, 311
Toulon	86, 230, 368
Track of Storm	13, 24
Tracks, Mosambique Channel . . .	49
Tracks of Cyclones	24, 115
Andaman Sea . . .	56
Andamans to Madras . .	54
Arabian Sea . . .	51
Average . .	25
Azores . . .	32
Banda, Celebes, and Sooloo Seas .	62
Bass' Straits . . .	64
Bay of Bengal . . .	54
Cape de Verd Islands . .	33
Cape of Good Hope, eastward, to Natal Coast .	37
China Sea	57

INDEX. 405

	PAGE
Tracks, Coast of Ceylon	54
Coast of North America and North Atlantic	28, 31
Coast of Van Diemen's Land	65
Curved	29
Diagram	27
Eastern Coast of South America	34
European Seas	85
Falkland Islands	35
of Storms defined	13, 24
Formosa Channel	58
General Theory of	28
in West Indies, Carribean Sea, Gulf of Mexico	28
Java and Anamba Sea	62
Madeira	33
Meridional	51
most Southerly known	26
Natal Coast to the Meridian of Madagascar	39
New Caledonia	370
New Zealand	78
Northern Pacific and Loo Choo Sea	65
of Antigua Hurricane	30
from the shift of wind	115, 119
Palawan Passage	59
Southern Atlantic, Cape Horn to Cape of Good Hope	36
on shore uncertain	26
Southern Indian Ocean to Coasts of Australia and Lat. 30° S.	42
Southern Ocean, Cape to Van Diemen's Land	40
Southern Pacific, tropical regions	73
South Coast of Australia	64
Straight-lined	28
Timor Sea	63
West Coast of Australia	50
Western Coast South America	81
North America	72
North America	28
between Trinidad and Tristan d'Acunha	37
Trade Wind	262
Transparent Clouds	266, 284
Trevandrum	53
Trident	35
Trincomalee	53

	PAGE
TRISTAN, Count de	98
Trombes	318
Tropical Tempest Simpiesometers	234
TROUGHTON and SIMS	234
True Briton, H. C. S.	156
Tudor	179
Tung Hai . . . *Note*	65
Turning and veering	13
Turon Harbour	59
Tuscan	144
Tweed, Steamer	189
Two Friends	75
Tyfoon, DAMPIER's description	286
Tyfoon, derivation of the word	362
Tyne, H. M. S.	32

U

ULLOA, Don Juan	2, 82, 84
Underoot	199
Unicorn	228
Use of the Horn Cards	132
Ushant	186, 187

V

VAILE, Mr.	289, 291, 294
VAN DELDEN, Captain S.	64
Van Diemen's Land	40, 79, 188
VAN WYCK, Captain	207
Vanguard, H. M. S.	88
Vautour	316
Vavao	75
Vectis	153
Vellore	263
Vernon	323
Vestal, Frigate	158
Ville de Paris	125
VILLENEUVE	87, 230, 368
VIRGIL	283
Virginie	273
Vitileva	73
Volcanoes and Volcanic centres, originating storms	22

INDEX. 407

	PAGE
Voss, Captain	323
Vulture, H. M. Steamer	59

W

WALES, Captain	148
WALKER, Mr. on atmospheric pressure	170
on Spout in the Bay of Naples	309
WALKER, Captain	363
WALTERSHAUSEN, W. S. Von	371
WATERMAN, Captain	6
Water Barometer, fluctuations	237
Waterspouts	309
in Cyclones	314
Wave, Storm	12, 359
Cyclonal	204
of Progression	204
WEBB, Captain	183
WEBSTER, Mr.	222
WELLER, Captain	148, 275, 285
West Coast of Australia	50
West Indies, Tracks	28, 29
Rates of Travelling	90
Whirls of Wind	20, 21
Whirlwinds and Waterspouts	297, 305
Land	302
at Cabul	298
in Europe	300
in fine weather	305
Noises with them	317
White Squall off New York, in 1852	306
WICKHAM, Commander	50, 51, 180, 242
WILKES, Commodore	78
William Wilson	341
WILLIAMS, Mr.	75
WILLIS, Mr.	238
WILLS, Captain	308
Willy-Waws	10
Winds classed	9, 10
Wind Points and Compass Points	106
Table	107
Windsor's Tornado Cyclone	364
WISE, Mr.	184

	PAGE
Woodbridge, U. S. A.	315
WOOLWARD, Capt. R.	307
Words, Explanation of	9
Works on Storms	6

Y

Yanaon	195
Yang-Tze-Kiang river	67
Yarkund	68

Z

Zaida	177
Zanzibar	49

G. NORMAN, PRINTER, MAIDEN LANE, COVENT GARDEN.

References to Chart No. II.—*The Southern Indian Ocean.*

		Authority.
I.	Track of the Rodriguez Cyclone,	Mr. Thom.
II.	Average Northern limit of Cyclones,	Mr. Thom.
III.	———— Southern limit,	Mr. Thom.
IV.	Track curving to the Southward about Mauritius; Blenheim, 1807,	Mr. Thom and Col. Reid.
V.	About Mauritius and Bourbon,	Mr. Thom.
VI.	Curve of the Culloden's Cyclone, March 1809,	Col. Reid.
VII.	H. M. S. Serpent, Feb. 1846,	H. Piddington.
VIII.	Futtle Rozack and other Ships, November 1843,	H. P. in 11th Memoir, Jour As. Soc.
IX.	Charles Heddle's Cyclone, March 1845,	H. P. 13th Memoir, J. A. S
X.	Fr. Frigate La Belle Poule, and Corvette Le Berceau, Dec. 1846, (See also O)	H. P.
XI.	H. M. S. Albion, Nov. 1808,	Mr. Thom.
XII.	H. C. S. Bridgewater, March 1830,	Col. Reid and Mr. Thom.
XIII.	H. C. S. Abercrombie, Jan. 1812,	Mr. Thom.
XIV.	Maguasha, Feb. 1843,	Mr. Thom.
XV.	H. C. S. Ceres, — 1839,	Mr. Thom.
XVI.	Timor and Rottee Cyclone, April 1843,	Mr. Thom.
XVII.	Malabar, Jan. 1840	Mr. Thom.
XVIII.	Boyne, Jan. 1835	Col. Reid.
a.	Mauritius, March 1811	Col. Reid.
b.	————, Feb. 1818,	Mr. Thom.
c.	————, Jan. 1819,	Mr. Thom.
d.	H. C. S. Dunira, Jan. 1825,	H. Piddington.*
e.	H.C.S. Princess Charlotte of Wales, Feb. 1826,	H. P.
f.	H. C. S. Orwell, Feb. 1827,	H. P.
g.	Thalia, April 1827,	H. P.

References to Chart No. III.—Cyclones of the Bay of Bengal, and Arabian Sea.

BAY OF BENGAL.

I.—Sand Heads,	.	June 1839.
II.—Coringa Cyclone,	.	Nov. 1839.
II. 2.—Cashmere Merchant,	.	Nov. 1839.
III.—Cuttack Ditto,	.	April, May 1840.
V.—Madras,	.	May 1841.
VII.—Calcutta,	.	June 1842.
VIII.—Madras and Arabian Seas,	.	Oct. Nov. 1842.
IX.—Pooree, Cuttack and Gya,	.	Oct. 1842.
X.—Madras and Masulipatam,	.	May 1843.
X.—Coringa Packet,	.	May 1843.
XI.—Bay of Bengal and Southern Indian Ocean,	.	Nov. Dec. 1842.
XII.—Briton and Runnimede,	.	Nov. 1845.
XIV.—Bay of Bengal and Arabian Sea Cyclone,	.	Nov. Dec. 1845.
XVIII.—October Cyclone,	.	Oct. 1849.

The above with Roman numerals refer to the Memoirs with the same number published in the Journal of the Asiatic Society of Bengal.

a.	Ongole Cyclone,	Oct. 1800.
b.	H. M. S. Centurion,	Dec. 1803.
c.	H. M. S. Sheerness,	Jan. 1805.
d.	Dover Frigate's,	May 1811.
e.	Kistnapatam; Palmers,	Mar. 1820.
f.	Burrisal and Backergunge,	June 1822.
g.	Liverpool and Oracabessa,	May 1823.
h.	Balasore and Cuttack,	Oct. 1831.
i.	London's,	Oct. 1832.
j.	Duke of York's,	May 1833.
k.	Calcutta,	Aug. 1835.
l.	Madras,	Oct. 1836.
m.	Protector's,	Oct. 1838.
n.	Kyook Phyoo,	May 1834.
o.	Kyook Phyoo; Siren and Java's,	Nov. 1838.
p.	Madras,	Oct. 1818.
q.	H. M. S. Cornwallis and Point Pedro,	Nov. 1815.
r.	H. C. S. Minerva,	Nov. 1797.
s.	Coromandel and Malabar Coasts,	May 1820.
t.	William Wilson,	Oct. 1836.

ARABIAN SEA.

A	East London's Cyclone, XVI. Memoir,	April 1847.
B	Buckinghamshire's, XVI. Memoir,	April 1847.
C	Higginson and Lucy Wright, VIII. Memoir	Oct. 1842.
D	H. C. S. Essex's, XVI. Memoir,	June 1811.
E	Bombay, XVI. Memoir,	June 1837.
F	Rajasthan's, } XIV. Memoir,	Dec. 1845.
G	Monarch's, }	Dec. 1845.

References to Chart No. IV.—*The China and Loochoo Seas and adjacent Pacific Ocean.*

No. of Track.	Dates.	Names.	Authority.
I.	17th July 1780	H. C. S. London	H. Piddington
II.	19th June 1797	H. C. S. Buccleugh	H. P.*
III.	20th to 23rd Sept. 1803	H. C. S. Coutts, Camden, &c.	H. P.
IV.	20th September 1803	H. C. S. Royal George, Warley, &c.	H. P.
V.	28th and 29th Sept. 1809	H. C. S. True Briton & Fleet	H. P.
VI.	29th and 30th Sept. 1810	H. C. S. Elphinstone & Fleet	H. P.
VI. 2,	1837	H. C. S. Vansittart	H. P.
VII.	8th and 9th Sept. 1812	H. M. S. Theban and Fleet	H. P.
VIII.	28th and 29th Oct. 1819	H. C. S. Minerva	H. P.
IX.	29th November 1820	H. C. S. Lord Castlereagh	H. P.
X.	18th and 19th Oct. 1821	H. C. S. General Kyd and General Harris	H. P.
XI.	14th and 15th Sept. 1822	H. C. S. Macqueen	H. P.
XII.	25th to 27th Sept. 1826	H. C. S. Castle Huntley	H. P.
XIII.	9th August 1829	H. C. S. Bridgewater	H. P.
XIV.	8th and 9th August 1829	H. C. S. Chas. Grant, Lady Melville, &c.	H. P.
XV.	23rd September 1831	At Canton	H. P.
XVI.	23rd October 1831	At Manilla and the Panama's first Storm	Mr. Redfield.
XVII.	24th and 25th Oct. 1831	Panama's second Storm	Mr. Redfield
XVIII.	25th and 26th Oct. 1831	Fort William	H. Piddington
XIX.	3rd August 1832	At Canton and Macao	H. P.
XX.	22nd to 25th Oct. 1832	Moffatt	H. P.
XXI.	28th August 1833	Briggs Virginia and Bee	H. P.
XXII.	12th October 1833	H. C. S. Lowther Castle	H. P.
XXIII.	3rd July 1835	Barque Troughton	H. P.
XXIV.	4th August 1835	H. M. S. Raleigh	Mr. Redf.
XXV.	16th to 22nd Nov. 1837	Ariel and Van...	H. P...

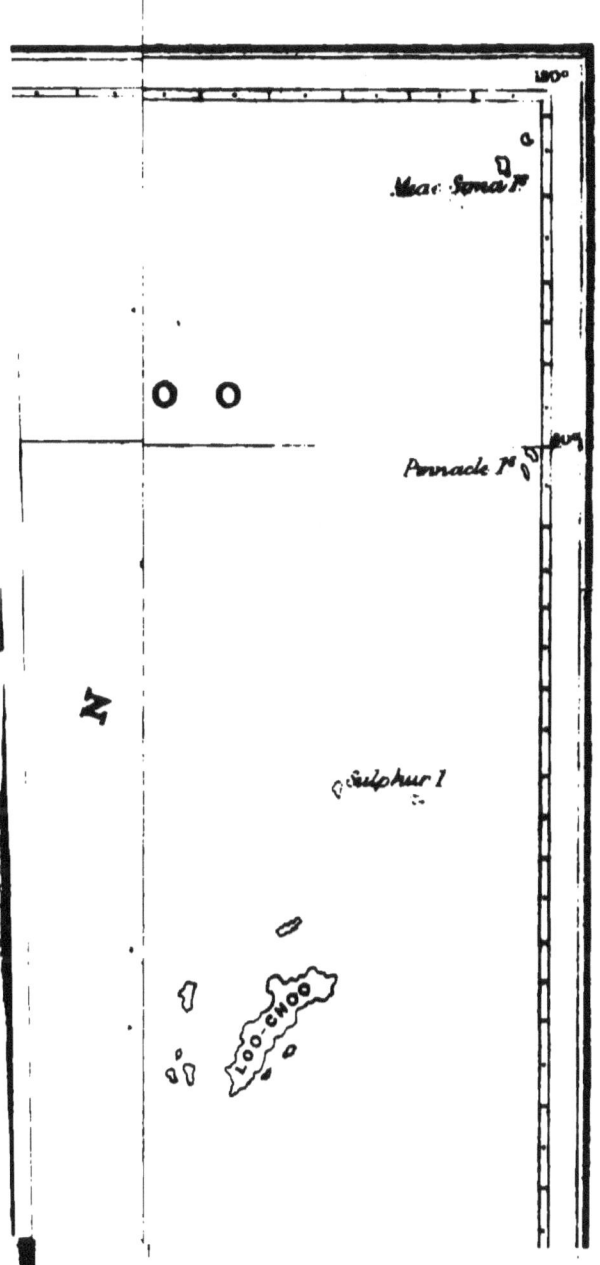